CARRYING THE WAR
TO THE ENEMY

C&C

CAMPAIGNS & COMMANDERS

GREGORY J. W. URWIN, SERIES EDITOR

Carrying the War to the Enemy

American Operational Art to 1945

Michael R. Matheny

University of Oklahoma Press : Norman

Library of Congress Cataloging-in-Publication Data
Matheny, Michael R., 1950–
Carrying the war to the enemy : American operational art to 1945 /
Michael R. Matheny.
p. cm. — Campaigns and Commanders ; v. 28)
Includes bibliographical references and index.
ISBN 978-0-8061-4156-5 (hardcover : alk. paper)
1. Military planning—United States—History. 2. Operational art
(Military science) 3. United States—History, Military. 4. Military
art and science—United States—History. 5. Tactics—History.
6. Strategy—History. 7. Unified operations (Military science)—
History. 8. Military education—United States—History. I. Title.
U153.M38 2011 355'.033573—dc22 2010030708

Carrying the War to the Enemy: American Operational Art to 1945
is Volume 28 in the Campaigns and Commanders series.

The paper in this book meets the guidelines for permanence and
durability of the Committee on Production Guidelines for Book
Longevity of the Council on Library Resources, Inc. ∞

1 2 3 4 5 6 7 8 9 10

FOR WENDY

JENNY

KEVIN

MEGAN

MICHAEL

Contents

Illustrations

Figures

Maps

Acknowledgments

For his patience, support, and mentorship, I am deeply indebted to Dr. Gregory Urwin. I would also like to acknowledge Dr. John Bonin, longtime friend, colleague, and fellow soldier. His interest in history and doctrine is both a source and a sounding board for many of my ideas. Additionally, I would like to thank Prof. Milan Vego, longtime mentor and friend, for his constant help and support. I am grateful to Capt. Steve Knott and Prof. Tom Hone for their comments on the chapter on seapower. I am also indebted to Dr. Phillip Meilinger for reviewing the chapter on airpower. I would like to thank Klemens "Van" Schmidt, cartographer extraordinaire, and Marsha Quesenberry, an expert in PowerPoint and all things technical. Dr. Evelyn Cherpak at the Naval War College Archives and all the good people at the Military History Institute in Carlisle, Pennsylvania, deserve my special thanks. I must acknowledge the support and constant encouragement of my wife, Gwen M. Matheny. As always, she has sustained me in everything I have done, or will do. Finally, I would like to acknowledge all those anonymous staff officers who soldier on despite the challenges and the sacrifices. During the interwar years, these challenges and sacrifices were substantial. They claim none of the glory or fame of their commanders, but they made that success possible. Their commitment to duty and their professionalism should remain an inspiration for those who follow. Finally, I must acknowledge the memory of my father, Farris R. Matheny. He served as an army private in the Pacific War, but he benefited greatly from the professionalism of his officers and never forgot the searing experience of war.

Introduction

A little more than three months after the Japanese attack at Pearl Harbor brought the United States into World War II, Army chief of staff George C. Marshall spoke to the mood of the nation when he asserted, "The time has come when we must proceed with the business of *carrying the war to the enemy*, not permitting the greater portion of our armed forces and our valuable material to be immobilized within the continental United States."[1] In World War II, carrying the war to the enemy would mean projecting, conducting, and sustaining large-scale military operations on a global scale. This is the story not only of growing American economic power but of the rise of a professional class of military officers capable of turning that power into military force. In its transformation from a frontier constabulary to a world military power, the U.S. military had to develop ways and means to apply armed force on the land and sea and in the air. The planning and execution of large-scale operations combining air, land, and sea forces form the essence of modern operational art. Operational art, that level of war between tactics and strategy, has, however, been largely overlooked by military historians.

Only a handful of books deal with the history of operational art.[2] Most military historians still view war within the framework of strategy and tactics. The scholars who have studied the operational level of war agree that it bridges strategy and tactics, but they rarely agree on the nature of operational art—that is, how commanders employ military force. Canadian historian B. J. C. McKercher refers to operational art "as the practice of generals for achieving operational success."[3] Another Canadian, John English, refers to it as the "art of campaigning."[4] The real divergence of opinion, however, emerges over what constitutes operational art. How do operational art and theory differ from the strategic or the tactical? Richard W. Harrison argues that the employment of army groups, consecutive operations, and deep operations characterized Russian operational art.[5] The Israeli military historian and theorist Shimon Naveh equates operational art with maneuver warfare that succeeds by defeating the enemy's organization through shock rather than attrition.[6] These historians, and most others who study operational art, are in agreement that the United States Army did not recognize operational art until the last decade of the Cold War. As Russell F. Weigley notes, "U.S. and British military thought before World War II neglected the operational art to focus instead on strategy and tactics."[7] I would argue, however, that although the American army did not officially recognize operational art as a third level of war, it did develop operational art during the interwar period, 1919–40, and practiced it to great effect during World War II.

Most military historians point to the interwar period as a time of stagnation, which they say accounts for the U.S. Army's lack of preparedness for World War II. David E. Johnson, who examines innovation in the army from 1917 to 1945 in *Fast Tanks and Heavy Bombers*, concludes that "the Army, in short, was responsible for its own unpreparedness. Tight budgets and an isolationist-minded Congress and public were powerful constraints, but the army would not have been ready even with adequate resources."[8] In *After the Trenches: The Transformation of U.S. Army Doctrine, 1918–1939*, William Odom agrees that the army was unprepared for World War II because following World War I "it failed to develop viable doctrine during the ensuing period of extended peace."[9] These historians and others may be quite right to point to the failure of the U.S. Army to modernize or develop a sophisticated combined arms doctrine during the interwar period, but in focusing on tactical doctrine or technology, they miss the evolution of U.S. military thinking at the operational level of war. It was at this level, particularly in dealing with logistically supportable joint and combined-phased operations, that senior American commanders did particularly well and laid a foundation for the Allied victory in World War II.[10] The best evidence for this maturing military thinking is found not in the published doctrine or scholarly works of American officers but in the military's educational system. It was at the Command and General Staff School in Fort Leavenworth, Kansas, and at the Army War College, then in Washington, D.C., that the rudimentary understanding of joint and combined operational art developed and was imparted to a generation of senior American officers.

Timothy K. Nenniger's *The Leavenworth Schools and the Old Army*, published in 1978, discusses the education and professionalism of the officer corps of the U.S. Army from 1881 to 1918. This is a groundbreaking study of the organization and the early influence of the Command and General Staff School up to 1918, but it does not cover the important interwar period, during which the army came to grips with the lessons of World War I. Likewise, T. R. Brereton's *Educating the U.S. Army: Arthur L. Wagner and Reform, 1875–1905* covers a formative period in the history of the Leavenworth schools and even the first years of the Army War College, but it does not address the later development or influence of these institutions. Harry P. Ball's *Of Responsible Command: A History of the U.S. Army War College* provides an excellent account of the organization and history of the college but does not detail the development of American military thought within the curriculum.

Historians who have studied military theory insist that operational art developed in Europe during the interwar period, but not in the United States. These historians consider the German or Soviet armies as the innovators and greatest practitioners of operational art. John English points to the Prussian and later the German Army as the pioneers in developing operational art. In "The Operational Art: Developments in the Theories of War," English traces operational concepts from Clausewitz through Helmuth von Moltke as the Germans attempted to meet the challenges of modern war.[11] Although English finds the roots of operational art in nineteenth-century Germany, real progress in this branch of military theory occurred during the interwar period. Certainly, this new conceptual framework was well in place by the end of the interwar period. In 1940, Col. Hermann Foertsch of the General Staff described the German concept of operations, emphasizing it as the link between tactics and strategy.[12] Jacob Kipp, David Glantz, and Shimon Naveh assert that the Soviet

Army pioneered operational art and brought it to its highest formula-
tion.[13] These historians credit Soviet military theorists Aleksandr A.
Svechin, Mikhail Tukhachevsky, and others with being among the
first specifically to analyze the third level of war and to develop its
theoretical foundation.

American military leaders also pondered the lessons of the Great
War. The changes in warfare and the need to move massive armies to
achieve strategic aims were no less apparent to competent U.S. Army
officers than to their European counterparts. Was there, then, no
comparable development of American operational thought? Most
historians, like English, believe that "contemporary American and
British interest in the operational level of war and the activity known
as operational art, of course, only dates back to the 1970s when the
U.S. Army sparked a renaissance in military thinking in the aftermath of the Vietnam War."[14] Contrary to this opinion, however, a
good deal of operational thought and synthesis was conducted in the
American armed forces during the interwar period. At Fort Leavenworth and the Army War College, the curriculum, lecture notes, and
even doctrine indicate increasingly sophisticated trends in American
military thought that forged a framework for operational art based on
experience, theory, and strategic requirements.

If historians have largely neglected operational art in general,
they are virtually silent on maritime operational art. The shelves
may be packed with the naval equivalent of drum-and-trumpet accounts of the clash of fleets in the modern era, but there is little real
analysis of the development of maritime operational art. *Professors of
War* by Ronald Spector and *Sailors and Scholars* by John B. Hattendorf, B. Mitchell Simpson, and John R. Wadleigh provide excellent
narratives of the development of the Naval War College, but like
similar studies of army schools, they do not specifically address operational art. Likewise, the role of airpower in operational art has been
ignored. Military history has largely focused on the strategic and tactical narrative of airpower. The story of modern American operational art is the story of joint operations—land, air, and sea. World
War I provided the stage for the introduction of airpower. It also presented the problems of modern mass warfare marked by an increased
complexity and scale that demanded study by the professionals destined to deal with it in coming conflicts.

The experience of World War I greatly influenced the officer education system established in the United States in the postwar period. The army reestablished the school system in 1919 to address many of the specific problems that emerged during the war, foremost among them the handling large armies in the field and preparing the nation for war. The School of the Line and the General Staff School at Fort Leavenworth prepared officers to staff and command large units; the Army War College, which initially reemerged in 1919 as the General Staff College in Washington, D.C., prepared them for duty with the Army General Staff.

At Fort Leavenworth, majors and captains attended the School of the Line. This course devoted one year to the study of brigade and division operations. Selected officers then went on to the General Staff School, which also had a one-year curriculum. In this second year, students studied corps and army operations. In 1922, these two schools were combined to form the Command and General Staff School, which remained at Fort Leavenworth. In the same year, the General Staff College in Washington, D.C., was redesignated as the Army War College. The curricula of these institutions during the interwar period evidence an appreciation for such key operational concepts as phased operations, center of gravity, and lines of operation, all of which became embedded in American military institutions and doctrine. The sophistication of interwar thought in the United States can be judged by the emphasis placed on theater structure, logistics, intelligence, joint operations, and combined warfare. The conducting and sustaining of large-scale operations was an evident strategic imperative. The army's requirement to defend the Philippines and fight a potential war in the Pacific against the Japanese mandated a broader understanding of modern expeditionary warfare. Operational art as taught and understood during the interwar years proved helpful to the American armed forces' preparations for the great challenge that loomed just over the horizon, World War II.

The lessons that the American military learned through two world wars were soon forgotten. The atomic bomb came to dominate military thought throughout most of the Cold War era. Large-unit conventional operations no longer seemed so relevant in the face of atomic deterrence or destruction. Operational art had little or no place in the "New Look" army of the 1950s. Only the bitter experience of Vietnam helped to launch a doctrinal renaissance, which led in 1982 to the rediscovery of operational art.[15] Now a centerpiece of American armed forces doctrine, modern American operational art is, however, rooted in the earlier renaissance that in fact occurred in the interwar years.

To avoid the confusion over the various definitions and descriptions of operational art as the third level of war, I have used the current U.S. military doctrinal concepts as a basis of discussion. At present, the doctrine of the United States military maintains that the operational level of war encompasses "the arrangement of battles and major operations to achieve military operational and strategic objectives."[16] At this level, commanders practice operational art by integrating ends, ways, and means across the levels of war.[17] At the heart of operational art is campaign planning. The campaign plan actually links tactics to strategy by determining where, when, how, and, most importantly, to what purpose military forces will engage the enemy. The campaign plan must balance these factors as it describes the employment of forces. In campaign planning current U.S. doctrine emphasizes several key theoretical concepts of operational design and operational functions. Among the key elements of operational design that can be traced back to the interwar period or even earlier are culmination, lines of operation, phasing, center of gravity, leverage, and linking tactical, operational, and strategic objectives.

Culmination refers to the point at which the attacker's strength no longer exceeds that of the defender because of a process of attrition or exhaustion. In campaign planning, the planner must be careful in offensive operations not to exceed the point of culmination. Lines of operation is a nineteenth-century concept that refers to the lines of approach to the enemy. Choosing the right lines of operation can be key to conducting and sustaining military operations. Since a single battle or major operation can rarely subdue a modern enemy, operations must be phased in a plan of campaign. Phasing recognizes the futility of the decisive battle and the need to sequence and think through all military actions necessary to achieve the strategic aim. This was perhaps the single most important innovation in military planning in the twentieth century. Finally, the concept of the center of gravity, which derives from Clausewitz, allows military planners to focus their efforts. The German military theorist defined the center of gravity as the "hub of all power and movement, on which everything depends. That is the point against which all our energies should be directed."[18]

The U.S. military currently recognizes several key functions of operational art. Among those that matured during the interwar years are maneuver, intelligence, sustainment (logistics), and command and control.[19] Those few scholars that discuss operational art usually focus on maneuver, but there is a great deal more to it than brilliant maneuvers that unhinge or quickly defeat the enemy. Intelligence has steadily increased in importance over the last century as the ability to acquire and manipulate information has advanced. Understanding the enemy and the environment combined with the ability to deceive the enemy are critical to the competent exercise of operational art. Perhaps most critical of all is the function of sustainment. Competent operators must organize and sustain the force as well as employ it. The growth of logistics as a staff function reflects its importance in operational art. This is particularly the case for expeditionary warfare of the kind the United States must conduct. Finally, getting command and control right as a function of operational art can greatly increase or decrease the chances of success.

Current U.S. doctrine mandates that the armed forces fight jointly—that is, the army, navy, and air force fight as a team.[20] The very essence of modern operational art is finding effective combinations of airpower, seapower, and landpower. Because the United States must often fight as a member of a coalition to achieve its strategic objectives, planning for joint and combined operations is conducted on a high order of complexity and sophistication, as compared to planning for unilateral warmaking.

Modern American operational art was forged on experience, theory, and strategic necessity. To examine the development of American military thought during the interwar period, this book will examine the curricula of the senior military schools at Leavenworth, the Army War College, and the Naval War College. American officers did not write much about military theory; there were no prolific American military writers such as Basil H. Liddell Hart, J. F. C. Fuller, or other German or Soviet military thinkers. However, the faculty and students of the senior American military schools labored on in workmanlike fashion to consider the problems and develop the solutions to modern warfare in order to meet the United States' specific strategic circumstances.

To understand the contribution of interwar military education to the actual planning and conduct of campaigns and major operations in World War II, this study considers four major operations, two each from the European and Pacific theaters: Torch, the invasion of North Africa in 1942; Overlord, the invasion of France in 1944; MacArthur's return to the Philippines in 1944; and Iceberg, the invasion of Okinawa in 1945. Operation Torch was the first major American operation of the war and demonstrates the level of American planning early in the conflict. Operation Overlord, arguably the most important major operation of the war, demonstrates perhaps the peak of Allied and combined-operations planning. The campaign of Gen. Douglas MacArthur in the South West Pacific theater was the largest in the Pacific war. Adm. Chester Nimitz's Operation Iceberg, the invasion of Okinawa, was the last major operation of the war. Iceberg, though controversial, represents the most developed American operational planning of the war.

In considering all these campaigns from an operational rather than tactical perspective, I place particular emphasis on the planning for these operations, in order to better understand the operational design and functions employed by American leaders. The U.S. military contemplated modern war during the interwar period as it considered the lessons of World War I and their implications for future warfare. Key to the senior military leaders' performance in World War II was their interwar education, which I highlight in the case studies. The development of a modern staff system and problem-solving process made possible the United States' ability to project, conduct, and sustain its forces in major operations. Campaign planning linking those major operations into a series of phased military actions to achieve strategic objectives in a theater of war was central to American operational art. The roots of modern U.S. operational art, which used all these key concepts to achieve victory, can be traced to the period between the world wars.

CARRYING THE WAR
TO THE ENEMY

1

The Roots of
Operational Art

"Civilization began, because the beginning of civilization is a military advantage." This assertion by Walter Bagehot overstates the case, but it does emphasize the close relationship between society and warfare.[1] As societies have evolved and become more complex, so has the warfare they wage. The art of war traditionally has two levels, strategy and tactics. Military strategy is the art of employing force to achieve political objectives.[2] Tactics is the art of placing and employing weapons and combat units on the battlefield. The evolution of strategy and tactics is determined by social organization, technology, the size of armies, and the scale of warfare. By the nineteenth century, Western civilization was sufficiently complex in social organization and technology to increase greatly the size of armies and the scale of warfare. This led to the development of a third level of war, operational art.

For much of military history, it was the role of strategy to choose the time and place to meet the enemy in battle with sufficient force to assure victory. Tactics then delivered victory by correctly employing weapons and units to maximum advantage. By the end of the nineteenth century, armies had become so large and the scale of war so vast that the traditional understanding of strategy and tactics no longer produced battlefield victories capable of yielding the desired political results. Large armies of more than a million men could not be defeated in a single decisive battle. In order to secure strategic objectives, a new, intermediate level of military operations had to be developed to link a series of battles in a theater of war. By the beginning of the twentieth century, some military theorists and practitioners began to recognize operational art as that third level in the art of war. The development and significance of operational art can be seen by briefly tracing the evolution of strategy and tactics.

The strategy and tactics of classical warfare were fairly simple, reflecting the resources and military capability of the city-states of ancient Greece. The Greeks concentrated land forces into a single army, normally numbering no more than twelve thousand warriors.[3] The army marched toward the enemy with the goal of either capturing territory or destroying the opposing army. Battles rarely lasted more than a single day and often proved decisive. In the centuries that followed, tactics and technology changed, but essentially the art of war still involved the concentration of forces into a single army to capture territory or destroy the opposition. The rise of the nation-state in the mid-seventeenth century brought significant changes. Geoffrey Parker argues that the West underwent a military revolution from 1500 to 1800 that allowed Europe to establish global empires. Parker maintains that the key elements of this military revolution were improvements in firepower, new types of fortifications, and an increase in the size of armies.[4] All these elements affected tactics and made European armies the most deadly in the world, but it was the dramatic increase in the size of armies that ultimately produced fundamental changes in the nature of strategy. The armies of the major European states increased tenfold between 1500 and 1700.[5]

By the end of the eighteenth century, the most powerful nation-states of Europe could field not only a single large army, but several such forces. Constant warfare and its importance to the state encouraged military writers to develop and popularize their ideas on how these forces should be used. Throughout the centuries-long decline of classical warfare and the rise of the West, military strategy—to the extent it was practiced—simply intended to bring the army into a favorable position for battle. Battle was the means to secure victory. In the wake of the devastation of the Thirty Years' War (1618–48), strategy shifted to emphasize maneuver rather than battle. Maurice de Saxe, marshal of France during the wars of Louis XIV, wrote, "I do not favor pitched battles, especially at the beginning of war. I am convinced that a skillful general could make war all his life without being forced into one."[6] How to accomplish this occupied many military writers during the period. Henry Lloyd, a British soldier of fortune, led a colorful life of military service with the French, Prussian, and Russian armies. His *History of the Late War in Germany* appeared in 1766. Lloyd was one of several writers who helped shift the interest of military theorists from the organization of armies to the conduct of operations.[7] His major contribution introduced the notion of the line of operation, which prescribed the movement of the army from its base to its objective.

Heinrich von Bulow was another writer of the late 1700s who helped shape changes in strategy. Bulow's *New Spirit of War*, written in 1799, was perhaps the best expression of the maneuver school. Adding the concept of base of operation to Lloyd's line of operation, Bulow argued, "The maneuvers of the field armies and the complex system of fortresses became the principal means of threatening the enemy's lines of operation while securing one's own, and replaced battle as the center of warfare."[8] The key to eighteenth-century warfare was to maneuver across the enemy's line of operation separating him from his base, thereby placing him at such disadvantage that he yielded the ground or the political objective. Even before Bulow's book was published, however, Napoleonic warfare broke upon Europe and again wrought fundamental changes in the objectives, organization, and employment of armies.

Napoléon dominated European warfare for twenty years, and his legacy it was dominant for another hundred years. Napoléon once again elevated battle as the means of victory. France fielded large armies swelled with conscripts and backed by the increasing resources of the centralized state. Napoléon maneuvered his army to give battle in the way most advantageous to the French; his object was nothing less than the destruction of the enemy army. Just as important, the organization of the larger armies into divisions and corps provided a new scope to how armies maneuvered to give battle. The French army adopted the division as a tactical unit in 1795 and grouped divisions into corps four years later. The division consisted of infantry and artillery. Divisions grouped together regiments and brigades into a force of four thousand to five thousand men. In turn, two to four divisions made up a corps. The corps also included a brigade or division of cavalry. The French Grand Army in 1799 consisted of seven corps.[9] The importance of these organizational innovations, largely required by the increased size of armies, was that large forces could now be articulated. Armies could maneuver over broad areas to ease supply and route congestion. In addition, corps were powerful enough to conduct independent operations, but they remained an integral part of an assembled army, normally only a day's march from the main body.

The advantages of the corps system can be seen in the Jena campaign of 1806. Alarmed by Napoléon's defeat of the Austrian and Russian armies of the Third Coalition in 1805, Frederick William III, king of Prussia, determined to make war on France in August 1806. Napoléon, never one to await events, concentrated his army in southern Germany and began his advance on Prussia in October. The French army of six corps moved in three parallel columns on a front of about thirty miles. Assuming the shape of a large square, French forces were prepared for a tactical thrust in any direction. Moving to the east of the Prussians, Napoléon concentrated four of his corps to attack what he believed was the main Prussian army, while sending the other two corps to move north and west to cut off the enemy's line of retreat. The Prussian army, which had grouped into two separate concentrations, had already begun to withdraw when it made contact with the French forces. Napoléon fell upon the smaller Prussian contingent with the main French army at Jena, while Marshal Louis-Nicolas Davout's corps of twenty-six thousand French soldiers engaged the main Prussian army of sixty-three thousand at Auerstadt. When the French First Corps, commanded by Marshal Jean Baptiste Bernadotte, finally arrived on the battlefield to the rear of the enemy forces, the Prussian army disintegrated. In one of his most decisive campaigns, Napoléon strategically deployed and tactically arranged his forces on the battlefield in such a way that a rout of the Prussian army was the result, leading to the occupation of Berlin.

The Napoleonic Wars engaged every major power in Europe. The increased manpower and resources available to the nation-states allowed them to raise several armies to conduct war simultaneously in different theaters. For example, as Napoléon planned the campaigns of 1805, he did not neglect his forces in Italy. In September 1805, Napoléon's chief of staff, Marshal Louis-Alexandre Berthier, summarized Napoléon's overall strategy in a letter to Gen. Gouvion Saint-Cyr, the commander of French forces in Italy, "The great blows will be struck in Germany, where the Emperor will go in person, and even if the operations of the Army of Italy are not successful, this must not influence your efforts because any success the enemy could obtain would only be of short duration. If the Emperor's operations are crowned with success, their first result will be to extricate the Army of Italy and send you the support you would need to throw coalition forces into the sea, recapture all of the country that you will have lost, and even to threaten Sicily."[10]

The need to coordinate operations in multiple theaters or even within a single theater certainly did not originate with the Napoleonic Wars. Coordination of forces at the strategic or tactical level is as old as the dispersion of forces on the battlefield or within a theater of war. As the British learned during the American Revolutionary War, managing forces from a great distance is never a simple task. For example, in the British campaign plan for 1777, three forces were to act in concert to seize the Hudson Valley, splitting the colonies. Lt. Gen. John Burgoyne was to move south from Canada by way of Lake Champlain while Gen. Sir William Howe was to move north from New York to join Burgoyne at Albany. Col. Barry St. Leger was to sweep down the Mohawk Valley, uniting with the other forces at Albany. This plan failed due to the inability of Lord George Germain, British Secretary of State for the Colonies, to coordinate these forces. General Howe's failure to cooperate with General Burgoyne helped to ensure the latter's defeat at the Battle of Saratoga. The forces involved in the British campaign of 1777 were comparatively small. Coordinating large armies and fleets such as those developed in the Napoleonic Wars for a coherent strategic purpose was a problem of much greater magnitude.

The organization of armies into divisions and corps as well as the size of armies and the scale of warfare assumed during the Napoleonic era were new in European warfare. Napoléon's stunning success led his adversaries to copy his methods and stimulated a renewed interest in the art of war. Military writers and professional officers studied Napoléon's campaigns intensively. The lessons they drew influenced Western warfare for the next century. In fact, many of the concepts developed by the military theorists of the nineteenth century continue to influence the conduct of modern war.

Perhaps the best known and one of the most influential interpreters of Napoleonic warfare was Henri de Jomini. Born in Switzerland in 1779, Jomini served as chief of staff to one of Napoléon's most famous corps commanders, Marshal Michel Ney. Eventually, Jomini deserted Napoléon's army and joined the Allies, serving as an aide and military adviser to Czar Alexander of Russia. A long-lived and prolific military author, Jomini published his most influential work, *The Summary of the Art of War*, in 1838. This book became immensely popular as military professionalism took hold in all the major armies of Europe. Ultimately, Jomini provided much of the vocabulary and many of the concepts for strategic theory in the nineteenth century. Building on the work of Lloyd and Bulow, he laid the foundation for European strategic theory that emphasized the base of operation, lines of operation, and lines of communication as the primary constructs for maneuvering armies in the field. Significantly, all modern forces still use these terms today.

Jomini asserted that the art of war consisted of five elements: strategy, grand tactics, logistics, engineering, and minor tactics. He intentionally omitted engineering and minor tactics from this study, but in summarizing the relationship between the other key elements in the art of war, he asserted, "Strategy decides where to act; logistics brings the troops to this point; grand tactics decides the manner of execution and the employment of the troops."[11] His later division of operations into strategy, grand tactics, and minor tactics is of particular interest. Jomini defined strategy as "the art of making war upon the map, and comprehends the whole theater of operations."[12] On the other hand, "grand tactics is the art of making good combinations preliminary to battles, as well as during their progress."[13] Jomini's division of the art of war into strategy, grand tactics, and minor tactics clearly suggests three rather than two levels of war. This emerging recognition of a third level of war marks the origin of operational art.

Jomini was not alone in discerning this new level of war. The other great military theorist of the nineteenth century, Carl von Clausewitz, also suggested an intermediate level between strategy and tactics. Clausewitz entered the Prussian army in 1792 as a twelve-year-old cadet. Within a year, he marched off to his first campaign in the War of the First Coalition against France, in 1793–94. He saw considerable action during the Napoleonic Wars, eventually serving as a corps chief of staff in the final campaigns of 1814 and 1815. Following the Napoleonic Wars, Clausewitz rose to the rank of major general and served as the director of Prussia's War Academy.

His major work, *On War*, written between 1824 and 1830, was described by Bernard Brodie, a twenty-first-century strategist, as "not simply the greatest but the only truly great book on war."[14] Clausewitz gave the clearest understanding of the three levels of war when he wrote about time and space.[15] As he noted, "The concepts characteristic of time—war, campaign and battle—are parallel to those of space—country, theater of operations, and position—and so bear the same relation to our subject."[16] Clausewitz associated campaign and theater of operations with a third level of war: "It is true that the term campaign is often used to denote all military events occurring in the course of a calendar year in all theaters of operations, but normally and more accurately it denotes the events occurring in a single theater of war."[17] He also indicated, "By theater of operation we mean, strictly speaking, a sector of the total war area which has protected boundaries and so a certain degree of independence."[18] Strategy for Clausewitz and for practically every military theorist and practitioner in the nineteenth and twentieth centuries was an inclusive term. It embraced both strategy at the highest level, what some would later call "grand strategy," and military strategy for the conduct of operations in a campaign within a theater of operation.[19] For the next century, theorists and soldiers struggled to find an appropriate way to differentiate between grand strategy, military strategy, and tactics.

Clausewitz defined strategy as "the use of engagement for the purpose of war. In other words, he [the general] will draft the plan of the war, and the aim will determine the series of actions intended to achieve it: he will, in fact, shape the individual campaigns and, within these, decide on the individual engagements."[20] Clausewitz used the term "engagements" to refer to battles. In an unfinished note presumably written in 1830, he gave the best description of strategy as what we now call the operational art: "The theory of major operations (strategy as it is called) presents extraordinary difficulties, and it is fair to say that very few people have clear ideas about its details—that is, ideas which logically derive from basic necessities."[21]

Though they called it strategy, both Jomini and Clausewitz wrote a good deal about the operational level of war. Along with the tactical level, it was the operational level that most interested the new class of professional warriors. The Napoleonic era gave birth to a new military professionalism. Britain's Royal Military College was established in 1799; the United States Military Academy was founded in 1802, and the Prussian War Academy opened in 1810. Long-serving professional officers looked to military writers as well as their own experience to further codify and develop their craft. This increasing professionalism in the nineteenth century naturally focused on a narrower interest in the employment of military forces in the field than the more encompassing considerations of grand strategy that might encroach on the prerogatives of their civilian or royal masters. Jomini's simple and prescriptive style was well suited to the needs of the new professional military institutions and their students. Riding the crest of interest in Napoléon and French military affairs, Jomini's *Art of War* was widely read and translated into several languages. Although Clausewitz was much more difficult to read, his influence would later rise with the prestige of Prussian arms following Prussia's success in the wars of German unification. The popularity and influence of both Jomini and Clausewitz, however, was due primarily to their ability to help others understand and deal with the growing complexity of war.

By the mid-nineteenth century the professionalization of the American officer corps was well under way.[22] In 1844, Henry Wager Halleck, a graduate of West Point, made an inspection tour of European fortifications. Two years later, he published *Elements of Military Art and Science*, a popular military text based on Jomini's *Art of War*. In 1862, Jomini's book was translated into English and went on to become a text at West Point and other American military institutions for many years. During the Civil War, Halleck rose to become the Union army's chief of staff in Washington, D.C. The Civil War was the first opportunity for Americans to employ mass armies on a grand Napoleonic scale. To employ these armies effectively, American officers learned through hard experience about the operational level of war.

The Anaconda Plan of 1861 established the North's initial grand strategy. Maj. Gen. Winfield Scott, commanding general of the United States Army in 1861, recommended blockading the rebellious Southern states and splitting the Confederacy by seizing the line of the Mississippi Valley. Eventually, the Northern strategy settled on a direct advance to Richmond, the Confederate capital, and the destruction of Southern armies. The organization and employment of large armies in the field required a new level of sophistication in the conduct of operations—the emerging level of war between grand strategy and tactics.

By 1862, both the North and the South had organized their armies into corps, but of greater significance was the organization and coordination of the armies in the field. The American Civil War saw several innovations in the development of operational art. The employment of several independent field armies distributed in the same theater of operations, the use of quasi–army group headquarters to control armies, the design of a distributed campaign plan, and deep strike operations all suggest a new and sophisticated level in the conduct of operations.[23]

In 1863, the North created the Military Division of the Mississippi under the command of Maj. Gen. Ulysses S Grant. This headquarters functioned essentially as an army group headquarters responsible for the Army of the Tennessee under Maj. Gen. William T. Sherman, the Army of the Cumberland under Maj. Gen. George H. Thomas, and later, the Army of the Ohio under Maj. Gen. John M. Schofield. These were independent armies within the same theater of war coordinated by a single commander for a common purpose. General Grant's campaign plan for 1864 presented an ideal example of the synchronization of forces for a unified strategic purpose. In a letter to General Sherman, he expressed his intent "to work all parts of the army together, and toward a common center."[24] Rather than independently pursuing the destruction of an enemy army or the capture of territory in the various theaters of operations, the Union high command had learned the value of synchronized operations.

In addition, the United States Army had begun in 1864 to employ large formations of cavalry in deep strike missions for operational objectives. Though both sides had sent their cavalry on such raids throughout the war, perhaps the most impressive was Maj. Gen. James H. Wilson's mounted thrust into Alabama and Georgia in 1865. General Sherman sent Wilson with a cavalry corps of more than thirteen thousand troops on a deep raid to deflect attention from Sherman's invasion of South Carolina.

By the end of the Civil War, the North was demonstrating considerable skill in the synchronization, command, and employment of large armies at the operational level of war. However, most of this expertise passed with this generation of officers. The security needs of the reunited nation called for small and isolated garrisons to police the western frontier. The professional army returned to constabulary duties and focused on small unit tactics rather than the study of the higher levels of war. Only the need to field large armies during World War I again forced American officers to think about the operational level of war, and by then, they would be learning from European professionals.

The German wars of unification from 1866 to 1871 again saw the collision of massive armies. The success of the Prussian army was built on the professional expertise of Helmuth von Moltke. Moltke, who was born in Denmark in 1800, served in the Danish army until joining the Prussian service in 1822. His career in the Prussian army was unspectacular until a series of staff appointments brought him into close contact with the royal family. In 1857, Moltke was appointed chief of the Prussian general staff, and over the next fifteen years he guided the Prussian army through a spectacular series of victories. In the Austro-Prussian War (1866), Moltke coordinated the deployment of three Prussian armies, concentrating the entire force on the battlefield of Koniggratz, decisively defeating the Austrian army. In the Franco-Prussian War (1870–71), he coordinated the deployment and operation of three armies consisting of more than 309,000 troops. In a series of battles culminating in the Battle of Sedan in September 1870, the Prussians and their German allies defeated the French army and brought down the French government. Continued French resistance led to the siege of Paris, but ultimately the Germans prevailed and the war ended with the proclamation of the German Empire at Versailles on January 18, 1871.

The key element in Moltke's conduct of operations was his efficient mobilization and concentration of large number of troops. The Prussians made the most of new technologies, particularly railroads and the telegraph, but it was the development of "the first deep-future oriented war planning system" that gave Prussians the critical advantage.[25] The Prussian general staff under Moltke's supervision became an extremely capable organization for the efficient planning and execution of the mobilization and deployment of mass armies. When combined with Moltke's competent generalship, these advantages overwhelmed the French.

Moltke described strategy as a "system of expedients."[26] He noted that "strategy governs the movements of the army for the planned battle; the manner of execution is the province of tactics."[27] Moltke used the term "operations" to describe the movement and deployment of troops prior to the decisive battle.[28] In fact, Lt. Gen. Hugo Freytag-Loringhoven noted, "In the German Army, then, starting in the general staff, the employment of the term 'strategic' has fallen more and more into disuse. We replace it, as a rule, by the term 'operations' and thereby define more simply and clearly the difference from everything that is referred to as tactical."[29] At the heart of Moltke's operational system was the pursuit of the decisive battle. In his "Instructions for Large Unit Commanders," written in 1869, Moltke observed that "war must attain the goal of the government's policy by force of arms. The battle is the great means to break the will of the opponent. This intent is the basis of all large and small engagements."[30] The main features of Moltke's operational art were the linking of mobilization to campaign planning and the ability to move and concentrate large armies for decisive battle.

European armies noted the lessons of the German wars of unification, specifically the importance of mobilization, the continuing relevance of the decisive battle, and most important, the need for even bigger armies. By 1914, the armies arrayed for the initial clashes of the Great War totaled 3.3 million men. Through four bloody years, the armies conducted operations aimed at achieving the decisive battle. The very size of the armies precluded their defeat in a single battle. Only the slow process of attrition proved capable of providing victory through the exhaustion of the enemy's resources and national will. As soon as the war came to an end, military thinkers began to ponder the new lessons of warfare. In the aftermath of the Great War, the professionals began to understand more completely the impact of the expanded battlefield, industrialization, and mass armies. In this process, the nature of modern operational art became more clearly defined in a new framework for the art of war.

In 1926, the British military theorist J. F. C. Fuller expressed a modern view of the three levels of war by dividing the art of war into grand strategy, grand tactics, and tactics. He wrote that "grand strategy secures the political object by directing all war-like resources— moral resources—moral, physical, and material—towards winning the war."[31] The duty of the grand tactician is to take "over the forces as they are distributed and arrange them according to the resistance they are likely to meet."[32] His fellow countryman and theorist Sir

Basil Henry Liddell Hart retained the use of the word strategy to describe operations, but he further refined grand strategy as "the policy governing its employment and combining it with other weapons: economic, political, psychological."[33]

Modern operational art found its clearest expression among Soviet theorists. The Soviet army struggled not only with the lessons of World War I but also with those of the Russian Civil War. The Soviet concepts of operational art evolved from the thinking of several men, A. A. Svechin and M. N. Tukhachevsky foremost among them. In 1923, Svechin proposed that operational art was the "totality of maneuvers and battles in a given part of a theater of military action directed toward the achievement of the common goal, set as final in the given period of the campaign."[34] Further, he established the relationship between operations, tactics, and strategy: "tactics makes the steps from which operational leaps are assembled; strategy points the way."[35]

Tukhachevsky's analysis of World War I led him to develop many key operational concepts. He recognized that technology expanded the battlefield, thus requiring successive and deep operations.[36] In order to penetrate enemy forces, particularly where those forces are arrayed in a continuous front, a series of battles combined with operations to strike deeply throughout the enemy's positions would be necessary. In fact, the concept of deep operations was the greatest achievement of Soviet interwar operational art. With the onset of Stalin's purges in 1936, however, innovative Soviet military thinking came to an abrupt halt.[37]

As war clouds gathered in the late 1930s, there emerged a fairly clear theoretical understanding of the three levels of war. Strategy or grand strategy dominated the highest direction of war and encompassed how the nation would organize and employ all its resources. The Germans classified the level below strategy as pertaining to operations. Likewise, Soviet theoreticians believed that the operational level of war involved the linking of major operations or battles to achieve strategic objectives. Since the beginning of military history, the understanding of the term tactics has been constant. Strategy, as we have seen, has had several meanings. Beginning with the introduction of large armies in Europe in the eighteenth century, strategy also came to be used to describe not only the higher direction of war but also the conduct of large-unit operations in the field. It was not until 1982 that the United States Army officially designated the third level of war as the operational level. Long before this official U.S. appreciation of the concept, however, American forces conducted large-unit operations with mass armies, with the roots of American operational art traceable to the Civil War. Modern American operational art dealing with mass industrial armies supported by airpower and mechanization, like that of European armies, was not developed until the period between the world wars.

By the end of the nineteenth century, the art of war could be divided into strategic art, operational art, and tactical art. Strategic art at the highest level has not changed much over the centuries. The Periclean strategy of defense, reliance on maritime strength, was as appropriate to Athens in the Peloponnesian War as to England in 1940, when Britain held out alone against Germany while hoping for American intervention. Tactical and operational art, on the other hand, have always been much more influenced by technology and military organization. Tactics, in fact, is how technology in the form of weapons can best be employed on the battlefield. Operational art, on the other hand, changes more slowly, evolving along with major shifts caused by technology by combining with advances in military doctrine, theory, and organization. Operational art is best defined as the use of force to achieve strategic objectives in a theater of operations or a theater of war. As the nature of the force and the means to employ it changes, so too does operational art. The maneuver and coordination of armies for a coherent strategic purpose evolved over the course of the nineteenth and twentieth centuries. The techniques of decisive battle, linking mobilization to deployment, and broad lateral maneuver by corps or armies were all a part of early operational art.

Modern operational art as it emerged at the end of the interwar period was characterized in theory by successive and deep operations, synchronized major operations and battles, and combined forces, including airpower. The American army's performance in World War II displayed all of these characteristics of modern operational art. Like their European counterparts, officers in the U.S. Army studied the lessons of World War I. In 1918, the United States sent more than 2 million troops to Europe. By the end of the war, the Americans had organized two armies and were in the process of forming an army group headquarters. As in the past, the peacetime American army experienced significant reductions; but unlike the army's leadership after the Civil War, U.S. officers who gained experience in World War I continued to think about and study modern war during the 1920s and 1930s. The origins of modern American operational art can thus be found in the military educational institutions of the interwar years.

2

OPERATIONAL ART IN THE AMERICAN ARMY BEFORE 1919

British historian Michael Howard has said that the First World War lies like a giant scar dividing our world from that which existed before 1914.[1] This is true not only for the general political, social, and economic development of western Europe, but for warfare as well. The face of war by 1918 was drastically different from the expectations and initial conditions of 1914. Another historian suggests that a battalion commander who marched off to war in August 1914 would barely recognize the routine of combat by 1918. The employment of aircraft and tanks, the dominance of the machine gun and artillery, and the conditions of prolonged trench warfare were completely foreign to his prewar training and education. By contrast, the battalion commander of 1918 would recognize the key features of combat some two decades later in 1939. Again massive European armies would be engaged, with merely better versions of the technologies introduced in World War I.[2] This transformation was achieved as the great European powers struggled to find new solutions to the military problems that confronted them in the Great War. For the Americans, who came late to this titanic struggle, it meant a wrenching transformation from an army that was at the outset little more than a nineteenth-century frontier constabulary to a modern military capable of challenging European armies on their own soil.

Theory and Practice

Theory and practice in the U.S. Army before World War I were shaped by the requirement to police newly acquired overseas possessions, the influence of European military trends, and a growing sense of professionalism. Altogether, the army's small size and its frontier and colonial missions slowed its progress in moving toward a modern conception of warfare. The reduction of the U.S. Army after 1865 had been dramatic. At the height of the war in 1865 federal troops numbered over a million officers and men, but within a year, shed of its volunteers, the Army counted only 57,000 regulars. The regular army continued to shrink. By 1877, Congress reduced it to 24,000 officers and men, and it rarely rose above 27,000 total troops until the outbreak of the Spanish-American War.[3] Until the end of the nineteenth century, this small army was scattered in small posts and absorbed in constabulary duties on the western frontier. After 1898, the army helped secure America's newly acquired overseas possessions.

In the last quarter of the nineteenth century, an increasing sense of professionalism gripped the American officer corps, despite its small size.[4] Journals, professional associations, and most important, a postgraduate school system sharpened American military professionalism. For practically all of its existence, the American army had served as a frontier constabulary. With the official close of the frontier in 1890, various officers began debating the army's mission in their new professional journals. They found a partial answer in the requirement to garrison and defend the United States' new overseas possessions resulting from victory in the Spanish-American War. The shift from a frontier constabulary to a colonial army required thinking about defending American possessions from modern foreign armies of potential aggressors. By 1899, the secretary of war, Elihu Root, determined "that the real object of having an Army is to provide for war."[5] Budgetary constraints and a traditional distrust of a peacetime military hampered leaders' efforts to prepare the army for modern warfare at the turn of the twentieth century.[6] Yet if the army could not maintain a large modern force, nor could it modernize the thinking of its officer corps through education and keep up with the latest military trends in the most advanced European armies.

There were many technologies at the end of the nineteenth century to catch the eye of military professionals: railroads, machine guns, quick-firing rifles, and artillery were among the most notable. Just as important, though, were the organizational advances necessary for managing the growing complexity of modern armies. Michael Howard notes that "the greatest military innovation of the nineteenth century was not technological, but rather the organizational institution of the general staff."[7] National general staffs and field staffs both arose in response to the increasing challenges of waging large-scale war in the nineteenth and early twentieth centuries. Competent staffs were crucial to the evolution of modern operational art. Moving, sustaining, and coordinating large armies in a theater of operations was complicated by the need to conduct joint and combined operations that integrated the new technologies. The Prussians led the way with the creation of a national general staff in 1806; this group evolved over the years into the Great German General Staff, a model much studied by other European militaries. After 1814, the Prussians established a general staff in Berlin and distributed other staff officers to the field staffs in corps and divisional commands. In July 1828, the Prussian army issued directions for the organization and responsibilities of the field staffs.

As the Prussian general Bronsart von Schellendorff observed, it was clear by the late nineteenth century that "the enormous numerical strength of modern armies, and the way they must be organized to meet the constantly changing requirements of war, render necessary great differences in carrying out the details of military operations. Consequently, the higher leaders and commanders require a regular staff of specially selected and trained officers."[8] Educating staff officers for large modern armies became a focus of military education in the nineteenth century. An imperial order on November 21, 1872, placed the Prussian Kriegsakademie under the Chief of Staff, and consequently the college was "regarded, to a certain extent, as a training establishment or school for the General Staff."[9]

All the major powers of Europe recognized the need to educate and train officers not only to lead the ever larger armies but to manage them. The British Staff College opened in 1858, and twenty years later the French established the École supérieure de guerre after the failure of French arms in the Franco-Prussian War. In America, the development of the national and field staffs, along with institutions to supply them with trained officers, evolved more slowly in the years before the country's entry into World War I.

From 1869 to 1883, Gen. William Tecumseh Sherman reigned as a dominating figure in the "Old Army." Sherman believed in small staffs and had little use for a chief of staff.[10] In 1881, however, Sherman directed the establishment of the School of Application for Infantry and Cavalry at Fort Leavenworth, Kansas. In its early years this school aimed at training and educating company grade officers for routine duty with troops. After problems in military administration became apparent during the war with Spain, the War Department recognized the need for significant improvements. Elihu Root, a successful New York lawyer, became secretary of war in August 1899 and during his tenure instituted important and long-lasting reforms. Among these reforms were the creation of a national general staff, replacement of the commanding general with a chief of staff to direct the national staff, and the establishment of a postgraduate military education system. This system included the General Service and Staff School at Fort Leavenworth and the Army War College in Washington, D.C.

By 1907 the army introduced a two-year course of instruction at Leavenworth to address its need for both competent tactical commanders and staff officers. The first-year course, the Infantry and Cavalry School, became the School of the Line. A second-year school, the Army Staff College, lasted until 1916. Following World War I, the Army Staff College was renamed the General Staff School, and in 1923 it became the Command and General Staff College. First year students studied troop leading and worked on practical problems in brigade and division operations. The faculty then selected about half of the students to go on to the second year in the staff course. In the staff course, the curriculum focused on general staff duties and corps operations. Competition for selection to the staff course was intense and considered an honor.[11] At the top of the postgraduate military system was the Army War College. The Army War College located in Washington, D.C., opened its doors to students in 1904. Its purpose was to become "a postgraduate course for the study of the greater problems of military science and national defense."[12]

Practice in the United States Army prior to U.S. entry into World War I was driven by the requirement to police the new overseas possessions gained in the war with Spain. Until 1908, 20 percent or more of the army served in the Philippines.[13] The rest of the army drilled and maintained itself in small stateside garrisons. The Punitive Expedition in 1916 focused the Army on Mexico while some of the largest battles in history were fought in France. Strategic theory, meanwhile, received little mention in the professional journals. From 1889 to 1905, only three articles dealing with strategic theory appeared in the monthly *United Service Journal*. There were discussions of the strategic and tactical lessons from the Boer and Russo-Japanese wars, but these articles offered little beyond a conventional analysis of lines of operation and observations on the relative power of offensive versus defensive operations.

As already noted, the army's new postgraduate military schools gave more attention to theory. American military thought before World War I focused on tactics and closely followed trends and developments in Europe, particularly in Germany. Likewise, the American understanding of military strategy or operational art followed European thought. The dominant operational paradigm in the nineteenth century was the pursuit of the decisive battle.[14] Driven by the image of Napoleonic warfare, the Moltkean examples of Koniggratz and Sedan, and the increasing geostrategic need for quick victory, European military professionals in the decades before 1914 still believed in the possibility of smashing enemy armies in battle. Some historians argue that European professionals did grasp many of the implications of the changing nature of warfare. For example, advances in weapons lethality and the potential of railroads to support and move the new mass armies were seen as means to expand the single battle of annihilation into decisive combat throughout the theater of operations. Field Marshal Alfred von Schlieffen, chief of the German General Staff from 1891 to 1905, referred to this concept as the *Gesamtschlacht*, or total battle. Schlieffen, the architect of the German plan to invade France in 1914, viewed combat throughout the theater of operations as part of one battle.[15] Germany's geostrategic dilemma of a two-front war against Russia and France still required that the expanded battle in either theater must be decisive to achieve quick victory. European operational art at the turn of the twentieth century consisted of maneuvering the army or armies in the theater of operations so that the conditions for decisive battle could be favorably achieved. To the extent that the theory of large-unit operations and theater strategy was studied at all in the U.S. Army at the beginning of the new century, it was studied in the postgraduate school system.

Two of the most influential figures in the development of the Leavenworth schools were Col. Arthur L. Wagner and Maj. Eben Swift. Wagner graduated from the United States Military Academy in 1875 and served in several campaigns against the Sioux and Ute Indians. In 1886, he joined the faculty of the Infantry and Cavalry School and, despite occasional interruptions, over the next seven years helped to shape the curriculum. As head of the Department of Military Art, strategy and tactics were his main concerns. Although chiefly focused on tactics, he gave some thought to the higher levels of war. In a lecture on strategy given to officers of the regular army and National Guard gathered for maneuvers at West Point, Kentucky, in 1904, he declared, "The art of war is broadly divided into the two subjects of strategy and tactics. Strategy is the art of moving an army in the theater of operations, with a view to placing it in such a position, relative to the enemy, as to increase the probability of victory, increase the consequences of victory or lessen the consequences of defeat."[16] Wagner went on to quote Jomini, Clausewitz, and a host of other European theorists.

Wagner's concept of strategy derived from a noted English military educator and author, Sir Edward Bruce Hamley. Hamley was the first professor of military history at the newly established British Army Staff College, where he taught from 1859 to 1865. The new military staff colleges needed texts, and these were often provided by the faculty. For his part, Hamley published *The Operations of War* in 1866, which became the sole official text for the British Staff College until 1894. For the next half century it served as a standard text in staff colleges around the world.[17] The demand for texts in English secured it a place in the curriculum of both the U.S. Military Academy and the Leavenworth schools for many years. Hamley's approach to military strategy was essentially Jominian. He illustrated lines of operation and communication as well as interior and exterior lines from Napoleonic and modern campaigns. The text stressed the importance of logistics and maneuver rather than battle in waging short, decisive campaigns.

The American concept of operational art prior to World War I was a mixture of Jomini and Clausewitz as distilled from European texts and articles and reflections on the Civil War experience. Wagner co-authored a text on strategy that drew heavily on Hamley's work but included more American campaigns. Assisted by Swift, Wagner wrote *Strategical Operations, Illustrated by Great Campaigns in Europe and America*, which was published in 1897. The authors admitted that "the basis of this work is that portion of Hamley's *Operations of War* relating to the subject of strategy. Some of the descriptions of campaigns are taken verbatim from Hamley; others are revised and rewritten either wholly or in part, and others again are entirely new."[18] As evident in Wagner's text, the American concept of military strategy was the movement of armies in a theater of operations. Still, Wagner does differ from Hamley in his emphasis on battle. Wagner insisted that "the enemy's main army is always the true objective, but there will often be intermediate objectives as necessary steps in reaching the ultimate objective."[19] Wagner also wrote another Leavenworth text, *Organization and Tactics*. This text also reaffirmed that "all strategical operations must terminate in battle."[20] Wagner read Clausewitz and shared his views on the importance of battle. One historian even described Wagner as an "early American disciple of Clausewitz."[21] The exposure of American military thinkers to Clausewitz became even more important in the interwar years.

American military thought continued to be heavily influenced by trends in Europe, but one of the more original American works from this period is John Bigelow's *Principles of Strategy*, published in 1891. Bigelow's purpose was to "discuss the subject of strategy in the light of American warfare, and thus furnish instruction for Americans, not only in the theory of this subject, but also in the military history and geography of their own country."[22] Bigelow used examples from the Civil War to describe political, regular, and tactical strategy. The author differentiated political and regular strategy based on the objectives. Bigelow used Sherman's Atlanta campaign to illustrate political strategy, which aims at the destruction or coercion of the opposing government as the military objective. Regular strategy has the enemy army as its objective and employs Jominian geometry and theory to get at it. This is certainly a more accurate reflection of the American Civil War experience, but also a recognition of the increasing trend toward total war in the late nineteenth century. In any event, foreign conceptions of warfare continued to dominate both the theory of war and the way officers were educated and trained.

The Leavenworth schools prior to the First World War also set the methods and provided many of the tools for officer education. While serving as Wagner's primary assistant in the Department of Military Art, Eben Swift introduced the applicatory method of instruction in 1894. Swift adopted this method of instruction from the Germans. This approach included war-gaming, staff rides, and map exercises. In addition, Swift introduced German troop-leading procedures including standard methods of issuing orders, such as the five-paragraph field order.[23] German influence increased at Leavenworth when Wagner's text *Organization and Tactics* was replaced in 1907 by Albert Buddecke's *Tactical Decisions and Orders*, Otto F. Griepenkerl's *Letters on Applied Tactics*, and von Schellendorff's *Duties of the General Staff*.[24] Despite the evolution of tactics and operations over the years, these German methods of instruction proved enduring. In fact, these methods remain the foundation of military instruction in the United States Army and have been extended to embrace the operational as well as the tactical level of war. The problem with the army's initial use of the applicatory method, however, was its preoccupation with practice rather than theory. Undoubtedly, the overriding concern was the need to train staff officers in the practical matters of management, but this did not leave much room for a systematic study of modern land warfare.[25]

In the last quarter of the nineteenth century and up to World War I, the Leavenworth schools varied their focus from time to time, but in the main they concentrated on tactics and the production of competent staff officers. Conducting and managing division and corps operations were the extent of large-unit study. There was no coherent view of large-unit or army operations beyond the Jominian legacy of battlefield geometry. The army conceived military strategy as the movement of an army in a theater of operations to achieve favorable conditions for decisive battle. Although logistics was understood as an important, indeed critical aspect of army operations, the Leavenworth schools developed no concept of phased operations or linking battles to achieve strategic objectives. Americans, like their European counterparts, still looked to the decisive battle as the means of achieving strategic success.

At the upper tier of the military's postgraduate system was the Army War College. At first, the Army did not intend to make the War College a military or service school, but rather a place where "problems involving military questions will be solved by groups of officers, [and] offensive and defensive plans will be worked out in a comprehensive way."[26] The War College functioned as an arm of the General Staff. These selected officers learned by working on real-world problems and plans. By 1890, both the army and the navy began to study and develop plans for a potential conflict with Great Britain.[27] Since its establishment in 1885, the Naval War College had been closely involved with the Department of the Navy in war planning. Likewise, the Army War College was charged with assisting the General Staff in studying contemporary strategic problems in which the officers selected to the War College served both as students and as apprentices to the General Staff.

Arthur Wagner and Eben Swift served in the War College much as they had at the Leavenworth schools. Wagner was appointed director of the War College for its first session in 1904, but death cut short his service. Swift arrived in 1906 fresh from the General Service and Staff School bringing with him the applicatory method of instruction. Swift, Maj. Gen. J. Franklin Bell (chief of staff, 1906–1910), and Maj. Gen. William Wotherspoon (president of the Army War College, 1909–1912) all served in senior positions in the Leavenworth schools. Together they changed the course of instruction at the War College to reflect the Leavenworth model with its increased emphasis on tactics, map problems, and staff rides. As General Wotherspoon remarked in his opening address to the War College class of 1911, the "course corresponds closely to that pursued in the Staff College at Leavenworth, and in Germany, which is the great model."[28] After 1908, the War College became more focused on instruction and less a functioning adjunct to the General Staff. There were, however, significant differences between the Army War College and the Staff School.

Unlike the Leavenworth schools, the War College was always intended to work closely with the Naval War College in the study of war planning and strategic problems.[29] Both colleges exchanged students and faculty and, on occasion, participated in joint studies. By 1917, the Army War College listed three naval officers and eleven Marine Corps officers among its graduates. Eventually, all the naval officers and seven of the marines achieved flag rank.[30] The requirement to protect overseas possessions underscored the need for service cooperation in projecting American power, and the War College recognized this need. As a memorandum drafted in late October 1909 put it: "The important subject of joint operations between the Army and Navy in oversea expeditions is discussed and studied in a series of lectures and practical problems involving the embarkation of expeditionary forces."[31]

On top of everything else, the War College worked with real war plans. Prior to 1890, American war planning typically only commenced with the initiation of hostilities. The example of the German General Staff suggested the need for future or contingency war planning in modern warfare. Although the army and navy began developing such plans in the 1890s, the experience of the Spanish-American War underscored the importance of joint cooperation in planning as well as execution. To meet this challenge, the Joint Board was organized in 1903 "for the purpose of conferring upon, discussing and reaching common conclusions regarding all matters calling for the cooperation of the two services."[32] The following year, the Joint Board approved a common list of designations for potential adversaries to be used in war planning. These designations referred to potential enemies by color: Red—Great Britain, Black—Germany, Orange—Japan, Green—Mexico. Gradually, this convention resulted in the color plans that dominated American war planning up to World War II. In the prewar period Orange, Red, and Green were the most frequently studied and exercised war plans at the War College.

The War College intended to prepare officers for "the higher duties of command" by studying "the tactical and strategical handling of troops, with special reference to those including and larger than a division." In actual fact, however, students received little instruction in large-unit operations. The pre–World War I army lacked a permanent corps structure. The land forces were divided into the mobile army and the coast artillery. The division became the basis of organization for the mobile army. When needed, divisions could be grouped into field armies. If several field armies operated in the same theater of war, they might be organized into full armies.[33] The War College maintained its tactical focus down to World War I.[34] The list of map problems for the 1909–10 course consisted of thirty-four division operations, five army operations, eight overseas operations, and five strategic problems. The same list indicated that only nine of the sixty-one map problems dealt with matters of supply.[35]

The exercises specifically designated as strategic map maneuvers, such as Problem No. 17 for the 1914–15 course, normally involved a field army consisting of four divisions. The problem consisted of how best to concentrate this corps-sized force in order to then conduct a movement to make contact with the enemy, followed by a battle.[36] Virtually all of the strategic map problems and exercises of the prewar period involved only field armies with four or fewer divisions; there was no evidence of phasing, just concentration, movement, and battle. There was also little attention paid to matters of supply. The map problems required students to perform a mission analysis, consisting of a statement of the mission, comparison of enemy and friendly forces, and development of courses of action and then to propose a decision on the proper course of action. In the course conducted in 1916, the last before America entered the Great War, the map problem dealing with the Red War Plan suggested the limited scale of U.S. exercises. In this case, the problem called for the army to field 196,000 men and 472 guns in seven divisions to seize key points in Canada.[37] By comparison, the British and French actually committed 750,000 men in three armies to the First Battle of the Somme, which raged from June to November that same year. The War College course did not change much in the last decade before the war. Although the War College sponsored lectures by U.S. observers to the European conflict, its map problems and exercises continued to focus on war plans for North America and on current tactical doctrine.

After witnessing three years of slaughter from afar, the United States was still unprepared for modern war, certainly in terms of equipment and training, and also in the education of its small officer corps.[38] The American army had made great progress in the decades before the war. The establishment by the War Department of a national general staff and the creation of a military educational system including a staff college and a war college all spoke to a greater sense of professionalism in the officer corps. Unfortunately, the educational system focused almost exclusively, and narrowly, on immediate American strategic and tactical requirements rather than a wider consideration of modern warfare. Without great peacetime armies to exercise, and with only fading memories of the Civil War and more recent expeditionary experiences, the officer corps depended on European conceptions of modern war.

Even though the Europeans proved capable of managing, moving, and sustaining vast armies, their conception of modern war failed to accomplish strategic decision. The Europeans studied and even conducted large-unit exercises prior to 1914, and they still got it all wrong. The United States Army had a long way to go even to conceive of the scale of modern war. At the Leavenworth schools and the War College alike, concepts of large-unit operations—armies and army groups—operational or strategic theory were little studied or exercised. The applicatory method embraced at both schools emphasized the practical and was best applied to the tactical level of war. Nevertheless, the U.S. Army had made a good start. The War College recognized the need for joint cooperation, and both schools provided a solid foundation for staff officers in the development of plans and the conduct of operations. Unfortunately, the realities of modern warfare would have to be learned in the hard school of experience. World War I would transform the American officer corps' understanding of war just as it transformed war itself.

The U.S. Army in the Crucible of Modern War

The United States declared war on Germany on April 6, 1917. Fifty-two days later Gen. John J. Pershing and a small staff of six officers stepped aboard the SS *Baltic* and headed for Europe. Pershing, the commander of the American Expeditionary Force (AEF), represented the American military experience and education of the preceding thirty years. Prior to 1917, he had fought against Indians in the American West, the Spanish in Cuba, and the Moros in the Philippines. A 1905 graduate of the Army War College, Pershing served as an observer in the Russo-Japanese War of 1905, and he led the Mexican border campaign in 1916–17. Now as the AEF commander, he faced his greatest challenge, as did the army that had produced him. Pershing fully grasped the urgency of the situation if America was going to make a significant contribution to winning the war.

Despite many years of war-gaming contingencies at the War College, the U.S. Army possessed no plan for how America might contribute to the Allies, how an expeditionary force might be organized, or even how the War Department itself might be expanded. The officers gathered around General Pershing were acutely conscious of the fact that the professional reputation of the American army would be tested on a world stage. American officers had long admired the efficiency of the German army. After years of using German tactical texts and methods, some were wary of taking on the kaiser's forces.[39] Whatever personal doubts Pershing's aides harbored, however, they quickly got down to developing an estimate of the situation. While aboard ship they concluded, "America must organize and put into France, armies, not divisions; a force of at least a million men should reach France within a year; guns and artillery ammunition for initial needs must be secured from the Allies."[40] From the beginning, Pershing envisioned large-unit operations with American formations under American command.

One of the first questions was where in France U.S. forces would be deployed, be organized, and fight. To a large extent, the answer would be determined by logistics as much as by strategy. British armies deployed to cover the channel ports while the French were arrayed in defense of Paris. The channel ports were fully committed to the support of British forces and French national needs. Only the ports in the southwest of France possessed the capacity to "supply the great forces deemed essential to win the war."[41] Adequate and clear lines of communication were critical to organizing and sustaining large-unit operations. The next question was what to do with the American forces once they arrived. The Allies were desperate for manpower and constantly clamored for the incorporation of U.S. troops and units directly into their own national forces. Secretary of War Newton D. Baker gave Pershing a mandate to cooperate but admonished him to keep "in view that the forces of the United States are a separate and distinct component of the combined forces, the identity of which must be preserved."[42] Pershing was firmly committed to the organization of an American army and to find an operational plan by which that army could be used decisively.

Building and organizing an American army capable of competing with European armies with already three years of war experience was no easy task. Shortly after Pershing's arrival in France, he tasked his G-3 Operations section to consult with the British and French staffs and propose an organization for American forces. A War Department panel known as the Baker Board was already in France working on the problem. Pershing's staff cooperated with the Baker Board and developed the General Organization Project. Published on July 10, 1917, this proposal became the basic blueprint for the American Expeditionary Force. The project asserted that "it is evident that a force of about one million is the smallest unit which in modern war will be a complete, well-balanced and independent fighting organization."[43] The plan called for the organization of five corps of six divisions each, with two divisions in each corps to be replacement divisions. The plan projected the total strength of the AEF as 1,328,488 in thirty divisions. The study noted that this force should reach France by 1918, but that at least three million men might eventually be required. This number included army, corps, and support troops.[44]

Pershing recognized the need for a modern staff to manage this large force. American general staff organization called for operations, intelligence, and administrative sections. After studying Allied staffs, Pershing decided to adopt the French army's system, which included staff sections for personnel (G-1), intelligence (G-2), operations and training (G-3), and supply (G-4). The difficulties encountered in training raw units shipped from the United States led to the creation of a separate section for training (G-5). Eventually, the G-5 supervised an entire military school system in France, providing training in virtually every critical skill from cooking to staff work. The general staff organization that was replicated in AEF corps and divisions remains the basis for modern staff organization. In addition to the general staff, the AEF formed a large technical staff to manage engineering, medical, transportation, aviation, tank, and other elements of modern warfare. The critical importance of railroads led to the creation of a transportation division headed by Brig. Gen. William Atterbury, a Pennsylvania railroad executive in civilian life.

The immense logistical challenge of shipping, deploying, moving, training, and sustaining a million men or more was characteristic of modern war. The General Organization Project of 1917 suggested that 20 percent of the American force in France would be dedicated to logistics, to maintain the lines of communication. In fact, it grew to 329,653 soldiers, or 35 percent of the total force.[45] Eventually, the organization charged with maintaining the lines of communication became known as the Services of Supply. Maj. Gen. James G. Harbord, a trusted confidant of General Pershing, commanded this organization. As commander of the Services of Supply, Harbord supervised the chiefs of procurement, transportation, supply, and construction. The logisticians divided the rear area into nine base sections to receive supplies from ports, an intermediate section in the center of the area for storage, classification, and transshipment of supplies, and an advance section for the distribution of supplies in the zone of operations.[46]

Modern war meant large armies that needed large staffs. The General Headquarters staff mushroomed from 186 officers and men in 1917 to 1,414 within one year, and with the addition of the supply and administrative departments it grew to three times that number: 4,271 officers and men.[47]

Modeled on European practice and experience, the AEF held the key to shifting the military balance to the Allies. Since 1914, the war had required unprecedented and ever increasing levels of military effort. The initial German plan to invade France, generally referred to as the Schlieffen Plan, involved seven armies comprising 2 million men. By 1918, the Germans had massed thirteen armies in four army groups on the western front, where they were opposed by three Allied army groups totaling thirteen armies as well. Both sides had reached the limits of their national resources. Germany's strategy depended on quickly ending the war in the East and hurling the remainder of its strength against the Allies in the West. The Allies had to hold on until the American army could organize itself into a modern effective fighting force. How this American force would be employed was a contentious issue for the Allies.

Shortly after General Pershing arrived in France, he directed the operations section of the newly formed General Headquarters (GHQ) to make a study for the future employment of the AEF. The section was to seek "a vital point where a quick telling blow could be struck, a blow which would strike against the whole German system on the Western Front."[48] By September 25, 1917, this small group of officers produced "A Strategical Study on the Employment of the AEF against the Imperial German Government." The report's authors identified two critical sectors on the western front: Saarbourg-Metz in northeastern France and Hirson-Lille in the northwestern portion of the country. It also offered this prescient assertion: "For a successful conclusion of the war, Germany must strike a decisive blow against the Western Front prior to the Fall of 1918."[49] The study further predicted that "unless internal disorders appear in Germany, it does not seem probable that the Allies can make a large offensive in 1918 with much chance of success."[50] The study anticipated only defensive or minor offensive operations in 1918 while the American army gained strength, organization, and experience. Decisive operations would take place in 1919.

The study recommended that the American army eventually take over part of the Saarbourg-Metz sector and use Nancy as a major base of operations. A detailed analysis of the rail system supporting the German forces in France suggested that seizing or destroying the two lateral railroads running behind the German lines in the vicinity of Metz would separate the right and left wings of the German army and "might well compel the evacuation of practically all territory West of the Rhine."[51] Striking in this region had the additional advantage of securing the valuable iron ore fields in the area. Metz, however, was heavily fortified, and so striking to either flank of the city was preferable. Regardless of where the Americans attacked in the direction of Metz, the St. Mihiel salient would have to be eliminated first to protect the flank of the attacking forces.[52]

Pershing's staff was a tight-knit group dominated by Leavenworth-trained officers.[53] Within four months of arriving in France his staff had studied the problem and recommended an operational objective that would guide the general's efforts to get American forces into the fight. A host of problems highlighted by logistics, a lack of organization, and German offensives hindered his ability to build an American army capable of achieving this operational objective.

In 1917, as America shipped troops to France and began the process of organizing and training its forces, the Germans knocked Russia out of the war. The German High Command, now led by Field Marshal Paul von Hindenburg and Gen. Erich Ludendorff, began shifting forces to the western front for decisive operations against the French and British. On March 21, 1918, the Germans opened a series of major attacks aimed at finishing the war before the American forces could tip the balance in the Allies' favor. The Germans committed three armies in a massive blow that collapsed the British Fifth Army and threatened the entire front. Four days after the beginning of the offensive, Pershing offered to delay the formation of U.S. corps and to provide Gen. Henri-Philippe Pétain, the commander of the French armies, with any serviceable divisions available.[54]

In this crisis, the Allies reached several significant agreements. Pershing's offer of American troops led to the early commitment of several U.S. divisions and supporting elements to combat in the next few months. Desperate for manpower, the French and British insisted that priority be given to American infantrymen and machine gunners in the shipment of forces to France. The result of this agreement postponed the organization of an American army, which required a host of supporting combat and service units, such as engineer, artillery, signal, medical, and quartermaster units. Pershing and Secretary of War Baker recommended approval as long as the Allies understood their intention to form an American army as soon as possible.[55] Finally, the crisis provided the impetus for a unified Allied command. On March 26 at Doullens, France, the British and French agreed to give Marshal Foch coordinating authority over their armies. Eight days later, the Allies better defined Foch's authority by giving him the title Commander in Chief of Allied Armies in France. The Allies specifically charged Foch with "strategic direction of military operations." The commanders of the national forces would have "tactical direction of their armies" as well as the right to appeal to their governments if they disagreed with Foch.[56] Pershing insisted the incipient American army be included in this agreement.[57] Under pressure from the German attacks, Foch became the theater and operational commander while Pershing, Field Marshal Sir Douglas Haig, and Pétain retained tactical direction of their armies.

Four more German offensives in the spring and early summer of 1918 drew American units into their baptism of fire, but no units larger than a division saw action. By the end of June, the AEF organized four corps headquarters, but not until July did the I Corps actually assume tactical direction of American divisions. On July 10, Pershing went to see Foch to secure his blessing for the formation of an American army and to lobby for a sector in the Château-Thierry region. Foch agreed in principle, but was vague on timing. In the meantime, the Third U.S. Division played a prominent role in stopping the final German offensive in front of the Marne by July 17. Ludendorff's great gamble to win the war failed and now the initiative passed to the Allies. Foch was soon planning counteroffensives to eliminate the German gains, and Pershing pressed to get Americans involved in the coming attacks. On July 18, the First and Second U.S. divisions under the French Sixth Army spearheaded the counteroffensive to pinch out the German salient near Château-Thierry. Later, six more U.S. divisions as a part of two U.S. corps operated under the French Sixth and Ninth armies. The success of this attack demonstrated the readiness of at least those American divisions and corps in the line.

On July 22, Foch agreed in writing to the organization of an American army in the vicinity of the St. Mihiel sector.[58] Two days later, the senior Allied commanders met at Bombon to set the course for the next series of offensive operations. Foch presented an outline for limited offensive operations that would set the stage for subsequent decisive battles. He was primarily interested in freeing Allied railways from German interference in three regions: the Paris–Avricourt railway in the Marne region, the Paris–Amiens railway, and the Paris–Avricourt railway in Commercy. This last objective required the reduction of the St. Mihiel salient and was assigned to the American army.[59]

On July 14, Lt. Col. George C. Marshall, until recently the G-3 Operations officer for the First Infantry Division, drove into AEF headquarters at Chaumont as a new member of the operations section of GHQ. The next morning, Brig. Gen. Fox Connor, G-3 of the AEF, walked into his new subordinate's room and told him to start planning for the reduction of the St. Mihiel salient. Soon to be promoted to full colonel, Marshall, was a graduate and a former instructor at the staff school. He now found full use for the time he spent at Leavenworth. Within days, Marshall became the operations officer for the First Army and began planning the first major American military operation of the war.[60] AEF headquarters issued the order to organize the First Army staff on July 24. Initially, just thirty-five officers and one hundred soldiers were available to fill out the staff, but it grew to more than six hundred officers over the next few months.[61] Organizing the First Army as a fighting force was a monumental task. American divisions were spread from Switzerland to the English Channel coast. Due to the decision to give priority to the shipment of infantry from American ports to France, there were few corps and army support units. In fact, only I Corps was organized and functioning at the end of July. The III, IV, and V Corps were organized but possessed no troops. The AEF Air Service had only six squadrons available and there were only three corps and army artillery brigades. Tank units were still in the process of organization and training. A good deal of the AEF's supporting artillery, aircraft, and tanks would have to be provided by the French.

The concentration of the American First Army in the St. Mihiel sector required significant logistics preparation. Engineers reconstructed more than forty-five miles of standard gauge and 250 miles of light railways. Nineteen railheads provided for daily supply and a stockpile of forty thousand tons of ammunition. The First Army communicated through telegraph and telephone lines, radio, and pigeons.[62] When the concentration was complete, the First Army included over six hundred thousand troops organized into four corps with sixteen available divisions. Combined French and American assets provided for 1,400 aircraft, 267 tanks, and 3,000 guns to support the attack.

This was the largest joint and combined operation conducted by the American army to date. The French provided 600 aircraft, 113 tanks, and much of the artillery. Pershing commanded the First Army but was under the direction of Gen. Henri-Philippe Pétain, the commander-in-chief of the French army. The First Army's instructions for the operation called for a main attack from the southern portion of the salient by I and IV Corps. A secondary attack would be made on the western side of the salient by the V Corps. The French Second Colonial Corps assigned to Pershing's First Army would attack the nose of the salient following the success of the main attack. The plan divided the main attack into four phases designating objectives for each echelon of the army over time. The first phase provided for an intermediate objective to which each division would reach as rapidly as possible without waiting for the advance of flanking units. Beyond this line, artillery could not provide a rolling barrage without moving forward. A second phase line was established for the corps. Each corps was to reach this line as fast as possible without waiting for units on its right or left. The third phase consisted of corps objectives for the second day. The final phase consisted of the drive to the army objective as directed by General Pershing. The secondary attack scheduled for the French Second Corps was also planned in two phases.[63] (See figure 1.)

The First Army G-3 issued the operations order for the St. Mihiel attack in the classic five-paragraph field order so often practiced in the Leavenworth schools' applicatory method. The order included annexes that described in detail the mission of the new weapons of war, aviation and tanks. The AEF Air Service divided aviation into pursuit, bombardment, and observation units. The pursuit groups were tasked with defending friendly observation assets and destroying hostile aviation to a depth of five kilometers behind enemy lines. Bombardment units were directed to attack railheads, command posts, enemy airfields, and bridges "at a medium distance from the zone of attack."[64] The army observation group provided photographic

Figure 1. First Army operations map for the St. Mihiel attack.
Source: Drum Papers, USAMHI.

and visual reconnaissance of enemy movements, concentrations, and
withdrawal. Unlike the particular concern shown for aviation, the
operations order simply allotted the tanks available for the attack to
the I and IV Corps.

Having occupied the salient for four years, the Germans had con-
structed a series of four or five defensive positions complete with a
dense network of barbed wire. Pershing described the German position as "practically a great field fortress."[65] The terrain in the western
edge of the salient, which ran along the eastern heights of the Meuse
River north of the town of St. Mihiel, was rugged and easily defended.
The southern face of the salient offered more promising open terrain,
which dictated the location for the main attack. At first, the Germans
committed eight divisions and one brigade to the defense of the salient. Concerned about the American buildup, Ludendorff then ordered the evacuation of the salient on September 8.[66]

At 1:00 A.M. on September 12, heavy artillery initiated the first
major operation by an American army in Europe. Four hours later, six
American divisions on the southern face of the salient went over the
top. The American troops advanced quickly, interrupting an orderly

enemy withdrawal and pushing the Germans out of the salient. By dawn on the thirteenth, the forces arrayed against the western and southern portions of the salient met, eliminating the salient altogether. Pershing ordered the attack to continue on to the army objective. The German troops' resistance stiffened as they withdrew behind the Hindenburg Line, a strong series of defensive positions. Although Pershing was convinced that a determined attack might penetrate the German defenses and open the way to Metz, Foch's directives committed the American general to an attack in the Meuse-Argonne sector.[67] The St. Mihiel attack, a major operation, had achieved its objective in Foch's theater strategy, but it was now time to turn to decisive operations along the entire western front. Even before the St. Mihiel operation was concluded, the American army began positioning for its attack in the Meuse-Argonne sector.

Lieutenant Colonel Marshall got the assignment to plan for the concentration of the army for this new operation. Overwhelmed with the responsibility, Marshall took a walk thinking "that I could not recall an incident in history where the fighting of one battle had been preceded by the plans for a later battle to be fought by the same army on a different front, and involving the issuing of orders for the movement of troops already destined to participate in the first battle, directing their transfer to the new field of action. There seemed no precedent for such a course, and therefore, no established method for carrying it out."[68] The Americans were thus introduced to an important facet of modern operational art. Foch was linking major operations in the theater to a single purpose, the preparation and conduct of decisive operations. repositioning portions of the American army still engaged in the St. Mihiel operation was certainly a challenge. It involved the "movement of approximately 500,000 men and over 2,000 guns, not to mention 900,000 tons of supplies and ammunition."[69]

The rapid concentration of the American First Army was made possible through the use of one of the new elements of modern war—motorization. Marshall arranged the concentration for each division by employing nine hundred trucks for the infantry and "by marching the artillery, motor supply trains and other vehicular transportation."[70] Marshall found that by using trucks and buses for transportation, he could move troops to the Meuse-Argonne concentration in a single night. Relying on horse-drawn transportation required anywhere from three to six days to move troops into the concentration

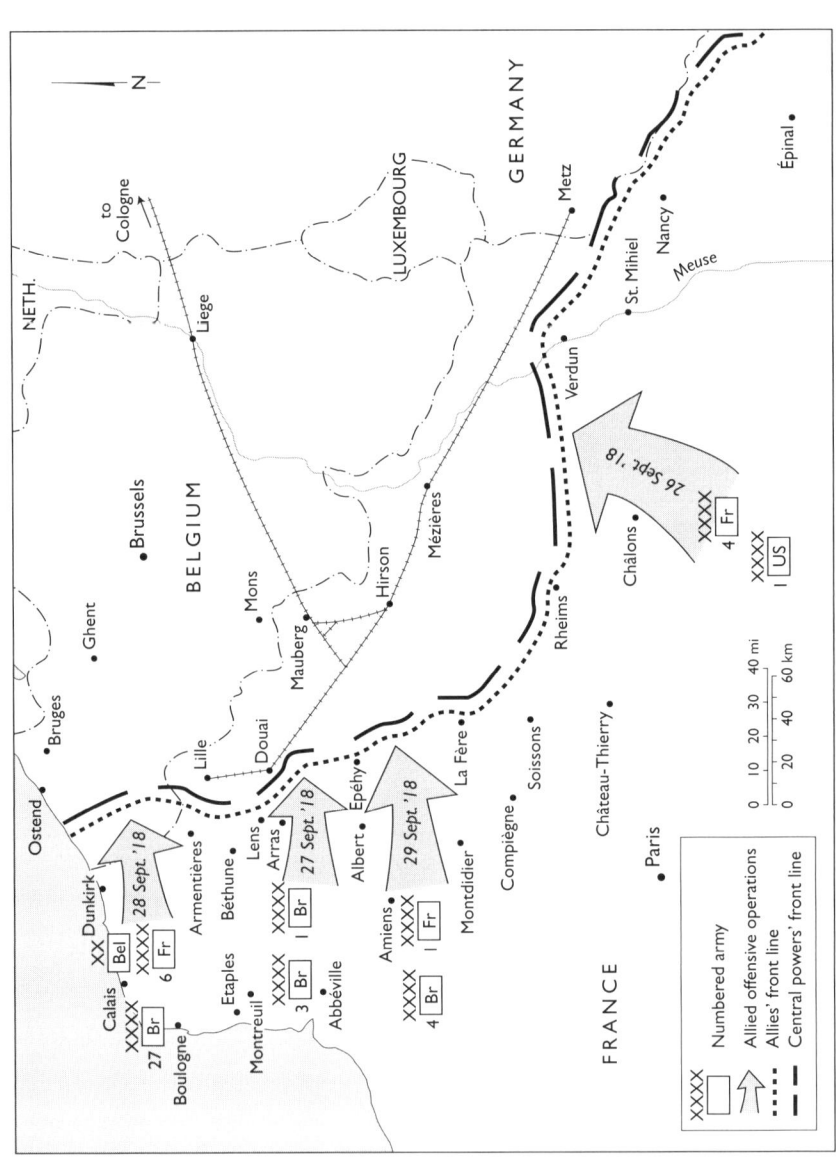

Map 1. The last Allied offensive of 1918.

area. In ten days, September 16–26, the American First Army muscled its way into position in the Meuse-Argonne sector.

Foch's plan for nearly simultaneous assaults along the western front called for four major operations launched by all the Allied armies. (See map 1.) The Allied commander-in-chief directed the British and Franco-American forces to make a large converging attack. The British Expeditionary Force (BEF) attacked toward Cambrai and St. Quentin, while the AEF in conjunction with French forces drove toward Mézières. These thrusts would seize the critical German lateral railroads as well as push out or bag several German armies. The specific mission assigned to the AEF was to attack northeast between the Meuse River on the east and the Argonne Forest in the west. The First Army's mission was to penetrate the Hindenburg Line and subsequently push toward the Stenay–le Chesne line.[71]

The Americans faced German Army Group von Gallwitz with eighteen divisions positioned along the front and twelve in reserve near Metz. Between the Meuse River and the Argonne Forest, the point of the American attack, the Germans had five divisions in the line. All the German divisions were greatly under strength and mostly of poor quality.[72] The Germans arrayed these divisions in a defensive zone consisting of four lines centered on dominating high ground. The Americans' zone lay astride the Meuse River valley, including the Argonne Forest on their left and the heights on both sides of the Meuse. The American First Army, which occupied sixty kilometers of front in this sector, had swollen to 890,000 men. According to Foch's plan, the French Fourth Army would attack alongside the AEF to the west of the Argonne Forest, also driving northeast as part of the general offensive.

The AEF had a total of fifteen U.S. divisions available and was given operational command of the French Second Colonial Corps and XVII Corps, an additional 120,000 troops. The French corps and elements of the U.S. III Corps occupied the lines east of the Meuse. By September 25, First Army's G-3 planners were prepared to employ this massive force in three operations. The first operation called for an advance of ten miles to force the enemy to abandon the Argonne Forest and connect with the French Fourth Army at Grandpre. The second operation called for a subsequent advance of ten miles to what Pershing called "the line Stenay to Le Chesne" in order "to outflank the enemy's position along the Aisne River in front of the French

Fourth Army and clear the way for our advance on Mezieres or Sedan."[73] The third operation was designed to clear the heights east of the Meuse River. The planners prepared this last operation with two branches or variations to be executed depending on the success of the main attack west of the river. (See figure 2.)

First Army Field Orders No. 20, dated September 20, 1918, again took the form of the now standard five-paragraph field order. The orders directed an assault with three corps on line. Each of the corps had three divisions in line and one in reserve. The army retained three additional divisions in reserve. As in the case of the St. Mihiel offensive, separate annexes covered the role of the tanks and aviation in detail. The tanks were "to destroy machine gun nests, strong points, and to exploit the success."[74] The First Army had only 189 light tanks (142 manned by Americans) available for the attack.

On September 26 at 5:30 A.M., the largest American army in history went over the top. Although Pershing hoped that the First Army might bull its way through the second German defensive position on the first day, the German defense tightened as the American attack lost its organization. The First Army pressed the attack and the Germans countered by throwing another six divisions by September 29 to reinforce their defenses. It became clear that new American divisions would be needed to resume the attack. Over the next two days, the First Army relieved and replaced three divisions in order to renew the general attack on October 4. Taking heavy flanking fire from artillery on the eastern heights of the Meuse River, Pershing directed the third operation attacking German defenses east of the Meuse. These attacks on October 8–10 made some progress but did not entirely eliminate the threat from the German artillery.

The First Army finally cleared the Argonne Forest on its left flank on October 10. Casualties had been heavy and progress slower than expected; Pershing reassessed the situation. The First Army required 90,000 replacements, but only 45,000 were on hand. Pershing, therefore, decided to break up arriving divisions to refill his depleted combat units. In addition, expanding the attack east of the Meuse and the growing strength of the AEF convinced Pershing to organize the Second Army under the command of Maj. Gen. Robert L. Bullard. Command of the First Army passed to Maj. Gen. Hunter Liggett, and Pershing became an army group commander on the same level as the other Allied senior commanders. Subsequent attacks finally got the

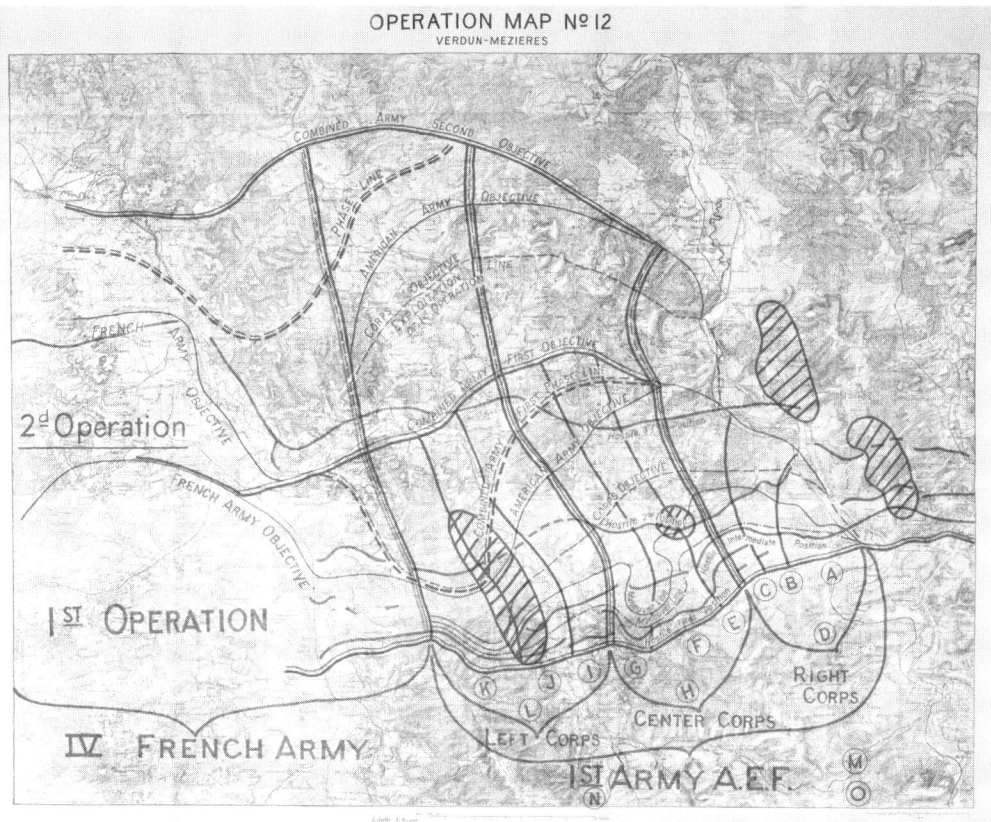

Figure 2. AEF operations map for the Meuse-Argonne offensive.
Source: Drum Papers, USAMHI.

American forces to their initial objectives by October 18. Major General Liggett scheduled the concluding attack for November 1 to coincide with an advance by the French Fourth Army. This powerful onslaught met with a German withdrawal and carried the American First Army up to the line of the Meuse from Sedan to Stenay.

All the Allied armies made rapid progress as the Germans continued to pull back along the entire front. The pressure of Foch's great counteroffensive achieved more than simply the seizure of critical German rail lines and the enemy's general withdrawal. It also cracked the German will to fight. Negotiations for an armistice had been under way since October 26. Finally on November 11, the guns fell silent as the Armistice took effect.

The Meuse-Argonne offensive was the last major operation of the U.S. First Army. It demonstrated not the AEF's mastery of modern war, but served as an initiation into modern large-unit operations. This operation by an American army included all the elements of modern war—the new technology of motorization, the airplane, and the tank. It required competent staffs and massive supply efforts to conduct combined and major operations to achieve the theater strategic objectives that ultimately proved decisive. The Meuse-Argonne operation lasted forty-seven days and cost 122,000 American casualties.[75] This seminal experience provided the Americans an understanding of the reality and the problems of modern operational art for the coming decades.

Assessment

No one associates World War I with creative generalship. Nineteenth-century operational art was unable to deliver stunning victories in the manner of Helmuth von Moltke in the German wars of unification. After the bold gamble of the German Schlieffen Plan failed, the new technologies and massive size of the armies prevented a repetition of those earlier victories. Moreover, the commanders' pursuit of victory through the elusive decisive battle invariably ended in stalemate and slaughter. Field Marshal Haig, commander of the BEF, conducted the Battle of the Somme in 1916 in the classic model of the Napoleonic decisive battle. Reflecting what he had been taught in the British Staff College, Haig conceived modern war as consisting of attempts to concentrate superior force against the enemy's principal army, engage the enemy on a wide front, wear him out, draw in his reserves, and then strike the decisive blow.[76] This tragic battle that cost 60,000 British casualties on the first day alone remains one of the enduring images of futile generalship in World War I.

Erich von Falkenhayn, chief of the German General Staff, was the first commander in the war to abandon the notion of decisive battle. He instead pursued a simple strategy of attrition in the West, such as at the Battle of Verdun in 1916. This encounter helped provoke the great French mutinies of 1917, but it bled the German army as well. The Germans then shifted forces to seek a decision in the East, where the greater space allowed for more maneuver. By the end of 1917, Russia, wracked by revolution, sued for peace. The Germans then shifted their offensive effort to the West to crush the Allies before American intervention could become decisive. The Allies braced for the heavy German blows sure to come in the spring of 1918.

The great tactical problem of World War I was how to achieve and sustain a penetration to an operational depth that would unhinge the defense and restore maneuver to the battlefield. Ludendorff became obsessed with the need for tactical penetration to such an extent that he lost sight of operational objectives. In a conference with his commanders, he announced, "I object to the word 'operation.' We will punch a hole into [their line]. For the rest, we shall see."[77] The German use of infiltration tactics in the 1918 offensives proved capable of tactical penetration, but Ludendorff was unable to make operational use of these successes. Foch parried the blows and counterattacked, at first reducing the German gains, then achieving limited objectives while preparing a decisive counteroffensive.

The American Expeditionary Force struggled to become a modern tactical and operational force. The divisions were fed into the Allied line to help stop the German offensives and later to reduce their gains. Only when the First Army was formed did the Americans at last have an operational force. Certainly, it was an imperfect instrument, lacking both experience and many of the corps and army troops necessary for effective operation. Needed troops and equipment it obtained from the French; experience it got the hard way. The First Army's reduction of the St. Mihiel salient gained two hundred square miles of French territory and allowed for improvement in lateral communications along the western front. The Meuse-Argonne offensive was a major operation critical in Foch's decisive counteroffensive. Foch's operational art in this war-ending counteroffensive relied not on attrition, but on seizing the enemy's critical lateral lines of communication to force the Germans to abandon northeastern France and much of Belgium. Much like the Allied strategy in 1944, Foch favored a broad-front strategy that involved all the Allied armies conducting nearly simultaneous major operations to fix and overwhelm the available German reserves. The pressure along the entire western front made manifest the Allies' superiority in men and materiel. Acknowledging the inevitable, the Germans sued for peace.

World War I was the first modern war. It involved massive armies and new technologies, all of which presented unique problems. During the war, the belligerents tried various solutions. Technological expedients in the form of tanks, poison gas, flame throwers, and other weapons could not overcome the advantages that technology also gave the defense. Doctrinal solutions, such as infiltration tactics, rolling barrages, limited attacks, and closer combined arms cooperation, could lead to tactical but not operational success. The old operational paradigm of the decisive battle was wholly inadequate. In the end, it was Foch's ability to bring successive major operations to bear at the right time in order to allow the full weight of Allied resources to convince the Germans to sue for peace.

In the years following the Great War, each of the armies pondered the lessons. All of them struggled to understand the full impact and potential of the new technologies—the airplane, the tank, and motorization. Soviet theorists attempted to overcome the barriers to operational penetration by matching new technologies with successive operations in what they called "deep battle." The Germans sought to restore mobility to the battlefield and perfect the battle of annihilation through motorization, armor, and the airplane. British theorists J. F. C. Fuller and Basil Liddell Hart also championed mechanization as the key to tactical and operational penetration. The French sought to impose greater control over the battlefield through the doctrine of "methodical battle."[78]

The Americans perceived different challenges. Their understanding of modern war was thoroughly shaped by the total-war experience of massive national mobilization and unprecedented scales of effort of the Great War. For the first time since the Revolutionary War, America went to war with allies. The nation's geostrategic position and interests meant modern war would be expeditionary, requiring joint and most likely combined operations. The essential questions for the American military in the coming decades at the operational level was how to project, conduct, and sustain military operations in a distant theater of war.

Just as in the period before World War I, the postgraduate military institutions shaped American theory and practice in this new modern warfare. Like their European counterparts, American officers pondered the operational implications of modern war. Over the next twenty years, the higher military schools educated and trained the generation of officers that would fight and win the next great war.

3

LANDPOWER

In the years following the Great War, the United States military sorted out the lessons from that massive conflict in a difficult period of fiscal constraint and public indifference. As in previous wars, the American army quickly melted away as the guns fell silent and the citizen soldiers were rapidly demobilized. By the end of June 1919, the army had discharged 2,736,218 officers and men. By 1920, the regular army had shrunk to 103,247 officers and men. Within seven years, it reached a rock-bottom aggregate strength of 134,829.[1]

As the nation settled into peace, congressional interest in funding the military faded. As early as 1922, Secretary of War John Weeks noted, "Economy has literally become the primary consideration in every departmental undertaking."[2] The following year, Secretary Weeks complained that the total expenditure for national defense, both army and navy, had declined steadily from around $11 billion in 1919 to just over $500 million in 1924.[3] Well before the Great Depression, the nation's armed forces contended with a shortage of money to train, modernize, or even maintain their authorized strength.

In addition, the pride so many Americans took in their country's victory in the war soon turned to disillusionment. Congress rejected President Wilson's attempt to involve the United States in the League of Nations as part of the peace settlement. Wilson's successor, Warren G. Harding, campaigned on the slogan of "Return to Normalcy." In practical terms, Harding's policy turned the nation toward isolationism and focused the public on the peaceful pursuit of happiness and business. America returned to its traditional antimilitary attitudes. In 1927, at the suggestion of France's foreign minister, the United States signed the Kellogg-Briand Pact, which outlawed war as an instrument of national policy altogether. In the following year, the pact was extended to a total of sixty-three nations. American society, both the elites and citizenry in general, showed signs of increasing disillusionment with war as the stupendous cost and sacrifice of World War I seemed incommensurate with the disappointing fruits of victory. At the beginning of the decade, John Don Passos helped inspire this mood with his popular novel, *Three Soldiers*. The 1920s witnessed the migration of a small "lost generation," as expatriate American intellectuals sought refuge and meaning in France. The widespread pacifism, disillusionment, and antiwar sentiment that blossomed in the twenties grew even greater in the thirties.

Regardless of the reductions in military strength, constant lack of funding, and public indifference or open hostility, U.S. Army officers of the interwar period soldiered on. For many of them, peace meant reduction in rank. Hugh Drum, wartime chief of staff of the First Army and a brigadier general, returned to instructor duty at Ft. Leavenworth as a major. Col. George C. Marshall reverted to major, while Dwight D. Eisenhower and George Patton reverted to captain. Drum and Marshall would regain their rank in the next few years, but most of the returning veterans who remained in uniform experienced a slow climb back up through the ranks. For new officers, it was worse. The West Point class of 1919, including future general Albert Wedemeyer, served as lieutenants for seventeen years.[4] Still, these officers, thoroughly imbued with a professional ethic, studied, trained, and attempted to sort out the lessons from the war.

The search for lessons from World War I began soon after the Armistice. General Pershing directed the organization of almost twenty boards to consider the particular lessons from the various branches and services of the AEF. The AEF Superior Board on Organization and Tactics convened on April 27, 1919, to review the findings of the subordinate boards and reach its own conclusions. Pershing chose the board members with an eye to both senior experience and reputation. Maj. Gen. J. T. Dickman, Maj. Gen. John Hines, Maj. Gen. William Lassiter, Brig. Gen. Hugh Drum, Brig. Gen. W. B. Burtt, Col. George

Spaulding, and Col. Parker Hitt, all senior officers with high-level combat and staff experience in the AEF, participated on the Superior Board.

As suggested by its title, the board focused on tactics and organization. It also reached some general conclusions confirming the AEF's overall experience in the war. The board confirmed the importance of the general staff system, specifically one that included an operational planning staff. It further recommended that the "division of staff duties thus defined should exist at the War Department and should extend down through all the tactical commands to include the battalion."[5] Undoubtedly reflecting on the Allied situation prior to Foch's rise to supreme command, the report noted, "No greater lesson can be drawn from the World War than that of unity of command is absolutely vital to the success of military operations."[6]

The board confirmed the importance of logistics, which it defined as everything that "embraces the supply of armies."[7] The board found that "the infantry must be recognized as the basic arm and all other arms must be organized and made subordinate to its needs, functions and methods."[8] This conviction heavily influenced the tactics and organization of the army in the interwar period. The board blamed the indecisive results in the earlier part of the war on limited-objective attacks in which the infantry was subordinated to the artillery. The report stressed the offensive in open warfare, in which all arms supported the infantry in reaching final objectives.[9] In summary, the board confirmed the AEF's staff organization, unity of command, importance of logistics, supremacy of the infantry, and the offensive spirit.

General Pershing became chief of staff of the U.S. Army in 1921, bringing with him all the prestige of a wartime commander as well as his wartime experience. He recognized that the experience of the war, the new military technologies, and the impact of the National Defense Act of 1920 required a postwar review of doctrine.[10] Having benefited greatly as commander of the AEF from the services of the Leavenworth graduates, and Pershing was a strong advocate of the military school system. As the War Department General Staff started the review of doctrine, Pershing encouraged them to work with the army schools. The task of revision was farmed out to the various schools and branches, then reviewed by the Training Division of the

G-3 Operations Division of the General Staff, and finally distributed in pamphlet form.[11] The Command and General Staff School at Fort Leavenworth began overhauling the *Field Service Regulations* (FSR), the capstone doctrinal manual, as early as 1920.

The commandant of the General Service Schools at Fort Leavenworth appointed a board of officers to revise the FSR. By 1922, the board submitted the *Manuscript for Training Regulations No. 15 (Field Service Regulations)* to the General Staff for review. In its attempt to extract lessons from World War I, the board considered the AEF Superior Board report, as well as many others. Its revised regulations confirmed the geometry of the battlefield in defining the theater of war as the "entire area of land and sea which is, or which may become, directly involved in the operation of war."[12] The board divided the theater of war into a zone of the interior and theaters of operations; the theater of war could comprise several theaters of operations depending on geography and the threat. The proposed FSR stated that war plans should consist of a detailed study of a particular theater, a plan of concentration, and "plans of operation for each major operation in the theater."[13]

The Leavenworth manual made critical assertions in discussing strategic principles and operations. It defined military strategy as "the art of moving armies in the theater of operations."[14] It concluded that "a plan of operation is a study of the exact lines of military activity proposed for a particular force during a phase of the campaign. Subordinate forces acting in conjunction therewith have separate but coordinated plans of operation."[15] The manual stressed the importance of seizing the initiative and taking the offensive. It listed piercing (penetration), frontal, envelopment, and turning operations as forms of strategic maneuver.[16] Further, the manual listed nine strategic principles of war: operational objective, concentration of effort, economy of force, strategic surprise, freedom of action, strategic security, strategic offensive, moral ascendancy, and strategic direction.[17] These strategic principles provided the basis for the principles of war adopted by the War Department in 1921.

The manual insisted that the "proposed FSR are distinctly American in the subject matter covered, [and] in the methods of execution of the underlying and controlling principles."[18] This was a wide-ranging document with sections on mobilization, training, and War Department functions. It included a great many good ideas—too many, in fact. Comments on the Leavenworth proposal indicated that it was a fine piece of military literature, but the army required a concise practical guide to service in the field.[19] The chief of staff appointed a committee to review and edit the manual. Eventually, the army published *Field Service Regulations 1923* as a more concise and focused manual. FSR 1923 became the capstone doctrinal publication for the army until it was replaced in 1939.

FSR 1923 kept much of the Leavenworth draft, specifically the geometry of the battlefield and the emphasis on initiative and offense. FSR 1923 omitted the Leavenworth manual's sections on mobilization, War Department functions, and strategic art. FSR 1923 boiled down the multivolume Leavenworth draft to 195 pages. FSR 1923 did contain sections on the employment of aircraft and tanks. These sections on the new military technology accurately reflected the AEF's wartime experience, but the strategic and operational lessons were left to the Staff School and War College instruction. Although FSR 1923 served as the foundation and official doctrine of the army for most of the interwar period, the Leavenworth manual remained influential. Much of the thinking that appeared in the Leavenworth draft found its way into the student texts and curriculum of the Staff School. Any evaluation of American operational art in the interwar period based solely on the official doctrine disregards the student texts, exercises, and curriculum of the postgraduate military schools that shaped and educated the future leaders of the army.

The Command and General Staff School

The Leavenworth schools reopened in the summer of 1919, and for three years it operated as the School of the Line and the Staff School. The School of the Line taught brigade and division tactics, while the Staff School provided instruction on corps and armies. In 1922, the two schools were merged into the Command and General Staff School. The merger allowed a greater throughput for officers since the time in school was reduced to one year. In 1928, the army reinstituted a two-year course to provide a more rigorous course of study. The two-year course continued until 1935 when once again the pressure for more officers to attend the school caused a reversion to a one-year curriculum.[20]

Officers highly prized selection to attend the Command and General Staff School. Selection and success in the course could make or break a career. The faculty rank-ordered the students according to merit. The commandant provided an individual efficiency report on each officer and selected those that were considered suitable for duty on the General Staff. The officers felt the pressure, as lectures, conferences, and problem solving filled the day from 8:30 A.M. until 5:00 P.M. five days a week. In conferences, the faculty divided the class into committees of eight to ten officers. Instructors assigned the committees topics for study, and a committee spokesman reported on the group's work.

The applicatory method in student exercises and staff rides still formed the fundamental approach to learning. Individual problem solving was the critical feature in grading and ranking the students. In 1926, the year Dwight D. Eisenhower attended the school, the curriculum called for seventy-eight map problems and terrain exercises that were worth a total of 1,000 points.[21] Eisenhower did well, graduating number one in his class. He was helped in part by George Patton, who had attended the year before and shared with Eisenhower his notes and the problems of the previous school year.[22] By all accounts, it was a rigorous year of instruction. Eisenhower later wrote an anonymous article for the *Infantry Journal* aimed at allaying anxiety about the course and providing practical advice on how to survive the ten months of intense instruction. He stressed the need for a positive attitude and good personal habits.[23] All the officers understood that attendance and success in the Command and General Staff School was important, indeed, critical to their careers. Competition among the students was keen, and hard work both in class and after class was characteristic of their year at Leavenworth.

Theory   Veterans of the AEF dominated the faculty through 1925 and remained a significant portion of the faculty throughout the decade.[24] The faculty used many texts and instructional materials employed by the AEF schools in that first year. The old German texts were tossed out, and instructors spent much of their time in the first two years writing new curricula to account for the new weapons of war, tanks and aircraft. Col. William K. Naylor, one of the returning officers, wrote *Principles of Strategy* as an American text to replace previously used European texts. Naylor had served as an instructor at the General Staff School from 1913 to 1915. Following the war, he returned as director of the school and taught the classes in strategy.

There was little that was new or original in Naylor's text; for the most part, the book is a compilation of his lectures based on the military thought gathered from Jomini, Victor Derrecagaix, Colmar von der Goltz, and others. From Derrecagaix's *Modern War*, Naylor adopted the Frenchman's discussion of "a project of operations."[25] In form and content, *Principles of Strategy* closely follows von der Goltz's *Conduct of War*. In fact, Naylor frequently paraphrases the German military writer.

Echoing von der Goltz, Naylor suggests that a campaign consists of a series of operations, all "connected by the bond of some common, fundamental idea."[26] He also borrows from the German author the idea that the first strategic principle is "to make the hostile main army the objective."[27] The focus on the main enemy army was a common theme in nineteenth-century military literature. Clausewitz states early in *On War* that the three broad objectives in war are the armed forces, the country, and the enemy's will. He suggests that these objectives follow in a natural sequence.[28] Later he discusses at length his concept of "center of gravity," which remains a fundamental element in modern operational art. Clausewitz maintained that in directing military operations, "one must keep the dominant characteristics of both belligerents in mind. Out of these characteristics a certain center of gravity develops, the hub of all power and movement, on which every thing depends. That is the point against which all our energies should be directed."[29] Based on his experience, he recommended the center of gravity normally might be found in the destruction of the enemy army, seizure of his capital, or a strike against his principal ally.[30] Naylor does not mention the center of gravity but simply accepts, as most nineteenth-century military theorists did, that the operational center of gravity is the enemy's main army.

Although Naylor's text was based on nineteenth-century military thought, some important new operational constructs found their way into American military curricula. One of the more important concepts was the Clausewitzian notion of the culminating point. Naylor cited Clausewitz, but most likely he absorbed Clausewitz through von der Goltz. Clausewitz noted that the strength of the attacker invariably diminishes until it reaches a point of culmination. This is the point where the attacker's strength no longer enjoys significant advantage over his opponent and poses significant risk to the attacker if he continues the attack.[31] Naylor, closely paraphrasing von der Goltz, notes, "Although originally superior to the enemy, and victorious in the past, troops may finally arrive, through an inevitable process of weakening, at a point which does not assure any future success, or, in other words, the point of culmination."[32] In operational art, "a general, with a correct estimation of the situation, should immediately recognize the arrival of this culmination . . . concluding a peace or else changing over to the defensive."[33]

The Command and General Staff School classes from 1923 to 1927 got a much stronger and more direct dose of Clausewitz from Lt. Col. Oliver P. Robinson. Robinson, who graduated from the staff school in 1915, served during World War I as the chief of staff of the Eighty-first Division and later as the chief of staff of the American Expeditionary Force to Siberia from 1918 to 1919. In 1923, Robinson was assigned to Leavenworth as an instructor and followed Colonel Naylor as the instructor of strategy. During the 1920s each class received ten hours of instruction in strategy toward the end of the school year. Robinson used the lectures in strategy to illustrate the principles of war, largely through a discussion of Clausewitz's *On War*.[34] Lieutenant Colonel Robinson believed that "Clausewitz's book on war, published in 1832, occupied about the same relationship to the study of the military profession as does the Bible to religious studies: "I have been unable to find a single proposition relating to strategy which Clausewitz did not cover in a broad general way."[35]

Robinson's lectures provided the students with an excellent view of operational art from a Clausewitzian perspective. He defined strategy as "that branch of the theory of war which has to do with the planning and effect of the various combinations, movements, and use of all the forces of a power or all of the forces in a given theater or theaters of operations. It takes into consideration tactics, logistics, material assets, the theater of war, psychology of the people and the national policy, both from the viewpoint of its own country and that of the enemy."[36] This definition clearly defines nineteenth-century strategy as twentieth-century operational art—all military activity within a theater of operations that takes into account a number of factors including logistics and national policy. Later Robinson updated Clausewitz's famous definition of strategy: "If, for the word battle in Clausewitz's definition, there be substituted, 'operations of war', which includes all those things which precede and lead up to the battle and the threat of battle, as well as the battle itself, there results the all inclusive definition: strategy is the use of the operations of war to gain the end of war."[37]

In mid-May 1926, with Eisenhower and four other future corps and army commanders in the audience, Robinson lectured on the "Principle of the Objective" by discussing at length the concept of the center of gravity.[38] Paraphrasing Clausewitz, Robinson insisted: "Therefore the first consideration under the principle of the objective is to determine the centers of gravity of the enemy's power. Then against this center of gravity the concentrated blow of all the forces must be directed."[39] Following Clausewitz's discussion of the center of gravity, Robinson noted that "the will of the people to carry on a war may be the real center of gravity of a nation, but in this situation the quickest way to reach that will is by a defeat of the hostile main forces."[40]

Later, Robinson lectured the class on the "Principle of the Offensive," emphasizing the operational concept of culmination. He told students it was critical that the commander "must make it his business that the culminating point will see the maximum result accomplished. He must stop his advance the moment he discovers that his strength would fail by undertaking more. Then he must pass to the strategic defensive conducted offensively, and thus as far as possible retain the initiative."[41] Robinson was, in effect, arguing that an operational pause must be taken before reaching culmination. His firm grasp of Clausewitz was reflected in virtually all of his lectures. From 1923 to 1927, Eisenhower, the future supreme commander and all six of the army commanders in World War II, plus twenty-five of the thirty-four corps commanders, sat through Robinson's lectures on strategy. He later compiled his lectures into *The Fundamentals of Strategy*, published in 1928. The book was used as a text and recommended for reading at Leavenworth, the Army War College, and the Naval War College throughout the rest of the interwar period.

The influence of Clausewitz on American military thinking during the interwar years can be debated. Clausewitz's *On War* covers the whole scope of war: theory, philosophy, strategy, operations, and tactics. To the extent American officers were exposed, directly or indirectly, to Clausewitz, they like other military professionals, undoubtedly read or understood him selectively. Eisenhower claimed he read *On War* three times and believed it was the most influential book besides the Bible he ever read.[42] Robinson noted that "in our library for instance—the first volume of Clausewitz's three is thumb marked, pencil marked, pages dirty and worn from use while Volumes II and III show very little evidence of use."[43] The term "center of gravity" was occasionally but not commonly used in student texts and exercises. What is clear is that there was a firm understanding of the three levels of war, of the need to focus combat power at the theater level, and of the concept of culmination of the offensive that required operational pauses.[44]

Doctrine   Many of these concepts did not find their way into the official doctrine of the U.S. Army, but they did appear in the student texts. The student text on *Tactical and Strategical Studies, Corps and Army*, originally published in 1922, went through five editions and was used throughout the 1920s. This text clearly establishes the three levels of war:

> In discussions of plans involving large forces there are utilized the terms project of operations, plan of campaign, and plan of an operation. The first relates to a national project prepared by the War Department for the execution of a war with a specific enemy, and may involve several campaigns. The second relates to the general conduct of forces in a single theater of operations and is the plan prepared by the commander thereof for the accomplishment of the mission assigned. It includes successive tactical operations. The third relates to a tactical phase of a campaign which generally involves several tactical operations before the mission is accomplished.[45]

In campaign planning the text describes the center of gravity as the "hostile decisive element" that is usually the enemy's main force, but goes on to explain that "under exceptional circumstances, the enemy's capital, his commerce, his industrial areas, or his resources may be the military objective. However, these objectives are generally secondary, in that they provide a means for the destruction of hostile forces."[46] The text further states that "the plan of campaign may also contemplate probable successive operations phases to continue the success of the primary operations, and consider steps to be taken contingent upon results different from those expected."[47] This is a clear expression of phasing and the need to develop branches and sequels in campaign planning.

A survey of the Leavenworth curriculum during the decade of the 1920s demonstrates several other key features of military education. In addition to the necessary focus on tactical instruction, military history and logistics made up a significant portion of the curriculum. In 1922, tactical instruction on corps and army made up 26 percent of the conferences, while military history absorbed 17 percent and logistics another 13 percent.[48] Military history presented as lectures and conferences on World War I and the Civil War provided a vehicle for deriving lessons from large-unit operations, campaign planning, and senior leadership. Logistics, both as a separate subject and as an integrated element in all exercises, pervaded tactical instruction. Consistently throughout the decade logistics specifically comprised 10–15 percent of the instruction. When compared to the 25–29 percent regularly dedicated to tactical instruction, logistics loomed large in the faculty's consideration.[49]

The Command and General Staff School at Leavenworth during the first half of the interwar period made great strides in sorting out the lessons from World War I. Its primary focus was on divisions, corps, and the army. The army was considered the strategic or operational echelon of maneuver, while corps and divisions were tactical units. The Staff School clearly recognized three levels of war and reflected several key elements of operational art in its instruction in campaign planning. These included the need for a clear focus for the application of combat power in the theater of war (a center of gravity or decisive hostile element), offensive culmination, and an extension of the concept of tactical phasing to operational phasing. In addition, instructors stressed logistics as an important and integral part of large-unit operations. The Staff School imparted doctrine through rigorous instruction and exercises. The study of the broader implications of World War I for strategy and large-unit operations was reserved for the army's senior educational institution, the Army War College.

The Army War College

The War Department reestablished the Army War College in 1919 with the mission to "train officers for high command and War Department General Staff Duty."[50] Initially called the General Staff College, the institution was renamed the War College in 1921 to avoid confusion with the Command and General Staff School at Fort Leaven-

The Army War College was instrumental in developing American operational art in the interwar period. Founded in 1903, the college was located in Washington, D.C., at the Washington Barracks, later renamed Fort McNair. Courtesy of USAMHI.

The Army War College's use of history to study war is highlighted by the use of historical staff rides. This staff ride, most probably of the Civil War battlefield at Fredericksburg, Virginia, took place during the interwar years. Courtesy of USAMHI.

worth. In 1922, the War Department further specified the mission of the War College to train officers in high command and general staff duty with units larger than corps. Maj. Gen. Hanson E. Ely, commandant from 1923 to 1927, used his first opening address to emphasize that the scope of the mission included strategy and logistics of all units larger than corps.[51]

Despite the efforts of the War Department to delineate clearly between the missions of the Staff School and the War College, there was overlap between the two institutions for most of the decade. Although the Staff School focused on division and corps, Leavenworth also taught army operations. With the addition of a second year of study at Leavenworth in 1928, the school provided even more instruction dealing with echelons above corps. As a result of the changes at the Command and General Staff School, the War Department altered the mission of the War College "to train officers in the conduct of Army and higher echelons; to instruct in those political, economic and social matters which influence the conduct of war; to train officers for joint operations of the Army and Navy; to instruct officers in the strategy, tactics, and logistics of large-unit operations in past wars, with special reference to the World War."[52]

There were significant similarities and differences in the instruction and the subjects covered at both schools. Both institutions used the same doctrine and the applicatory method, employed military history, and emphasized logistics in the curriculum. Leavenworth was legendary for its rigorous methods and adherence to the school solution. Col. H. B. Crosby, assistant commandant of the War College, noted in his orientation lecture in 1924 the difference between the methodology of the two schools. "I believe I speak the truth when I say that no one helps his rating by blindly accepting the views of the faculty on any subject," Crosby declared. "This is distinctly a college —where we learn from an exchange of ideas and not by accepting unquestioned either the views of the faculty or the views of the student. At Leavenworth we accepted and should have accepted the principles and doctrines laid down by the faculty of that school. Here we reach our own conclusions, faculty and student, following a full and free discussion of the subject."[53] Leavenworth was about training; the War College was about education.

There were other significant differences between the curricula of the two schools. The War College began as an adjunct to the War

Department's General Staff to assist in the preparation of war plans. Unlike the General Staff School, the War College worked with real war plan scenarios. Virtually all of the War College map exercises dealt with the color plans. Throughout the interwar period, the College most frequently exercised the Plan Green (war with Mexico), Plan Red (war with Britain), and especially Plan Orange (war with Japan). As intended from its inception, the War College was always concerned about joint operations between the army and the navy. The national war plans, generated by the Joint Board, inherently involved joint operations. The broader perspective of the War College in preparing the nation for war and in conducting it called for the study of both strategy and operations. The college considered operations within a broader political, economic, and social context.

The War Department charged the War College with producing officers capable of serving on the General Staff and as commanders or staff officers of armies and army groups. In 1922, the War College faculty and curriculum changed to mirror the organization of the General Staff. The G-1, G-2, G-3, and G-4 courses taught lessons in personnel, intelligence, operations, and supply, respectively. Courses in war plans and command rounded out the early curriculum.

Like the Command and General Staff School, the War College made extensive use of military history for instruction. Unlike students at the Staff School, officers at the War College studied a broad array of campaigns, war plans, and great commanders to draw their own lessons. The faculty and students particularly examined World War I to discover the lessons of modern war. From these analytical studies, the students reached conclusions on campaign planning and increased their understanding of the operational level of war.

Joint Operations   Throughout the 1920s, the students routinely studied the campaigns of World War I. Criticisms of German and Allied operational art consistently pointed to poor command and control, lack of joint planning, and insufficient emphasis on logistics to account for failures on the part of either side. A committee in 1923 studying the Schlieffen Plan of 1914 found that "the plan failed to provide sufficiently for cooperation in the field, in the armies of the wings and center with distinct tasks did not have group commanders to coordinate their activities."[54] The report concluded "there should be a plan of campaign or an outline of proposed operations, simple in conception, stating clearly the objective, based on the principle of offensive, and of movement, but not encroaching on the initiative [of] the commander. This part of the plan will also designate the theater of operations, [and] list the troops required. Locate or define the initial concentration areas, prescribe the organization of the command and indicate the cooperation of the Navy (which the German plan did not do)."[55] Significantly, the committee criticized the Germans for the failure not only to coordinate properly their armies within the theater of operations but also to include the navy in their planning.

The lack of German joint planning was featured in most of the committee reports reviewing the opening campaign. A committee in 1927 reviewing both German and British naval plans of 1914 insisted that "in naval plans, the singular feature is their lack of coordination with military ones, and in Germany their domination by military plans."[56] The committee concluded that "naval cooperation is essential in any major effort, with the service having paramount interest in the operation in control. All war plans should be a result of studies by both services working together in their preparation."[57] Another committee reporting the previous year on the same subject argued that if the German Imperial Navy had interfered with the movement of the British Expeditionary Force from Britain to France, the Germans would have had sufficient force to overcome the French and win in the opening campaign.[58] This committee drew the lesson that joint planning agencies are necessary to coordinate between the two services.

From the beginning, the War Department charged the Army War College with encouraging joint training and education. As early as 1920, the commandant of the college suggested an exchange of students with the Naval War College. By 1927, three naval officers and three Marine Corps officers annually attended the Army War College. The college also added two navy officers to its faculty. Both as faculty and as students these officers contributed to improvements in joint planning.

Joint war games between the services' war colleges began in 1923, with an exercise of the defense of the Philippine Islands. Joint games were held again the next year, and the majority of the War College class was participating by 1925. The students and faculty maintained communication between Washington Barracks (the Army War College) and Newport, Rhode Island (the Naval War College), by telegram.[59]

Joint exercises were not confined to the map. In 1925, the chief of staff, Maj. Gen. John Hines, lectured the War College class on the recent Army-Navy exercises in Hawaii. He noted that 50,000 officers and men had participated. He raised the issue of joint staffs instead of liaison officers. Finally, he noted that the only real problem was lack of coordination between army and navy air forces.[60] Two years later, the commandant, Major General Ely, involved the War College in a joint exercise in New England. The students prepared course of action briefings, estimates, and incredibly detailed plans for a two-corps assault on the New England coast. The 182-page series of orders included administrative and field orders for embarkation, debarkation, and naval fire support, a communications plan, and numerous appendixes. The Commandant and six other army officers boarded the flag ship to supervise the exercise.[61] Lt. Col. Charles Keller made this key point in 1926 while addressing the War College: "The real importance of the annual Joint Exercises is but little realized at the present time. Our geographical location alone would appear to dictate the necessity for this class of training."[62] The exercises and the faculty together drove home the importance of joint operations in future warfare; the only question remaining was who would command them.

During the interwar period, the Joint Board was responsible for joint planning and establishing the means for army-navy cooperation. In 1926, the board established two methods of joint coordination in military operations: paramount interest and unity of command. Under the principle of paramount interest, the service whose function and requirements are of greater importance maintained authority and responsibility for coordination. In this arrangement the service with paramount interest could give operational missions to the other service. Under the unity of command principle, forces of one service were assigned to a commander who was empowered to coordinate the services by the "organization of task forces, the assignment of missions, the designation of objectives, and the provision of logistic support; and to exercise control during the progress of operations to insure the effective effort toward the accomplishment of the mission."[63]

In their studies, the students preferred the principle of unity of command. In 1928, a committee of students charged with developing lessons from the study of British and German naval plans during

World War I recommended: "That in all major joint expeditionary forces a single supreme commander with a suitable joint staff be designated for control of the entire campaign."[64] Another subcommittee report in 1928, chaired by Maj. Simon B. Buckner, the future commander of the Tenth Army during the invasion of Okinawa in World War II, emphasized the importance of unity of command. Buckner's committee further noted that the army was deficient in training for landing operations and suggested that it might adopt some of the methods employed by the Marine Corps. Buckner and his fellow students recognized that in any future wars close teamwork with the navy would be required in the formation of broad strategic plans, as well as joint land, sea, and air operations.[65] The students anticipated that future wars would require projecting significant combat power overseas.

The students were right. The importance of joint command and joint operations would play a prominent role in the operational art exercised by American commanders in World War II. These officers recognized that in operational art all combat power, naval and land power alike, must be brought to bear in the theater to achieve strategic objectives. One of the key features of modern operational art is the ability of the theater commander to bring all the various capabilities at his disposal to bear. In their study of joint operations, the students of the War College recognized the need to leverage the capabilities of each of the services in achieving strategic objectives. The third element of national military power, airpower, had been born over the trenches in World War I. The promise and role of that new force forged in the Great War was a subject of much debate between its most fervent advocates and the practitioners of landpower.

Logistics American officers understood that the mass armies of modern war required massive logistics to support them. At both Leavenworth and the War College, an interest in logistics pervaded the curriculum. In 1929, a student committee at the War College determined that the German failure in 1914 was due to the weakening of its right wing, a lack of control between the General Headquarters and the various armies, and finally rigidity and inflexibility of Germany's supply system.[66] The students concluded that the solution was mass, formation of army group headquarters to coordinate armies in the theater, and a greater emphasis on logistics in theater planning.[67]

There was a general perception that German attention to logistics was lacking during the war. Beginning in 1925, the college required each student to write an individual study that took the form of a staff memorandum in each of the major staff areas. Maj. H. S. Grier's staff memoranda in 1926 on German logistics in World War I found the German supply organization too rigid and too complex.[68] The lesson was obvious, Grier concluded: "Every war plan and every plan for operations should have included in it a plan of supply the scope of which must be broad and contain the requirements, the procurement and distribution plans and policies for the situation to be met."[69] In a similar vein, fellow student Maj. James W. Barber expressed the belief "that the U.S. General Staff Organization for supply is far superior to that of the Germans."[70]

One of the great lessons of the war was the need for total mobilization of the nation's industry and manpower to create and sustain the mass armies required in modern warfare. In fact, the War Department established the Army Industrial College in 1924 to address mobilization problems that had been encountered in World War I. The War College gave mobilization planning special attention in the curriculum. How to logistically sustain the mass armies provided by total mobilization formed a significant part of the curriculum as well. In the 1920s the G-4 course covered various aspects of strategic and operational logistics. In 1926, the course included lectures on ship-to-shore supply, rail movement, and studies dealing with campaign analysis from a logistics point of view.[71] Students examining the impact of logistics on modern strategy recognized "that as armies have increased in size, the necessity for the protecting of their lines of communications has increased the influence of logistics upon operations."[72]

As one student committee noted, "Military strategy is realized by means of maneuver and logistics."[73] The importance of logistics increases with large-unit operations. Deployment, concentration, and sustainment are at the very core of operational maneuver. It is clear from the curriculum that both faculty and students recognized that logistics at the operational level is not just a matter for logisticians. Most of the students at the War College came from the combat arms—infantry, artillery, and cavalry—yet they were required to write detailed logistic annexes during exercises. All students were required to understand how supply works at the higher echelons of army organization. Brig. Gen. Fox Conner lectured the 1925 class: "Material (supply) has enormously increased in importance, so much so that the G-3 is more concerned with the possibilities of supply than with anything else or with everything else put together."[74]

The army understood and emphasized logistics. It had long been a practice to exchange faculty and students between the Army and Naval War Colleges. Adm. W. V. Pratt, president of the Naval War College, during an address to the Army War College in 1926 noted that the army officers serving as faculty at the Naval War College helped establish that institution's course in logistics: "Without their earnest effort we never would have been able to get as far as we have in what we call our Logistic Course which is, after all, supply personnel, materiel, matters of the sort which naval men as a rule do not appreciate because we carry with us our ninety days supply."[75]

Even in the theory of war, American officers did not leave out logistics. In 1928, a committee reporting on "War and Its Principles and Methods and Doctrine" quoted Clausewitz, while adding some distinctly American observations. The committee included Maj. Dwight Eisenhower and Lt. Col. Oliver Robinson, who had lectured Eisenhower's General Staff School class on Clausewitz just the previous year. Its report, probably written by one of these officers, is thoroughly Clausewitzian in its discussion of the theory of war, even though Clausewitz himself never discussed logistics. The committee report nonetheless concluded that logistics that involves the entire complex mechanism of organization and administration is part of the preparation for war, and of the constant replenishment of the means to wage it, but most officers consider logistics to be part of the conduct of war itself.[76]

The relationship between operational maneuver and logistics was clear to these American officers. The faculty drove home this lesson: "Every phase of military operations, every strategical or tactical conception is inextricably interlocked to a greater or lesser extent with some phase of supply and transportation."[77] Logistics determines "the art of the possible" for the operational commander, and its role in campaign planning is crucial.

Campaign Planning In 1928, Brig. Gen. Frank Parker, the Army General Staff G-3, lectured the class on the application of strategy. He criticized the strategy of the Germans in 1918: "The offensive maneuvers of the German Armies on the western front beginning March 21, 1918 and ending with the attack of July 15 in Champagne, seems to have had no definite strategic inspiration. Their tactical successes in Flanders, on the Somme, and on the Aisne, at this time seems to have no part in the continuity of a strategic idea involving the whole front. The only result of these attacks was to establish dangerous salients at long intervals of time and space."[78] The next year, faculty instructor Lt. Col. A. D. Chaffin similarly lectured the class on the strategy of the Central powers and criticized the Germans for adopting a strategy of attrition in 1915 and for having no strategy or operational intent in the offensives of 1918. He offered this rhetorical question: "What were Ludendorff's objectives in those attacks; not what did he want, but what did he expect and plan for?"[79] Both Parker and Chaffin emphasized the notion that, in strategy or operational art, combat operations must be coordinated and guided by the commander to achieve a strategic objective in the theater of operations.

Campaign planning is the primary means through which the commander exercises operational art. Students in a 1926 committee report, noting the nature of modern campaigning and the role of the commander, observed: "After the War of 1870, the commander tended to become a Director of Operations and was seldom seen on the field of battle. Von Moltke, Oyama, Foch, Hindenburg and Pershing were Directors of Operations."[80]

The theater commander, frequently responsible for several armies, no longer managed a single decisive battle but directed operations that might involve several battles all linked to a common strategic purpose. Linking several battles in a theater called for phasing. In an orientation lecture to the class of 1925, Col. C. M. Bundel, director of the War Plans Division, advised the students:

> It is becoming apparent that the whole of the war effort is not a rigid, indivisible affair that must be handled as such. In fact, an analysis shows quite clearly that it is divided into several distinct steps or phases which, while inherently distinct, nevertheless are interdependent and in some cases overlapping. It is believed that the differentiation of these phases is essential to clear understanding and correct solution of the many problems involved.[81]

The students at the War College practiced campaign planning through exercises based on actual war plans. In 1919, the Joint Board was reorganized and charged with coordinating matters of mutual interest between the army and the navy—joint exercises, joint procedures, and war plans. The Joint Board was primarily an attempt to develop a national planning system. To this end, it formed a Joint Planning Committee, consisting of officers from each service's respective war plans division, to prepare estimates and plans for its review. By 1925, the college was teaching that there were four types of plans: the joint plan, army strategic plan, General Headquarters (GHQ) plan, and the theater of operations plan. The Joint Planning Committee developed the joint plan stating the national objectives, summarizing the situation, and prescribing missions to the services. The General Staff then developed the army strategic plan, which was essentially a directive from the secretary of war allocating forces and directing mobilization. The War Plans Division (WPD) of the General Staff wrote the GHQ plan. In theory the WPD would form the staff of the general headquarters established in a theater of war. This plan organized the theaters of operations, allocated forces, and gave broad missions to subordinate commands. Finally, the theater commander developed the operation plan.[82]

The joint plan was the capstone plan; the others were supporting plans, all linked in their support of objectives to the higher plan. The War College settled on the five-paragraph field order as the format for all the plans.[83] The significance of this national military planning system for operational art lies in the fact that campaign plans would be nested in a series of national and joint plans that were intended to ensure joint cooperation in the theater to pursue strategic objectives. The operational commander would pursue not just victory over the enemy in the field but strategic objectives with political purposes.

In 1926, Lt. Col. Charles Keller of the War Department General Staff informed the class that a strategic plan might include the following: an estimate of the situation, a general concept of the war (from the joint basic plan), phases of the war, missions assigned by the joint plan, strategic concentration, theaters of operations, administration and supply, and plans for operations.[84] A few months after Lieutenant Colonel Keller's lecture, the faculty expanded on their view of the process in a memorandum of what the plan of campaign in this planning system might look like. They combined the doctrine found in the Leavenworth student text *Tactical and Strategical Studies, Corps and Army* with their understanding of the national planning system. The memorandum on campaign planning quoted the Leavenworth text by describing the campaign plan as a guide to operations that may contemplate probable successive phases in operation. The joint basic plan, the army strategic plan, and the theater operations plan all made up a planning sequence that established objectives, missions, and guidance from national strategic objectives down to operational objectives in a theater of operations or war. The General Headquarters plan provided the campaign plan for the theater. According to the faculty, the campaign plan must include an estimate of the situation that considers military, geographic, topographic, political, and economic conditions. The plan should include the objective, general plan of operations, method and location of concentration, and general supply policy.[85]

War College students not only studied this planning system; they exercised it. In 1925, Map Problem No. 1 in the War Plans course required the students to draft a campaign plan for an army in a war against a Red-Orange alliance. Immediately following World War I, the only two powers capable of challenging U.S. interests were Great Britain and Japan. As early as 1919, the Joint Board considered the possibility of conflict with these two powers. As unlikely as a war against the British Empire might seem, it provided significant training value in planning for large-unit operations. The plan called for the ultimate mobilization of nine American armies to seize Canada while the navy concentrated in the Atlantic to defend the eastern seaboard and wage war on British commerce.[86] The campaign plan devised by the students followed doctrine closely, but it did not include phasing. Of more interest is the basic war plan provided by the college. In this scenario, the United States is involved in a war against Great Britain in the Atlantic theater and Japan in the Pacific theater. The basic plan called for three phases: mobilization and concentration, joint army-navy operations to capture Nova Scotia, and finally, destruction of Red and Crimson (Canadian) forces wherever found. The strategy was "to seek a favorable decision with Red while offering the maximum practical resistance to Orange in the Pacific Ocean. After the defeat of Red, to undertake a general offensive against Orange in the Pacific."[87] This "war game" strategy for a two-ocean war involving Japan and a European ally required prioritization of the theaters just as in reality some fifteen years later.

The focus of the War College increasingly shifted to war planning. In 1924, the assistant commandant informed the class in his orientation lecture that the entire course was based on the preparation of war plans.[88] The next year, the director of the War Plans Division assured the students that "almost without exception everything undertaken during the year has its application, either direct or indirect, to the preparation of war plans."[89] From 1920 to 1929, the college exercised War Plan Red, Red-Orange, or Orange every year. War Plan Red allowed the exercise of joint large-unit operations. The entire class normally conducted a field reconnaissance in the northeastern United States to validate its plans with the actual terrain. The prominence of War Plan Orange recognized the very real threat Japan posed to the Philippines and U.S. interests in the Pacific. This war plan provided concrete and specific challenges to military planning. In the long run, War Plan Orange stimulated the most productive American thought on operational art.

Toward the end of the decade, in September 1929, Maj. Gen. W. D. Connor, commandant of the War College, went to Europe to visit other institutions of higher military education. To his surprise, and satisfaction, he found the Army War College to be virtually unique. The British Staff College was more comparable to the Command and General Staff School. The British College of Imperial Defense included members from other government offices, such as the Foreign Office and the Treasury. Perhaps because of the presence of so many civilian officials, the Imperial Defense College did not war-game or tackle any detailed military planning.

The Versailles Treaty abolished the German General Staff and the Kriegsakademie. Training and education for the "leader staff" that functioned as a general staff took place in divisional schools. The two-year staff training course in the divisional schools included historical studies and division- or corps-level war games. The German War Department School did extensive war-gaming, mostly at the corps level, but the student body consisted of only thirty junior officers. Major General Connor found the French Center of High Military Studies most comparable to the War College. He noted, however, that this institution did not war-game or prepare officers for service on higher staffs. Connor concluded that "all in all, I came back home with a greater feeling of satisfaction with our course at the Army War College than I had when I left."[90] The War College seemed uniquely positioned to study the relationship of strategy and operations.

THE THIRTIES

The 1930s was a dark decade for America. On October 29, 1929, the stock market crash initiated a national economic depression that spread throughout the world. During the 1920s the United States produced more manufactured goods than all the other six great powers combined, but the value of U.S. manufactured goods in 1933 was less than one-quarter what it was in 1929. By 1933, around 15 million workers had lost their jobs and were without means of support.[91] For the American military, the postwar budgetary cutbacks that began in 1919 and continued throughout the 1920s became even worse in the following decade.

In 1932, Gen. Douglas MacArthur, chief of staff of the U.S. Army, noted in his annual report to the secretary of war "the universal and inescapable influence" of the economic depression. According to MacArthur, retrenchment was the dominant factor shaping military policy. The strength of the army had dipped to 12,180 officers and 119,888 enlisted men, which MacArthur described as "below the point of safety."[92] There was barely enough money to maintain a skeleton military force, let alone funds to train and exercise it or to acquire modern technology. Only when clear potential threats arose in Europe and the Far East near the end of the decade did Congress demonstrate interest in increasing military funding. Although the Japanese invasion of Manchuria in 1931 and the Nazi rise to power in Germany in 1933 had foreshadowed trouble in the years ahead, not until Imperial Japan assaulted China in 1937 and Germany marched into the Sudetenland and Austria in 1938 did the level of international tensions and potential threat even attract congressional notice. Congress authorized an increase in army strength to 165,000 for fiscal year 1938. In that year, Congress also finally appropriated money for the Vinson Naval Parity Act, which authorized the navy to acquire 100 ships and 1,000 airplanes over the next five years. President Roosevelt proposed a new rearmament program in November 1938 that included $500 million for 10,000 airplanes. Some historians cite 1938 as the beginning of American rearmament. If so, the navy and the Army Air Corps were the chief beneficiaries.[93]

Despite the increased congressional support for the military at the end of the decade, public pacifism was reflected in a strong isolationist mood. The popular 1930 movie *All Quiet on the Western Front* was symptomatic of the growing antiwar movement at the beginning of the decade. The use of regular troops to evict World War I veterans of the Bonus Army from Washington, D.C., in 1932 further diminished the army's prestige.[94] In 1934, Senator Gerald Nye of North Dakota chaired a committee investigating the charge that munitions manufacturers and financiers conspired to involve the United States in World War I. The following year, retired Marine Corps Major General Smedley Butler published *War Is a Racket*, which *Reader's Digest* later condensed and popularized. Butler toured the country for two years sponsored by the League against War and Fascism. His status as a long-serving veteran of foreign wars, twice decorated with the Congressional Medal of Honor, gave credibility to the views of the antiwar, isolationist movement.[95]

Isolationism and the antiwar sentiment also found a powerful political voice. Former president Herbert Hoover, Senators Robert Taft, Robert La Follette, William Borah, and Hiram Johnson, and Nye, and the *Chicago Tribune*, *New York Daily News*, and Hearst newspaper chain represented a powerful isolationist bloc both in and out of government. Isolationism and antiwar rhetoric helped to create a national mood hostile toward military preparedness. Despite the German conquest of Poland and the shocking defeat of France in June 1940, Congress passed by only a single vote an extension of the conscription act in August 1940.

The Great Depression and the public's prevailing antiwar mood affected not only America's military institutions but also the officers who led them. Low pay, a slow rate of promotion, and civilian indifference or hostility made pursuing a military career personally difficult for these men. Future general Maxwell Taylor graduated from West Point in 1922, but he did not attain the rank of captain until 1935. In his memoirs, he recalled the interwar period as a challenging time during which many officers resigned their commissions to pursue other careers. Taylor, who chose to stay in the army, noted that those who remained "in these doldrums, were saved by some inner feelings of the importance of their profession, reinforced by the influence of the Army school system."[96] He enjoyed the rigor of the Command and General Staff School at Leavenworth, and "the year at the War College was a time for mature reflection on the broadest problems of the military profession in company with congenial fellow professionals."[97]

Future General of the Army Dwight Eisenhower turned down a job with the Hearst newspaper chain that offered triple his military pay. This during a period in which the Hoover administration mandated a 10 percent across-the-board pay cut for the army in 1930.[98] Robert L. Eichelberger, future commander of the Eighth Army in World War II, was one of Eisenhower's Leavenworth classmates. As Eichelberger recalled, "There are ampler ways of making a living than a professional military career. But I guess it must have its satisfactions; despite business offers I never seriously considered getting out of the Army."[99] Similarly, but earlier in his career, future Fleet Admiral Chester Nimitz declined an offer of $40,000 per year to work for a commercial company when he was making $2,880 as a junior naval officer.[100] The professionals stuck it out in a period of increased financial retrenchment and public indifference and even hostility. For most of the interwar period, the Army could not afford large formations, new technology, or even adequate manpower levels. With plenty of time on their hands, few units to command, and little modern equipment with which to train, education became the central focus in preparing the military for war. Only in the military's postgraduate schools could any real preparation for war be made in the minds of the officers who would fight it.

Command and General Staff School

From 1930 to 1936, the Command and General Staff School at Fort Leavenworth, Kansas, continued its two-year program. Maj. J. Lawton Collins reported to the Command and General Staff School on August 29, 1931. He attended both first and second years, recalling that the second year was chiefly devoted to the logistics of large units, corps, armies, and army groups.[101] The 1933–34 second-year course devoted thirty-one of fifty-two conferences to army-level operations. The course included instruction on tanks, mechanized cavalry, and joint operations. The course included five lessons on overseas expeditions. Interestingly, there was as much instruction on the employment of the air corps as the field artillery in army operations.[102] There were frequent lectures on military history to illustrate large-unit operations. The army reduced the Leavenworth course to one year beginning in 1936. The Command and General Staff School then concentrated on tactics and logistics at the division level and below.

By far, the most remarkable document to come out of Leavenworth in the thirties was *Principles of Strategy for an Independent Corps or Army in a Theater of Operations*. Written in 1936 by anonymous members of the faculty, this text was remarkable because of the obvious influence of Clausewitz, its clarity in expressing operational concepts, and its analysis of the impact of modern warfare on operations within a theater.

The influence of Clausewitz permeated *Principles of Strategy*. The text stressed the importance of history in the study of campaign planning. Underscoring the school's belief in military history, the text asserted that "only a leader well versed in military history will possess those qualities which are found among the great captains of history."[103] Citing another famous Clausewitzian observation, the text affirmed that the role of chance meant that "the issue of battle is always uncertain."[104] To overcome this uncertainty, or "fog of war," the commander needed special qualities of character and determination. Such observations can be found in *On War*, where Clausewitz discussed them at length.[105]

Clausewitz's influence was even more evident in the text's discussion of mass and the strategy of annihilation. All other things being equal, mass—numerical superiority—decided the issue. In fact, the fundamental law of strategy is: "Be stronger at the decisive point."[106] *Principles of Strategy* strongly embraced the battle of annihilation and concluded that only the wide envelopment could achieve it.[107]

The operational concepts found in earlier Leavenworth texts were presented more clearly and forcefully in 1936. The three types of military art were reaffirmed as the conduct of war, strategy, and tactics. The conduct of war related to employing not only the armed forces but also political and economic measures in achieving national aims in war. Strategy was defined as "the art of concentrating superior combat power in a theater of war," which would defeat the enemy in battle. The text conceived of tactics as "the art of executing the strategic movements prior to battle and of employing combat power on the field of battle."[108]

This framework of military art allowed for other operational concepts included from earlier texts. In regards to successive or phased operations, it was noted that the commander "must look further into the future and must see beyond the battle itself."[109] Indeed, modern conditions meant that "final victory will be achieved only through a succession of operations or phases."[110] The notion of a culmination point was also discussed.

Principles of Strategy also included a new analysis of the changing nature of warfare and its impact on operations within a theater. It noted that perfection of road networks, plus the telegraph, radio and airplane, had "decidedly modified the art of war."[111] The new technology allowed for a greater distribution of force in a theater providing for more secure operations on exterior lines. The increase in range and firing rates of modern weapons made frontal attacks more hazardous and "as a result the envelopment from one or both sides is the object of modern strategists."[112]

Principles of Strategy insisted that mechanization, motorization, and airpower mandated wide envelopments to move around the flanks of the enemy. The mission of motorized units and tanks "must be to attack the flank and rear of the enemy, and to prevent the hostile withdrawal. Aviation and tanks must disrupt the lines of communication far in the rear."[113] The text recognized that modern warfare increased the importance of logistics. "Any enemy interference with an army's supply system has far reaching consequences. The larger the force the greater will be the consequences."[114] Wide envelopments made the most of the new mobility, targeted the enemy's lines of communication, and by preventing enemy withdrawal could be made more decisive. Although frontal attacks were discouraged, if a penetration was necessary, it should be done: "By massing a preponderance of force while economizing elsewhere, the commander plans to achieve an advance deep into the hostile formation. If this operation is successful, it is frequently decisive. It has for its object the separation of the enemy's forces into two parts and then the envelopment of the separated flanks in detail."[115]

This analysis compares favorably with the writing of the most prominent military theorists of the day. In fact, it could have been written by the German tank advocate Heinz Guderian, or the Soviet general Mikhail Tukhachevsky. Curiously, many of Tukhachevsky's ideas received official sanction when the Red Army published them as the *Field Service Regulations of the Soviet Union, 1936*. The main difference lay in the fact that Tukhachevsky saw mechanization as providing the means of deep operations, his preferred maneuver. While the Russians favored penetration leading to envelopment, the Americans leaned toward the German solution of wide envelopment.

Published the same year, *Principles of Strategy* went beyond this analysis to consider new approaches to strategy. It made a key assumption by stating, "Strategy is concerned with making an indirect approach accompanied by movements intended to mystify, mislead, and surprise the enemy."[116] The text went so far as to assert that if two armies confronted each other with their lines of communication secure, all their combat power present, and without being surprised, no strategy had been used at all."[117] This logically led to the emphasis on the enemy's flanks and rear and to wide envelopments.

The great British theorist Sir Basil H. Liddell Hart first proposed his thesis of the indirect approach in *The Decisive Wars of History*, published in 1929. He would later expand this operational construct into a strategic prescription in his *Strategy of the Indirect Approach*, which was not released until 1941. Original or not, *Principles of Strategy for an Independent Corps or Army in a Theater of Operations* was remarkable for its synthesis of modern thought, combining Clausewitz, the indirect approach, and modern technology.

How influential was *Principles of Strategy*? The Command and General Staff School hammered home the doctrine to such an extent that the Army Staff took issue with the emphasis on wide envelopments. The staff's objections were hotly debated in the War College.[118] Regardless of the debate, the text was quoted in lectures given by senior faculty at both the Naval War College and Army War College.[119]

After 1936, the Command and General Staff School reverted to a one-year curriculum. The army's desire for more graduates encouraged this move, which led to an inevitable decline in the amount of time available for the study of operational art. The one-year course became focused exclusively on division and corps tactics. Military history and logistics were still emphasized, but large-unit operations and army operations became the sole responsibility of the Army War College. The overlap between the two institutions in operational art virtually ceased with the introduction of the regular one-year course at Leavenworth. The Army War College continued to use the doctrine and student texts developed at Leavenworth with regard to large-unit operations, but the college persisted on its own path of educating students through exercises, lectures, and conferences on joint and combined operations.

The Army War College

On September 1, 1937, Col. Ned B. Rehkopf, assistant commandant of the Army War College, stood before the incoming class and delivered the general orientation lecture. He noted that the mission of the college was to train officers "for the conduct of field operations of the Army and higher echelons; to instruct officers in those political, economic and social matters which influence the conduct of war, to instruct officers in the duties of the War Department General Staff, train officers for joint operations of the Army and Navy; and to instruct officers in the strategy, tactics and logistics of large operations in the past."[120] The War College curriculum now began with a Preliminary Command Course (PCC) to pick up the instruction no longer covered at the General Staff School. This course covered the organization, tactics, and logistics of the army, army group, and GHQ. The remainder of the year consisted of the Preparation for War and Conduct of War courses.

The Preparation for War course began with the general staff or G courses. About one month was devoted to each staff section, but by the end of the decade, G-4 received more days of instruction than any other. The G-2 course surveyed the political, economic, social, and military features of the major world powers. This course provided the greater strategic context for the subsequent military planning in the Conduct of War course. Complementing the G-2 course, each week the students discussed and evaluated a foreign news article. Analytical Studies followed the G courses. This subcourse used history to examine leadership, high command, joint operations, and national defense organizations. Analytical Studies had long been a part of the curriculum, and depending on the commandant, the subcourse used military history to discover principles and search for trends, or as a laboratory for operational research.[121] Over 20 percent of the committee reports through these years dealt with joint or coalition subjects.[122]

Following Analytical Studies, the Preparation of War Plans subcourse allowed students to prepare plans from the national to theater level. The students did not work with actual war plans but used non-country-specific color plan scenarios: Green for a minor effort, Red for a major effort, and Orange for an overseas effort. The faculty advised students: "Do not concern yourselves with the probability or improbability of such wars nor in the useless conjecture as to whether we would or would not fight alone or associate with allies. Our task is to teach ourselves how to formulate war plans and in doing so we select different types to cover as many situations as time and personnel permit. However, the plans we do make should be as complete, logical and accurate as possible under the governing circumstances."[123] War planning and war-gaming at the War College were not meant to test war plans but to train and educate competent staff officers and future commanders in the art of war.

The Conduct of War course lasted until the end of the school year and consisted of map exercises and maneuvers. From 1928 to 1932, this course consisted of four two-sided map maneuvers. These map maneuvers included operations in Puget Sound on the Pacific coast, a defense of the Chesapeake and Delaware bays on the Atlantic coast, a campaign in the Allegheny Mountains, and the recapture of Luzon in the Philippines. Beginning in 1933, the college replaced free-play map maneuvers with faculty-directed map problems in which student staff groups solved operational situations. The faculty and fellow students critiqued the solutions, students then were presented with new requirements and situations.[124] The course included a strategic reconnaissance in the Northeast or in the Delaware Bay area, as well as a command-post exercise, when funds were available. Throughout the year, lectures supported instruction on a variety of subjects from international economics to history to specific military and naval topics—all provided by the best expertise available. The college's mission and the curriculum that supported it covered a good deal, but clearly the focus on large-unit operations, joint operations, logistics, and military planning in the theater provided American officers an education in operational art.

The faculty divided the students into committees or staff planning groups. The committees reported to the entire class the results of their study or briefed their solutions to operational problems. Classes met every day with morning and afternoon sessions except for Wednesday and Saturday afternoons. Despite the Saturday morning classes, for the students, living was easy. Omar N. Bradley, a future army group commander in World War II, attended the War College in 1933–34. Bradley was impressed by the difference between the War College and Leavenworth. At the War College, he recalled, "there was very little pressure. We were not graded on our work; there was no class standing to be achieved, no one of importance to impress."[125] In his spare time, Bradley organized a baseball team with Jonathan Wainwright as umpire, William F. Halsey as shortstop, and himself as pitcher.[126]

By the end of the decade, not much had changed. Maj. Maxwell D. Taylor entered the last War College class before the European war began in 1939. He agreed with Bradley that "there was none of the individual competition among the students which characterized Leavenworth; the year at the War College was a time for mature reflection on the broadest problems of the military profession in company with congenial fellow professionals, most of whom were destined for senior assignments in the approaching global war."[127] Taylor also noted that the focus of instruction was on the "military problems of the theater of operations and at the seat of government."[128]

Just as in the 1920s when many future senior commanders passed through the Staff School at Leavenworth, many of the same men attended the Army War College in the decade preceding World War II. All of America's future army group commanders and all but two of the nine future army commanders were students at the War College in the 1930s. The other two army commanders, Simon B. Buckner and Walter Krueger, had attended in the 1920s. These men would fight battles and wage campaigns all over the world in the coming war. In the decade preceding the great struggle, however, they waged battles and campaigns in exercises and on maps to learn their craft and develop the organizational and staff skills to achieve victory in modern war.

The lessons of World War I, as distilled in the curriculum of the War College in the twenties, continued to be studied and taught into the thirties. The emphasis on logistics remained evident in virtually all of the exercises and map maneuvers. The scale, scope, and detail in campaign planning became more refined and more sophisticated, particularly as war clouds gathered at the end of the decade. More than anything else, the specificity of war planning, particularly the War Plan Orange scenario, helped to develop meaningful and modern solutions to problems in operational art.

Logistics   "You need very few Napoléon Bonapartes in war, but you need a lot of superb G-4s."[129] That was the opinion of Maj. Gen. Fox Connor, given in a lecture to the War College class in 1931. General Connor spoke with some authority on the issue, having served as the G-3, deputy chief of staff for operations in the AEF. Here was the senior operations officer in the AEF unequivocally underscoring the importance of logistics. He noted that "since the war we have paid more attention in our schools to matters of personnel, intelligence and supply. But we are not yet anywhere strong enough in our attention to logistics, the details of supply and the technique of G-4 work."[130]

Connor was widely respected in the army, not only for his specific service with the AEF but also as an officer of wide learning and professional competence. George Patton introduced Dwight Eisenhower to General Connor, and subsequently Connor became Eisenhower's mentor. Eisenhower greatly admired Connor and remembered him as "the ablest man I ever knew."[131] While Major Eisenhower served with Fox Connor in Panama, the general impressed upon him the value of military history and insisted he should read Clausewitz's *On War*.[132]

The faculty increasingly recognized that one of the key differences between tactical and operational art is logistics. In a lecture on the "Strategy of Supply," Maj. Gen. William D. Connor, former commandant of the War College, specifically made the point: "In this summary that I have hastily sketched, we have seen the functions of the leader, the Commander in Chief, change materially. Originally, they pertained mainly to tactics and did not need to take supply matters into account. Conditions have changed so that today tactics have passed entirely out of his list of functions and supply matters have come to occupy a predominating position in those functions."[133] The War College continued to emphasize the connection between operational art and logistics. As early as 1931, the faculty had begun providing fewer situations to the student groups so that "more time was given for the careful working out of the logistic features of operations."[134] By 1937, the Army War College was even teaching naval logistics.[135]

In exercises, the faculty demanded the students pay attention to logistics. In his critique of the Command Post Exercise in 1938, Lt. Col. B. Q. Jones criticized the students serving as army commanders because "not enough stress was laid upon logistics, the es-

tablishments and operation of the line of communication in the rear."
He further echoed Maj. Gen. William Connor by emphasizing that in
the conduct of modern war, "the commanding general once his plan
has been launched, becomes a logistics general and he must delegate
naturally, without losing touch, the conduct of actual combat to his
subordinates, [and] that the success of operations from then on depends upon careful attention and certainty that the logistics arrangements shall be maintained in spite of all operations."[136]

The emphasis on logistics came naturally to U.S. officers raised
in an industrial society that outproduced most of Europe. Americans
were good at it. Arguably, the army's frontier tradition also contributed to its attention to supply. The army projected force for sustained
periods of time in remote frontier areas. In these situations, logistics,
even if on a much smaller scale, could mean the difference between
success or failure. More important, in modern warfare, the requirement for an overseas expeditionary force demanded it. Projecting the
kind of power needed to wage mass industrial warfare across oceans
required attention to detail and a thorough understanding that logistics determined the art of the possible. For the army students, in their
studies, their planning, and their exercises, logistics always held a
central place in their understanding of operational art.

Campaign Planning   One of the distinguishing features of American campaign planning in the interwar period was the sensitivity
to national policy. Unlike the German General Staff prior to World
War I, the American officers were not purely military technicians.
They were very aware of the political, economic, and even social
dimensions of strategy in which they must plan campaigns. Some
historians have criticized U.S. war planning in the interwar period as
unrealistic.[137] The fact is the United States government had no single
institution to bring together national security planning into a single
forum. The Joint Board occasionally solicited input from the Department of State, but without success.[138] Although the American military established a planning system from the national level down to
the theater level of operations, there was no way effectively to coordinate with the other federal agencies or departments. In the absence
of guidance from above, the military planners simply developed their
own national objectives, as well as their appreciation of the economic, political, and even social strategic context. Inasmuch as oper-

ational art must bridge the tactical and strategic levels of war, this was a critical step in campaign planning.

Many, of course, recognized the problem with this approach to strategic planning. A committee of students at the Army War College charged with reporting on war planning in 1931 noted: "It is unwise to accept the statement of the national objectives made by the Joint Board, a purely military body. The statement of the national objectives, political, commercial, and economic, lacks the support and concurrence of those agencies of the U.S. charged with the determination of policies governing those objectives."[139] In the question-and-answer period following the presentation, an instructor, Col. Leon B. Kromer, noted that even if the Department of State participated in Joint Board planning, "no representative of the State Department could commit the government to a definite political line. After all, circumstances will determine just what the political objectives will be, because of the fact that we have no one enemy and our principal objectives therefore will be subject to the combinations that come at the time."[140]

Still, the Joint Board made war plans. Certainly, some of the American war plans maintained and updated from the twenties to the thirties were unrealistic—War Plan Red (war with Britain), for example—but War Plan Green (Mexico) or War Plan Orange (Japan) were well within the realm of the possible. Regardless of their likelihood, the joint estimates demonstrated that military planning flowed from national objectives and required an understanding of political, economic, and social context. The joint estimate for War Plan Orange was a model of detail and insight. In another question-and-answer period following a discussion of joint plans in September 1932, the War College commandant, Maj. Gen. George S. Simonds, insisted: "While I agree with what the committee put over about the desirability of calling all the other people in, the Joint Planning Committee hasn't always fallen down on taking into account the things they spoke of. I have in mind the joint estimate of the situation of the Orange Plan. The one I have in mind is perhaps out of date now, but they did go into that with a great deal of detail as to the political, economic, and social conditions of the United States and Japan. They got up a fine document."[141]

Since practically all of the planning scenarios used in the War College for exercises and map maneuvers were based on existing war plans, the students were not allowed to use or consult the real plans.[142] Nonetheless, the detailed student plans reflected this appreciation for strategic context. In briefing their solutions, students frequently began with a statement of national objectives. Student joint plans included a diplomatic and economic annex, as well as measures requiring cooperation with other government departments.[143] The sophistication in campaign planning increased significantly in the thirties and included for the first time coalition (allied) as well as joint planning.

In 1934, Major General Simonds reestablished a faculty war plans division that included his air corps and navy instructors. He also introduced into the War Plans course a problem dealing with coalition warfare. For the next six years until the War College ceased classes in 1940, at least one student committee developed a war plan called "Participation with Allies." The first of these student plans dealt with a coalition composed of the Soviet Union, Britain, China, and the United States pitted against the Japanese Empire. U.S. Navy captain William F. Halsey prepared the plans offered as options open to the allies. Col. Jonathan Wainwright served as chief of staff for the student group and prepared the war aims of each ally. The plans generated by this committee incorporated much of the operational design developed in earlier years. In the resulting scenario, Japan was involved in major ground operations against the Soviets in Manchuria and threatened U.S. and British possessions in the Pacific. The center of gravity of the campaign was determined to be the Japanese army and fleet. The Soviets were to remain on the defensive until the combined British and U.S. campaign provided an opportunity for a crushing allied counteroffensive.

The plan envisioned four phases, which brought the Allied (British and American) main effort up from the south. In the first phase, British and Chinese land and air forces from Hong Kong operated against the Japanese forces in Fujian (Fukien) Province. In the second phase, the allied fleet with a corps of U.S. ground troops penetrated Japan's Pacific defense line and conducted joint operations against the Japanese in Shandong (Shan-tung) Province. In the third phase, the allied air forces isolated the Japanese in Korea by bombing their lines of communication. Joint operations then secured Korea and coalition forces advanced on Mukden (present-day Shenyang). At this time, the Soviets began their counteroffensive, which resulted in a massive allied envelopment of enemy forces on the mainland. The final phase called for operations against the Japanese home islands to end the war.[144] The operational employment of airpower, phased joint and combined operations, and even operational envelopment are all evident in this concept of operations.

In April 1934, during the question-and-answer period following the students' presentation of their plan, the commandant observed: "Although there is a comparative calm in that region right now compared to what it was a short time ago, if you have been following the world situation, as I hope you have, it must be very apparent that there are conflicting interests which are moving toward inevitable conflict of various sorts, which may end up in armed conflict. In that theater it is quite evident that probably the first actors to come on the stage will be Orange (Japan) and Pink (Soviet Union)."[145] Three years after the commandant's prophetic remarks, Japan invaded China. Within another two years, the Soviet Union clashed with Japanese forces in the massive battle at Nomonhan along the boundary between Manchuria and Mongolia. In 1941, Great Britain and the United States joined China in the war against Japan. Three years later, the Soviet Union joined the coalition against Japan.

Subsequent committees developed coalition plans against a Nazi confederation in 1935, a central coalition of Germany, Italy, Austria, and Hungary in 1936, and in 1937 against an enemy coalition of Germany and Japan. These were theater strategic plans in which the students did not produce any detailed plans for theater operations. That sort of work was reserved for routine class problems dealing with the color war-plan scenarios. Typical was the requirement for student groups working the Green and Orange scenarios in 1936, which mandated they produce a joint estimate, a joint plan, an army strategic plan with three prescribed annexes, and a theater of operations plan.[146] The kind of detailed campaign planning called for in classwork built on progress made during the twenties, specifically phasing, joint operations, unity of command, and as already discussed, logistics.

Committees charged with examining the problems of war planning in joint operations were consistent in their conclusions. In terms of format, the five-paragraph field order remained the recommended template for theater operations plans. Improving upon planning formats from the previous decade, the operations paragraph now listed major subordinate forces with assigned tasks by phase.[147] The students generally used this approach in their classwork. Like their predecessors in the 1920s, the students continued to favor unity of command in joint operations. The official doctrine still allowed for either unity of command or command to the service with paramount interest. Expressing the common view, a committee in 1931 recommended that the principle of paramount interest be discontinued and that "the system of unity of command, as defined in the pamphlet, be used in coordination of all combined operations."[148]

Toward the end of the decade, the creation of joint staffs to serve the commander of a joint force was also noted. In a lecture on joint operations by Maj. Charles Bolte, the question-and-answer period highlighted the need not only for unity of command in joint operations but also for joint staffs. When asked if joint staffs were necessary, the instructor replied they were "absolutely essential." The commandant, Maj. Gen. John L. DeWitt, confirmed, "Joint staffs, I think are necessary and if they work personally together, if their personalities fit in, your chances for success are increased a hundred fold."[149] These discussions invariably arose as the students developed plans for expeditionary warfare involving land, naval, and air power. Students worked hard and wrote incredibly detailed plans using fictional or even unrealistic scenarios, such as war with the British Empire. These exercises still provided valuable staff training, but the most important progress in American operational art was made with the most demanding, realistic, and pressing problem of expeditionary warfare in a potential war with Japan.

The Impact of War Plan Orange   War Plan Orange originated in 1906 amid anti-Japanese rioting and expanding U.S. interests in the Pacific following the Spanish-American War. Tensions with Japan arose in that year over treatment of Japanese immigrants on the U.S. West Coast. Continuing immigration from Japan combined with American racial bigotry sparked anti-Japanese riots in California in 1907. Concerned about the increasing tensions, President Theodore Roosevelt asked the Navy if it was studying how to fight Japan.[150] For the next three decades, both the navy and the army studied hard to devise a plan to defeat Japan. Through most of its existence, War Plan Orange considered three phases: Japanese forces overrun U.S. possessions in the Pacific, the U.S. Navy and an expeditionary force advance to the Far East to recover them, and finally, U.S. forces establish an economic blockade that would force Japan to capitulate. Throughout the interwar period, there was a good deal of debate about how the second phase should be accomplished.

Initially, the intent was to build a base that would secure the Philippines and support the American fleet in offensive operations. The U.S. government refused to put money into the project, and in 1922 it traded away this option for mutual arms reductions in the Washington Conference. That left American war planners only two other options: a short war scenario in which the fleet rushes to the Philippines in order to save the army garrison and force an early decisive battle with the Japanese fleet, or a long-war scenario involving a step-by-step advance across the Pacific. By 1934, the step by step advance across the mandated islands became the approved solution. War Plan Orange was the most realistic, most studied, and most productive of all the planning scenarios devised in the interwar period. The plan helped to drive organization, doctrine, and the American approach to operational art.

Any drive across the vast expanse of the Pacific naturally held great ramifications for American seapower. Logistics drove the plan. If the navy could not count on an existing base in the Far East, then it must be prepared to seize bases along the way. Recognizing the importance of airpower, the navy would have to bring its own airpower along with it. This encouraged the navy's acceptance and support for naval aviation—aircraft carriers. Seizing islands also required troops trained for amphibious assault. The adoption by the Marine Corps of amphibious operations as a primary mission was not only the result of the desire for an institutional niche in modern war; it was an operational necessity for the navy.[151] For the army, the requirement to defend or retake the Philippines and provide an expeditionary force for the advance across the Pacific was uppermost on planners' minds. In every case, this expeditionary warfare required combining all elements of military power in a particularly challenging theater.

Scenarios for War Plan Orange were studied in one form or another every year at the Army War College during the interwar period. More than any other planning scenario, this one forced the students to consider all the facets of modern operational art: phasing, unity of command, joint operations, logistics, and how to combine seapower, airpower, and landpower in the theater. Phasing was evident in every solution. In 1934, Maj. Edward Almond provided a typical phasing construct in an oral presentation of his solution: "You may recall the successive steps or operations contemplated in the plan after the capture of the island of Luzon were in the following order: Formosa, Okinawa group, Jakishima Group, Amanmioshia Group." These islands, according to Almond, would provide the land air bases necessary to bomb the Orange homeland. Like most planners, Almond believed that if the home islands could be bombed or blockaded into submission, "land invasion is of last resort."[152]

In 1936, Lt. Col. Orlando Ward presented his group's army plan that consisted of the step-by-step approach to retaking the Philippines:

> After conference with the Navy it was decided that the scheme of operations would initially have three phases; the first phase to include the seizure and securing of adequate fleet bases, initially in the Mandate islands and subsequently in the Southern Philippines; the second phase to include the progressive occupation of the southern and central Philippines and the assembly there of a force for the capture of the entire archipelago. The next phase is to include the capture and preparation of Luzon as a base from which the main Orange forces and strategic areas could be subject to attack by Blue [American] ground and air forces, supported by the Blue fleet.[153]

Each of the services had a vital role to play in the plan, and the students paid due respect to the importance of airpower. The students serving as naval planners in Ward's group insisted that basing in the southern Philippines was necessary until air superiority was achieved. They also noted that "the magnitude of the logistics involved [in a Pacific war] is appalling."[154]

The 1936 student plan for War Plan Orange was incredibly detailed. The plan filled twelve volumes that included a joint estimate, an army strategic plan, the navy basic plan, a Western Pacific theater joint plan, army and navy operations plans, even a joint basing plan. The joint estimate included a thorough diplomatic and economic appraisal, establishing national objectives for both the United States and Japan. The estimate contained the classic military decision-making process comparing courses of action and reaching a recommendation. The students planned for the necessary mobilization and concentration of the means in the basic army and navy plans. The Western Pacific theater joint plan called for unity of command in which the "commander will control and coordinate the operations and logistics of the two services."[155] Providing for the joint commander to direct logistics as well as operations was a significant innovation. Of the twenty-four officers participating in this planning group, nineteen would become generals, including the briefer, Lt. Col. Orlando Ward.[156]

The following year, another group briefed the inevitability of a long war with Japan. They decided to "seek the Blue victory primarily by a measured westward advance, with the main effort in the direction: Hawaii—Orange mandated islands—Brown (Philippines)—Orange homeland, coupled with economic and diplomatic pressure against exterior sources of Orange support." The group characterized their concept of operation as a "step by step steam roller advance."[157] They projected a five-phase sequential operation that envisioned the capture of the Marshall Islands, the Caroline Islands, bases in the southern Philippines, Luzon, and subsequent operations against Orange. They paid particularly close attention to timing the phases so that forces are concentrated in Hawaii to "exploit any success attained in the early operations."[158]

This plan highlighted a consistent feature of Pacific plans: the imperative for joint command. "In wars where joint action of the major services is a rarity, the parallel channels of command . . . will, as proven by history satisfy the normal need. But in a war in the Pacific, where joint action obtains almost continuously, this method is likely to fail. Unity of command is prescribed for all operations and the forces will be referred to as the Expeditionary Force and will include army and navy contingents."[159] (See figure 3.)

One of the more interesting and detailed student presentations took place in 1938. In this solution, the group considered basically three approaches to Japan: from the north by way of the Aleutians, through the Central Pacific, and up from the South Pacific by way of the East Indian Barrier. They rejected the northern approach as impractical. They rejected the Central Pacific approach as a frontal assault that would run into the strength of Japanese defenses. Instead, they chose an indirect southern approach combined with a diversionary operation threatening the central mandates. (See map 2.) The

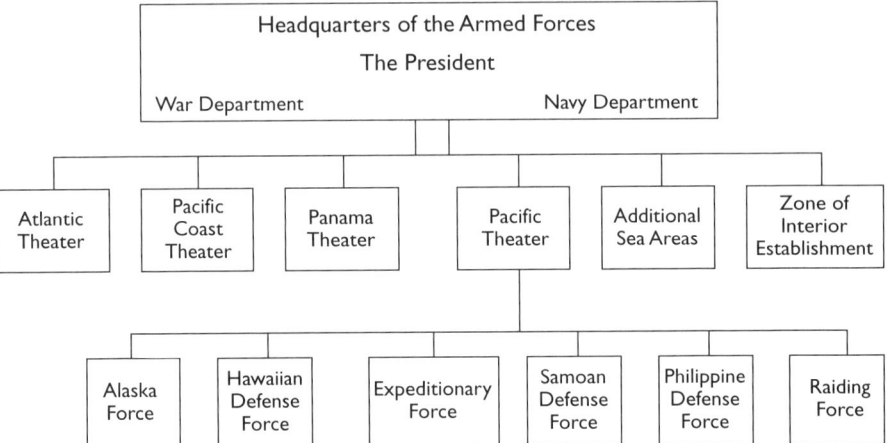

Figure 3. Army War College student solution for command in War Plan Orange exercise in 1937. Source: U.S. Army War College File, USAMHI.

students referred to their operational solution as an envelopment and were convinced this approach would be more decisive. They believed a quick strike via the southern approach would make for a shorter war. In defense of their course of action, the students asserted that the American people would not tolerate the high casualties of a long war required in the step-by-step approach through the Central Pacific.

Capt. Edward J. Foy, the naval instructor, soundly criticized the group not only for its psychological assessment of public will but also for attempting to shoehorn the fleet through the Torres Strait. From the navy's perspective, "the game is played over and over again at Newport and this particular solution invariably loses. I don't believe that anybody who has any conception whatsoever of war would attempt any such weird method of winning as this."[160] The army students were clearly more willing to accept operational risk to the fleet than was Captain Foy, but they admitted the "weakest element in the plan is logistics, because of the distance from home bases without the existence of intermediate bases."[161] The students vigorously defended their solution and provoked debates on joint command, the role of airpower, and the development of branch or alternate plans based on enemy actions. Regardless of the validity of this course of action, of more importance were the details in the planning.

The plan called for appointment of a single theater commander, the commander in chief of the U.S. fleet, who would assign missions

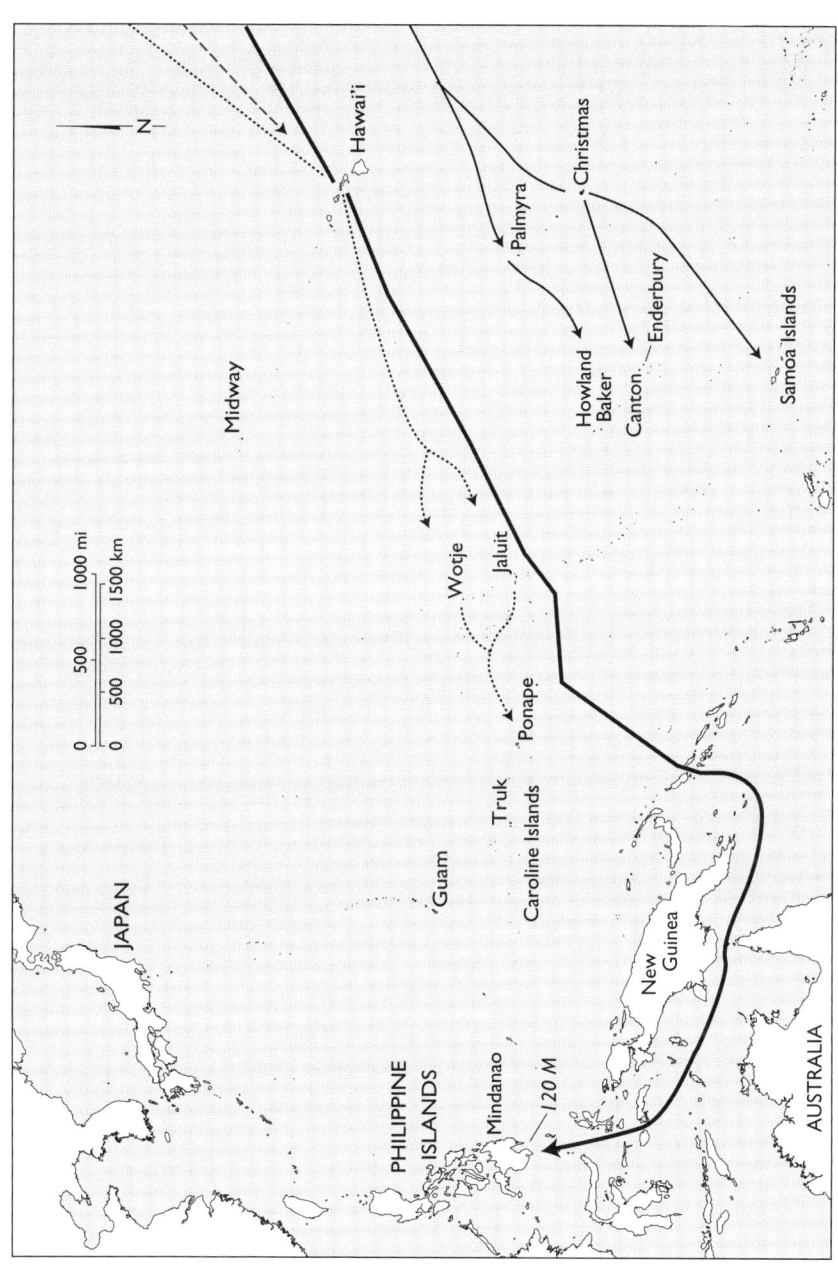

Map 2. Army War College student solution for War Plan Orange exercise in 1938.
Source: U.S. Army War College File 5-1938-21/1, USAMHI.

to the army expeditionary commander. The plan included sequels and branches providing options to the commander in the event operations went differently than expected. The inclusion of joint command, phasing, an indirect approach, detailed planning for logistics, even deception were marks of student sophistication in campaign planning. The commandant commended the group for their innovative approach and offered this final observation: "What we have had this morning is something we have wanted and which we want all the time: thoughtful, careful preparation, a full presentation, and a confident defense."[162]

The problems in planning for a war against Japan were significant. As it turned out, the navy pursued a deliberate island-by-island campaign across the Central Pacific, much like that envisioned in the later War Plan Orange. The army, led by Douglas MacArthur, pursued its own campaign along the southern approach from Australia up through New Guinea to retake the Philippines.

Of course, not every student solution was a model of detailed planning. The staff planning procedures, however, such as the military decision-making process comparing courses of action, developing estimates, and reaching a decision, were the same. Even those students whose groups did not war-game War Plan Orange benefited from the class discussion and debate on student solutions. As Omar Bradley noted many years later of his War College work, "Another group wargamed the operation of a half-million-man field army in the Philippines. I found that lecture valuable background later in the European war."[163]

The most significant American preparation for war in the interwar period occurred in the military's postgraduate school system. Particularly at the Army War College, this generation of military leaders developed their understanding of operational art, and it was this understanding that shaped the campaigns they would lead and ultimately contributed to their success.

ASSESSMENT

In the 1920s the army sorted out the lessons from World War I. These lessons focused initially on tactics and organization as derived from the many boards convened immediately following hostilities. The larger lessons in modern operational art were studied, discussed, and taught in the army's postgraduate school system. Both the General Staff School and the War College recognized the third level of war, the operational level. The War College particularly focused on large-unit operations. Modern warfare meant total mobilization and mass armies led by expanded staffs capable of planning and supervising vast operations. The geostrategic position of the United States required the projection of combat power sustained by enormous logistical efforts over vast distances. As a result, American officers learned to recognize the critical role of logistics in operational art.

The Command and General Staff School emphasized combined arms—cooperation between the branches of the army—the infantry, artillery, and cavalry. The War College emphasized joint operations—cooperation between the services. The increasing capability of airpower brought added importance to the projecting of combat power in the theater in order to achieve strategic objectives. Airpower, in particular, expanded the commander's ability to shape and influence operations in the theater. The theoretical elements of center of gravity or the hostile decisive element helped to focus that combat power. The national planning system helped to ensure that combat operations in the theater would be tied to national objectives.

In their study of World War I, U.S. officers extrapolated from the tactical phasing characteristic of American battles in that war to the need for phasing operations in the entire theater in future wars. Logically thinking their way through the application of force in the theater from deployment to concentration, to the maneuver of armies and army groups, they recognized the changing nature of decisive battle. Decisive battles might change the course of a campaign but could not decide the outcome of a war in an afternoon, a day, or (by World War I standards) perhaps even months. Modern war meant marshalling all the combat power available in the theater, seapower and airpower alike, to project, maneuver, and sustain large units in a planned sequence of operations.

Operational art as the employment of military force in a theater of operations or a theater of war to achieve strategic objectives was well understood by the American military by the end of the 1930s. This understanding grew largely from the traumatic experience of World War I. American confidence and optimism that the United States could meet any challenge in 1917 was tempered by the sober postwar realization among professional officers that their country had been unprepared for the sheer scale of modern warfare. Experience, theory, and strategic requirements continued to shape American operational art in the final decade before the next world war.

The employment of military power in distant theaters of operations called for an operational art that combined airpower, seapower, and landpower in expeditionary warfare. The solutions to this problem developed in the service schools included phasing, emphasis on staff skills, and recognition of the importance of airpower, logistics, and joint operations.

The need in World War I to coordinate infantry and artillery introduced American officers to tactical phasing. The development of operational phasing in the interwar period was driven by the need to plan the linking of battles and coordination with other services. The vast distances of the Pacific so challenging to logistics demanded basing to extend operational reach. This required phasing and attention to detail not only to project power but also to sustain it. Seizing bases as a way to extend and sustain airpower and seapower further encouraged a necessary phase-by-phase approach to campaign planning: Landpower could be combined with seapower to secure air bases, which in turn would allow airpower to support and extend the operational reach of all services. The army, the navy, and air corps could not each fight its own separate war against the enemy. They must fight the same war leveraging their capabilities against the enemy's center of gravity. This recognition of joint operations was fundamental in providing the flexibility required to win campaigns in the next war.

The U.S. Army by 1940 had come a long way from the days when General Sherman expressed such fierce disdain for staffs. In the interwar period, the army perfected the military decision-making process and staff procedures. The G-2 intelligence and G-4 logistics staff sections became planning staffs equal to that of the operations staff, responsible for handling the complexity of modern war. These staffs understood well by the end of the decade the demands of modern operational art. American operational culture at the end of the interwar period can best be characterized as a preference for expeditionary warfare conducted through joint operations dependent on mass and backed by an expertise in logistics. In the coming decade, facing the greatest war in history, U.S. officers were well served by their interwar preparation in the postgraduate military school system.

In response to the threat of world war in 1940, and to the pressing need for experienced officers to manage America's expanding services, the military closed down most of its postgraduate system. The army replaced the Staff School regular course with a series of abbreviated staff training courses. The War College shuttered its doors in the summer of 1941. The ultimate test for the army's leadership, the students of the interwar period, was less than six months away.

4

AIRPOWER

Airpower was born in America. On a windy day in December 1903 in Kitty Hawk, North Carolina, two bicycle mechanics launched powered heavier-than-air flight. Orville and Wilbur Wright labored long and hard pursuing their interest in flying. They may have believed that flying machines would benefit mankind by improving commerce and transportation, but they looked to the United States Army as their first customer. By 1905, they felt confident enough to write the secretary of war that "flying has been brought to a point where it can be made of great practical use in various ways, one of which is that of scouting and carrying messages in time of war."[1] After several attempts, the Wright brothers got the attention of the military authorities and in 1909 sold the army a Wright flyer for $30,000. The aircraft's obvious and intended use in reconnaissance and in communication initially consigned U.S. military aviation to the army's Signal Corps.

By 1914, the army addressed the role of aviation in military doctrine. Field Service Regulations stipulated that its primary role was either tactical or strategic reconnaissance. Strategic reconnaissance by "aeroplane is effective within a radius of 150 miles from the starting point and is for the purpose of determining the position, strength, and direction of advance of the large elements of an enemy's force and also the character of the roads, railroads, streams and the general military topography of the theater of operations."[2] Interestingly, the regulations also mention that the aero squadron should prevent hostile aerial reconnaissance, but do not say how this might be done.

The only practical experience American airmen gained prior to the U.S. entry into World War I was in connection with the failed Mexican Punitive Expedition in 1916. The First Aero Squadron equipped with eight old unarmed biplanes struggled to assist Brig. Gen. John J. Pershing to chase down Pancho Villa. Plagued by maintenance problems and difficult field conditions, the initial employment of American airpower promised much but delivered little.[3] By contrast, in Europe at the same time the Great War raged in the skies as well as upon land and sea. Airpower may have been born in the United States, but it matured in Europe.

THE WAR IN THE AIR

The military potential of powered flight drew plenty of interest from European armies already engaged in an arms race prior to World War I. Orville Wright traveled to Europe to drum up interest as early as 1910. In that year he secured a contract from the German army for five Wright planes.[4] Soon other aircraft companies, such as Bleriot, Farman, and Latham, sprang up and competed for military contracts. There was also a good deal of interest in the military utility of lighter-than-air ships. The Germans ranked first in airships under the leadership of Count Ferdinand von Zeppelin. At the beginning of World War I, the great airships known as zeppelins seemed to possess significant advantages over the airplane. In range, lift, and endurance the airship far exceeded the early airplanes in providing the potential not only for long-range reconnaissance but perhaps operational or strategic bombing. Eventually, however, airships proved to be a technological dead end. Rapid and steady advances in aviation technology favored the airplane. Within a decade of the Wright brothers' first powered flight, all the major powers of Europe had incorporated the winged aircraft into their military establishments.

By 1914, the major powers accepted the utility of airpower at least in reconnaissance, if not potentially as a supporting combatant arm. European air services were organized into units with specific missions and were well on their way as an autonomous arm. The British established the Royal Flying Corps in 1912. The French followed suit two years later, creating the Direction de l'aeronautique militaire.

By 1913, the Germans organized five aviation battalions and formed a separate inspectorate for flying troops. The first German aviation doctrinal manual written in that year specified strategic and tactical reconnaissance, artillery observations, dropping bombs, and fighting other aircraft as missions.[5] At the outbreak of World War I, the Germans fielded 232 airplanes. Russia began the war with 190 aircraft, and France had 162.[6]

At the beginning of the war, perhaps airpower's greatest contribution was, in fact, at the operational level. As expected, aircraft greatly expanded the ability of army commanders to see the battlefield. In 1914, as German armies swept through northern France, French aircraft helped to keep the French high command informed of enemy movements. The information provided by aerial reconnaissance in part helped to confirm Gen. Joseph Joffre's decision to fight the Battle of the Marne. Even more directly, as General Hindenburg maneuvered German armies to smash the Russians on the eastern front at the Battle of Tannenberg, he noted, "Without the airmen, no Tannenberg."[7] As the war of maneuver settled into trench warfare, particularly on the western front, the belligerents looked at all the possible ways to exploit the new form of warfare.

Within three years, the war in the air expanded significantly. By the time the United States entered the war in 1917, the basic outline of the tactical, operational, and strategic employment of airpower was apparent. At the tactical level, aircraft spotted for artillery, engaged in close air support of troops, and contested for local air superiority. At the strategic level, the Germans began bombing London using airships as early as 1915. As aviation technology advanced, the influence of airpower at all the levels of war grew along with its increased capabilities. Although in the grand scheme of the war, airpower's overall influence may have been peripheral, it was certainly suggestive for the future. By 1917, giant German Gotha bombers replaced the zeppelins in bombing London and Paris. At the tactical level, bombers became an extension of the artillery. As capabilities improved, aircraft extended their attacks to operational depths targeting troop concentrations, railway junctions, depots, and supply lines. By the time the United States entered the war, airpower had demonstrated its ability to influence operations at the tactical, operational, and strategic levels. The Europeans had developed organizations and a fair amount of doctrine, if not theory, on the employment of airpower based on a good deal of experience. Into this revolution in warfare, the United States Army's aviation detachment of the Signal Corps brought sixty-five officers, eleven hundred men, and two hundred obsolete airplanes.[8]

In the summer of 1917, the French premier, Alexandre Ribot, cabled Washington asking for forty-five hundred airplanes, five thousand pilots, and fifty thousand mechanics for the 1918 campaign. This request formed the initial basis for the American air program. Unfortunately, the French premier's faith in American mobilization was not matched by the War Department's efficiency or preparedness. Officers were soon dispatched, however, to seek out designs and aircraft that might be quickly produced in the United States. The American officer who eventually became one of the most controversial and influential advocates of airpower was already in France. Lt. Col. William "Billy" Mitchell was sent to France as an observer, arriving there four days after the U.S. declaration of war. Mitchell was a regular officer commissioned from the ranks. Born in 1879 the son of a wealthy U.S. senator from Wisconsin, he grew up in Milwaukee and enlisted in the army as a private during the Spanish-American War. He served in Cuba and the Philippines, eventually obtaining a commission in the Signal Corps. After attending the staff college at Leavenworth, Mitchell served on the General Staff. Fascinated by flying, he learned to fly at his own expense and transferred to the aviation section of the Signal Corps. In the spring of 1917 he found himself the senior American aviator in France.

Mitchell sought out senior French and British air officers and began to shape his own ideas about how aircraft should be employed. In May, Mitchell met with Maj. Gen. Hugh Trenchard, head of the Royal Flying Corps in France. Trenchard insisted to Mitchell that "the only way to handle airpower is to unify it all under one command." The British air commander also impressed upon the American the need to mass airpower and use it offensively. Mitchell noted that the British were in the process of building up an independent air force "designed to attack the interior cities of Germany."[9] Trenchard touched upon some of the most critical questions for the employment of airpower, among which were who would command it, and how would it be employed.

In the summer of 1917 General Pershing moved the AEF Air Service from the Signal Corps and established it as a separate branch. By

October, Mitchell drafted the "General Principles Underlying the Use of the Air Service in the Zone of Advance." Following the practice in other Allied armies, Mitchell divided aviation into two categories: tactical and strategic. Tactical aviation included observation, tactical bombardment (within twenty-three kilometers of the front line), and pursuit—air-to-air combat. The mission of strategic aviation was to destroy enemy depots, factories, lines of communication, and personnel. Daytime bombardment also provided long-range reconnaissance and encouraged "hostile airplanes to rise and accept combat." Night bombardment simply aimed to destroy distant enemy targets.[10] The potential for strategic bombing also caught the eye of another officer, Maj. Edgar S. Gorrell.

Major Gorrell accompanied an American team to Europe to determine what aircraft the United States should produce. Gorrell met Giani Caproni, an Italian aircraft engineer and a good friend of Guilo Douhet. Douhet is often cited as one of the foremost theorists of airpower. His influence on American air theory, however, appears to be more indirect than direct.[11] Caproni handed Gorrell a pamphlet entitled "Let Us Kill the War, Let Us Aim at the Heart of the Enemy" by Nino Salvaneschi, an Italian journalist. The pamphlet reflected many of Douhet's ideas on the potential of airpower, particularly the need to attack the enemy's economic and industrial means of war, rather than the enemy army. Gorrell passed the pamphlet around the air service and in November 1917 wrote a memorandum for a U.S. strategic bombing plan for 1918.

Gorrell's memorandum called for a strategic bombing effort to "drop aerial bombs upon the commercial centers and the lines of communication in such quantities as will wreck the points aimed at and cut off the necessary supplies without which the armies in the field cannot exist."[12] Gorrell outlined four industrial target groups for around-the-clock bombing at a range of 150–200 miles from air bases. Brig. Gen. Benjamin Foulois, the chief of the air service at that time, promptly approved the plan and assigned Gorrell as the head of Strategical Aviation. In December Gorrell attended a meeting of Allied air chiefs to discuss the strategic bombing plan for 1918. The British were well on their way to establishing their own independent bombing force specifically organized for this kind of strategic bombing. Pershing insisted that the U.S. air arm remain tied to the needs of the ground force. To make sure everyone understood, the name Strategical Aviation was changed to GHQ Air Service Reserve. In any event, the difficulties in American industrial mobilization limited production to only one strategic bombardment squadron, fielded two days before the Armistice.

The problems in management and production in the United States meant that the air service, like U.S. ground forces, would largely be equipped by the Allies. Getting the AEF Air Service off the ground posed large problems not only in equipment and training, but command as well. Brig. Gen. William Kenly had replaced Mitchell as the chief of air service in September 1917. Kenly soon returned home, however, and Foulois took command. Foulois was one of the first American aviators, but Pershing was unhappy that after a year of effort the air service seemed to have made little progress. He turned to an old West Point classmate, Brig. Gen. Mason M. Patrick. Pershing told him, "In all of this Army there is but one thing which is causing me real anxiety. And that is the Air Service. In it there are a lot of good men, but they are running around in circles. Somebody has got to make them go straight. I want you to do the job."[13] Patrick, an engineer, had no experience of aviation, but he was a fine administrator. Patrick finally established a workable chain of command with himself as the AEF air chief concerned primarily with training, equipping, and providing air units to Pershing's ground forces. Foulois became Patrick's assistant overseeing air service support functions. Billy Mitchell became the chief of air service for the First Army responsible for the employment of the air arm.

The American air units worked for ground commanders. Corps commanders could depend on observation squadrons to help see the battlefield, spot for their artillery, and track the progress of their infantry. In keeping with his views, Mitchell successfully kept the majority of the air units centralized at the army level. As the chief of air service for the First Army, Mitchell developed and executed the air plans to support the major battles of the American Expeditionary Force. His first opportunity to show what airpower could do was the battle for the St. Mihiel salient.

In September 1918, Mitchell was able to mass 1,481 aircraft to support the American attack on the St. Mihiel salient. This air armada included 701 pursuit, 366 observation, and 414 bomber aircraft. Less than half were manned by Americans; the British, French, and Italians contributed the rest. Mitchell described his mission to "pro-

vide accurate information for the infantry and adjustment of fire for
the artillery of the ground troops; second, to hold off the enemy air
forces from interfering with either our air or ground troops; third, to
bomb the back areas so as to stop the supplies for the enemy and hold
up any movement along his roads."[14] Mitchell intended to employ
mass air strikes against one side of the salient and then the other.
Deep raids against the enemy rear would focus attention away from
the front lines, and quickly allow the Allied force to achieve air superiority. As the American First Army pushed the retreating Germans
out of the salient, the enemy could muster only 243 aircraft in the
sector by the third day of the attack.[15] Mitchell's force quickly established and maintained air superiority. Pershing commended Mitchell
for the air effort, and they both turned immediately to the next major
challenge, the Meuse-Argonne campaign.

The St. Mihiel salient was the American First Army's baptism of
fire, but the Americans met only light resistance as they hurried the
retreating Germans out of the salient. By contrast, the battle for the
Meuse-Argonne was a meat grinder. This forty-seven-day battle cost
122,000 U.S. casualties and sorely tested the American army in modern warfare. The Meuse-Argonne was part of the final offensives
planned by Foch to exhaust and crush the German army. In the fall of
1918, the Allies massed a total of 5,000 aircraft to seize control of the
air and hammer the German army. The Germans mustered only
3,000 airplanes of all types to oppose allied air strength.[16] As Foch
launched his near simultaneous offensives, the Germans proved unable to achieve permanent air superiority in any sector on the western front. The Meuse-Argonne sector protected the vital railroads
running parallel to the front and guarded the key approach to Sedan.
Unlike in the St. Mihiel battle, the Germans fought tenaciously with
some of their best troops to protect this vital area.

In support of the American attack, Mitchell controlled a total of
842 aircraft. Americans manned 604 aircraft, the remainder being
provided by the Allies. The plan for air support consisted of four
phases: advance preparation, during the artillery preparation, during
the attack, and exploitation. In each phase, the aviation units performed tactical tasks, but the bombardment and observation units
significantly extended the army's reach.

During the artillery preparation, the bombardment aviation was
"to harass the enemy by attacking his troop concentrations, convoys,

stations, command posts and dumps, to hinder his movement of troops and to destroy his aviation on the ground."[17] The air units targeted troop concentrations and convoys beyond the range of artillery, approximately ten to thirty kilometers behind the lines. During the attack, the air objective was "to prevent the arrival of reserves and to break up counterattacks."[18] In the last major operation of the war, aviation practiced its classic missions of achieving air superiority, engaging in reconnaissance, and interdicting enemy movements on the battlefield. In these missions lay aviation's future contribution to modern operational art.

American infantrymen went over the top and on the offensive on September 26, just ten days after the conclusion of the St. Mihiel battle. On the first day, the Germans opposed Mitchell's air force with 302 aircraft. As the other allied attacks began, the Allies withdrew some of their aviation units to support their own offensives. By mid-October, Mitchell's force dwindled somewhat to 756 aircraft, while the Germans reinforced their air units in the sector to a total of 504 aircraft. By November, Mitchell had 697 planes to oppose 486 German aircraft.[19]

Consistent with his views, Mitchell continued to mass air units and send them on deep raids. During the tough fighting, Pershing reined in Mitchell to focus on the close tactical fight. As the infantry battle stalled against the German defense, Pershing wanted to focus as much combat power as possible to help get his troops forward. He believed that "the tendency of our air force at first was to attach too much significance to flights beyond the enemy's lines in an endeavor to interrupt his communications. However, this was of secondary importance during the battle, as aviators were then expected to protect and assist our ground forces."[20] Cooperation with ground forces needed to be improved in order to be effective, and soon Mitchell had fighter pilots visiting the trenches and infantrymen visiting airfields to participate in actual flights. Following Pershing's orders, Mitchell kept his air force tied closely to the immediate battle. The issues of who should control air assets and how they should be used would dominate discussions on airpower in the postwar era and into the next war. Clearly, at this stage the ground commanders insisted that airpower's purpose was to support the ground forces and their immediate requirements to see and influence the battlefield.

Even though Pershing made sure the AEF Air Service remained in close support of the infantry, Mitchell did seize opportunities to demonstrate the potential of airpower in the larger role of battlefield interdiction. On October 9 in the woods near Damvillers on the east bank of the Meuse River, the Germans concentrated a considerable body of troops to threaten the right flank of the American III Corps. Mitchell assembled 200 bombers and 110 fighters to strike the German force. In two huge formations, the bombers unloaded 39 tons of bombs, disrupting the counterattack.[21] This was the most striking example of the war of the potential for air units to extend the operational reach of the army not only through reconnaissance but by projecting combat power.

The Meuse-Argonne campaign was the U.S. Army's most meaningful experience of modern warfare in large-unit operations during World War I. The First Army and later Second Army functioned both tactically and operationally as part of the Allied war-winning multi-offensive assault on the western front. The lessons drawn from this experience shaped the American understanding of modern war for years to come. The tactical role of airpower seemed clear enough, but its strategic and operational role appeared to be limited by the state of technology. The experience of World War I, however, at the strategic and operational levels was suggestive of the potential of airpower. The primary problem during the war was the inability of landpower to achieve a penetration of enemy defenses to an operational depth. For its part, airpower demonstrated an ability to strike at operational, even strategic depths. The open question was how effectively airpower might be able to do this in the future and how this might change the dynamics of modern warfare. As aviation technology improved, the promise of airpower ignited a debate that began almost immediately in the postwar period and lasted throughout the interwar period and beyond.

The Twenties

Perhaps the clearest transformation in warfare wrought by World War I was the advent of airpower. Airpower's inherent range and flexibility brought a capability to influence operations throughout the theater. Thinking through the new impact of airpower on warfare, the air service's more fervent advocates and the established authorities for landpower frequently reached different conclusions. To the advocates of airpower, the experience of World War I showed a great deal of promise; however, the role of airpower in modern warfare according to American doctrine in the interwar period was determined not only by experience but also by technology, the budget, and the bureaucracy.

The army documented the official lessons of the war with regard to aviation in the AEF Superior Board's report on aviation. The report's conclusion did not look to the future, but simply confirmed that during the war aviation's major contribution was in reconnaissance and ground attack rather than distant bombing. The report added, "Nothing so far brought out in the war shows that aerial activities can be carried on independently of ground troops, to such an extent as to materially affect the conduct of the war as a whole."[22]

Brig. Gen. Billy Mitchell was one of the most public and fierce advocates of airpower. After the war, Mitchell became the director of military aeronautics and gathered around him veteran airmen whose ideas about airpower were often contrary to the official views of the War Department. He lobbied for an independent air force and constantly advanced the claims of airpower's new role in warfare. During a congressional hearing on an appropriations bill in 1921, Mitchell challenged the navy to permit a bombing test of warships. The subsequent test that sank the captured German battleship *Ostfriesland* and the cruiser *Frankfurt* may have helped prove Mitchell's point, but it made him few friends in the military establishment. Eventually, Mitchell's criticism of the administration of aviation affairs following the crash of the navy dirigible *Shenandoah* led to his court-martial and subsequent conviction for conduct prejudicial to the service. He resigned from the army, but remained a public figure for some time. In fact, he is better known as a publicist for airpower in the twenties and thirties than for any detailed theories about its employment. In his best-known work, *Winged Defense* (1925), he argued for the creation of a Department of National Defense that would include three separate services: the U.S. Army, U.S. Navy, and a Department of Aeronautics.[23]

Despite Mitchell's concern for the independence of the air service, the new combat arm made steady institutional progress in the twenties. The National Defense Act of 1920 officially made the air service a branch of the army. In 1926, the Air Service was redesignated the air corps and given a separate assistant secretary of war responsible for aviation matters. The relationship of the air corps to the army was now analogous to the that of the Marine Corps to the navy. The Air Service Tactical School at Langley Air Station was likewise redesignated the Air Corps Tactical School (ACTS) in that year. At the end of the decade, the school moved to its final home, Maxwell Field, near Montgomery, Alabama. From its inception to its closure in 1940, the ACTS was the source of American air theory.

In the early years the air service intended the school to prepare field-grade officers for higher command. The shortage of air service field-grade officers quickly changed the focus of the curriculum. Company-grade officers made up most of the twenty to thirty students attending the nine month course throughout the 1920s. Since there was little material on the actual employment of airpower, the faculty necessarily included a good deal of tactics and subject matter covered at the General Staff School at Leavenworth. In fact, Maj. Ira C. Eaker, future commander of the Eighth U.S. Air Force, believed the curriculum had two purposes: first "as a preparatory course to the Command and General Staff School at Leavenworth"; and "second for the education of the Air Corps officer in his own arm—the Air Force."[24]

The course certainly followed the familiar applicatory method used at Leavenworth. Conferences on a variety of subjects were followed by an illustrative problem discussed by the entire class. Faculty then issued the students individual problems, often map problems, that they then graded using school-developed solutions. The problems required the students to act as group commanders and write paragraphs two (mission) and three (execution) of the standard five-paragraph field order. In addition to instruction on the other combat arms, specific classes on airpower included each of the arms of aviation: bombardment, attack, pursuit, and observation. Although each of the air service students were qualified pilots, a practical flying course was added in 1923. More and more of the curriculum throughout the twenties was focused on airpower as the faculty began to develop their own texts and an increasing confidence in the potential of military aviation. Throughout the interwar period, the vision of airmen at the school grew along with the promise of advancing technology, but there was a growing divergence between what was being taught at the school and what made it into official doctrine.

One of the officers very much involved in the development of air doctrine in the twenties was Maj. William C. Sherman. Major Sherman was more of a moderate among the early group of American military aviators and one of the lesser-known pioneers of American air theory and doctrine. Born in Augusta, Georgia, in 1888, Sherman graduated from West Point in 1910. He served in various staff duties during World War I, but eventually returning to aviation as the chief of staff of the First Army Air Service in 1918. Following the war, he was assigned to ACTS, where he literally wrote the book on airpower.

In 1921, Sherman authored a text on air tactics for use by the school. Two years later, the air service issued this document as *Training Regulations No. 440-15: Fundamental Principles for the Employment of the Air Service.*[25] After review by the Command and General Staff School, the Army War College, and the army staff, TR No. 440-15 became official U.S. Army doctrine in 1926.[26] Not surprisingly, the approved regulations adhered to the army position that organization and training of members of the air service is based on "the fundamental doctrine that their mission is to aid the ground forces to gain decisive success."[27] Observation units were assigned to divisions, corps, and armies; attack and pursuit units were assigned to field armies, while an Air Service General Headquarters (GHQ Air Force) maintained a reserve of attack and bombardment units.

The GHQ Air Force was an operational tool for the theater commander during wartime. According to doctrine, the theater commander organized the GHQ Air Force into large units, highly mobile and capable of effective action within the theater of operations or against distant objectives. The regulation stressed that upon the outbreak of war, the first priority was to secure control of the air. Air units then served the double purpose of assisting the ground forces directly by joining in the ground battle and indirectly by operating against hostile lines of communication.[28] The air service employed attack aviation only to the depth of the enemy's rear corps area. Bombardment aviation might be used directly to assist ground troops, but it was best employed in the combat zone attacking enemy communications centers, ammunition and supply depots, troop concentrations, and transportation lines. Although granting its subordinate role to the theater commander, the doctrine did allow that bombardment units could operate "deep into hostile territory beyond the combat zone against targets which may be far removed from the field of battle, with the object of destroying sources of military supply, main lines of communications, mobilization, concentration, and military industrial centers."[29]

As technological improvements provided faster and more capable bombers in the coming years, strategic bombing became a central feature of instruction at ACTS, but not in army doctrine. Strategic bombing offered airpower the promise of institutional independence and a greater, perhaps even decisive, role in warfare. Officially, airpower, like the other army branches, assisted the infantry. Operationally, airpower's range and flexibility made it an ideal force for shaping and influencing the theater by attacking the enemy's lines of communication.

The Command and General Staff School and the War College taught official air doctrine rather than the unproved theories of strategic bombing that became increasingly popular at ACTS. In exercises and conferences, instructors stressed the role of airpower in assisting ground forces. In a conference on the tactics and techniques of air corps bombardment at Leavenworth in 1928, Maj. Oscar Westover, former commandant of ACTS, still stressed that the primary role for bombardment was in striking the enemy's industrial centers in accordance with a definite strategic plan. If these objectives were beyond the range of the bombers, then the lines of communication were the next best set of targets.[30]

The faculty and students at Leavenworth were less concerned about the strategic employment of airpower than they were about the ability of airpower to influence operations in the theater. In exercises, the faculty required the students to plan for the employment of an air division in attacks against the hostile lines of communication. A problem presented to students in 1928 asked them to plan for the employment of a Blue air division against Red forces attacking as part of a fictional country located in the central United States. The scenario located the Red capital in St. Louis with a reinforcing Red fleet moving up the Mississippi River. The options available to the Blue commander for using his air assets included attacking the capital, industrial centers, supply bases, bridges across the Mississippi, or the hostile fleet moving up the river. The approved solution called for a hard initial strike at enemy air bases, then switching to the enemy fleet, and subsequently concentrating on attacks against movements of hostile reinforcements and lines of communication.[31] The clear preference here was the need to shape the theater operationally rather than strike strategically deep into enemy territory.

Likewise at the War College, the students planned for the employment of airpower in accordance with official doctrine. In 1925, a committee of students prepared a study on the employment of air units that was eventually approved by the War Department for use at the college pending the publication of *Training Regulations 440-15*. It used much of Major Sherman's earlier Air Service text. The study asserted that the mission of the air service was to assist the ground forces to gain strategic and tactical success by destroying enemy air forces, then attacking enemy ground forces and objectives on land and sea.[32] The students noted that "while strategical bombardment does not involve direct cooperation with ground troops, it is always so employed as an integral part of the broad plans of operation of the military force, as to have a direct bearing on their success."[33] In 1927, Maj. H. C. Pratt lectured the War College class that most large-scale operations aim at cutting the hostile lines of communication. He acknowledged that airpower might not do this as effectively as ground forces, but that it could cut a line of communications at a vital point such as a river or along railroad lines.[34]

In 1926, Major Sherman collected his notes from lectures at both the ACTS and the Command and General Staff School and published them in a book entitled *Air Warfare*. This book represents the best expression of American thought on airpower during the twenties. Sherman emphasized the human dimension of warfare, that is, psychological and moral factors. He discussed at length the principles of war as related to aerial warfare. He established a priority for the employment of bombers that became the centerpiece of the air corps' internal doctrine in the coming decade. Sherman argues that bombardment should target population centers, the enemy system of supply, and fortifications, as well as provide for coastal defense. Specifically, "the long range of the bomber should be utilized to the full, and every sensitive point and nerve center of the system put under pressure in an effort to paralyze the whole."[35] This concept included attacks on the lines of communication.

Sherman asserted the other tenets of airpower: namely, centralized command, employment in mass, and the requirement for air superiority. The book contains a good deal of tactical and technical information, but he managed to cover practically every aspect of air

The Air Corps Tactical School at Maxwell Field in Alabama developed American theory and doctrine not only on strategic bombing during the twenties and thirties but also on the operational employment of airpower. Courtesy of the AFHRA.

This classroom at the Air Corps Tactical School during the 1930s demonstrates the emphasis on exercises and the influence of the applicatory method practiced by the army at the Command and General Staff School. Courtesy of the Air Force Historical Research Agency.

The employment of land-based airpower to support and coordinate efforts with naval forces was exercised at the Air Corps Tactical School, as depicted here. Courtesy of Air Force Historical Research Agency.

Maj. Gen. Henry H. Arnold (center), chief of the Army Air Forces, confers with several of the airpower theorists and future practitioners on the air staff in 1941. From left to right: Lt. Col. Edgar P. Sorenson; Lt. Col. Harold L. George; Brig. Gen. Carl Spaatz, chief of staff; Arnold; Maj. Haywood S. Hansell; Jr.; Brig. Gen. Martin F. Scanlon; and Lt. Col. Arthur W. Vanaman. All were graduates of the Command and General Staff School. Lt. Col. George and Maj. Hansell were particularly influential in developing U.S. views on the employment of airpower at the Air Corps Tactical School. Courtesy of Air Force Historical Research Agency.

warfare. He pointed out the operational impact of aviation in recon-naissance. Observation aircraft greatly extended the eyes of the oper-ational commander and could limit the enemy's operational move-ment. He noted how practically every army in the later years of World War I was constrained to move only at night to avoid observation.[36] Sherman also thought about naval air operations and logistics. He devotes a chapter to naval aviation, forecasting that aircraft carriers would replace the battleship and come to dominate maritime opera-tions. His chapter on air logistics reflects a fundamental American recognition, perhaps even preoccupation, with the demands of supply in modern war.

The Thirties

The trends in the theory and doctrine of the employment of airpower that originated in the twenties continued into the thirties—as did the disagreements. U.S. airpower continued to benefit from its own in-stitutional representation, fierce advocates, advances in technology, and Americans' fascination with flight. Suspecting corruption in the air transport companies, President Franklin Roosevelt ordered the Army Air Corps to fly the airmail starting February 19, 1934. With little time to prepare and with inadequate organization and equip-ment, the Air Corps quickly experienced fifty-seven accidents and twelve fatalities. Negative publicity and public concern prodded the War Department to convene a special committee to investigate the inadequacies of the air service that April. Chaired by former secretary of war Newton D. Baker, the Baker Board recommended the estab-lishment of a permanent peacetime General Headquarters (GHQ) of the Air Force.[37] The War Department accepted many of the Baker Board's recommendations and established the GHQ Air Force on March 1, 1935, giving it responsibility for all Air Corps tactical units. This headquarters reported to the Chief of Staff of the Army in peace-time and to the theater commander in time of war.

Unlike tanks, over which institutional advocacy was split be-tween infantry and cavalry, airpower was well represented by the GHQ Air Force, the Air Corps Tactical School, and active profes-sional and public interest. The trend among airpower advocates to emphasize the role of strategic bombardment increased with the in-troduction in the early thirties of such aircraft as the Boeing B-9 and

Martin B-10 bombers, which were capable of speeds in excess of 200 miles per hour and a ceiling of 21,000 feet. By 1935, their long-range performance was significantly increased with the introduction of Boeing's B-17. These aircraft gave the United States the most capable strategic bombers in the world. At the same time, the development of fighter aircraft lagged, which only added weight to the enthusiasts' claims for strategic bombing.

The disagreements, however, between the Air Corps and the War Department on the employment of this growing capability remained. Chief of Staff General Douglas MacArthur directed the army staff to prepare a statement on the employment of airpower. The War Plans Division released a revised War Department TR 440-15, *Employment of the Air Forces of the Army*. The draft of this regulation insisted the "land campaign and battle was the decisive factor in war."[38] After the faculty of the ACTS reviewed this document, it insisted that the principal mission of airpower "is the attack of those vital objectives in a nation's economic structure which will tend to paralyze that nation's ability to wage war and thus contribute to the attainment of the ultimate objective of war, namely the disintegration of the will to resist."[39]

The revised TR 440-15, dated October 15, 1935, was, to a certain extent, a compromise. The War Department was willing to concede some strategic role for the Air Corps, but was most concerned about support to ground forces. The regulation noted that the functions of the GHQ Air Force consisted of "operations beyond the sphere of influence of the ground forces, immediate support of the ground forces, and in coastal frontier defense and in other joint Army and Navy operations."[40] Operations beyond the ground force's sphere of influence included enemy air forces, war industry, critical points along the lines of communication, and troop concentrations.[41] The regulations indicated that the theater commander would provide the GHQ Air Force commander with broad general missions in which the air commander would select targets. The theater commander might also direct special missions and designate the major objectives. Although it noted the range and versatility of airpower, the regulation recognized the importance of basing. "Air bases suitably located are essential to the operations of air forces. If necessary, localities suitable for use as air bases must be seized and held in the prospective theater of operations after the outbreak of hostilities."[42]

Official doctrine captured the views of the ground commanders, modified somewhat for the benefit of the increasingly confident airpower advocates. Leavenworth and the War College faithfully taught the tenets of the employment of airpower contained in TR 440-15. The Air Corps Tactical School, however, preached a more strident view of airpower. In the last decade before World War II, the ACTS developed theories on the employment of airpower that both supported the official doctrine and exceeded it. Of all Air Corps generals in World War II, 82 percent passed through the doors of the Air Corps Tactical School during the interwar period.[43] They absorbed a philosophy, theory, and doctrine on airpower that described its role at the operational and strategic level. The lines between the operational and strategic role of airpower sometimes blurred, but the promise of the prestige and institutional independence inherent in the strategic potential of airpower was a powerful lure to the faculty and eager students.

The Air Corps Tactical School provided a broad military education in the thirties. About half the curriculum was devoted to air subjects, one-fourth to ground subjects, and one-fourth to subjects common to both.[44] To teach these subjects, the academic departments by 1934 consisted of air tactics, basic and special instruction, flying instruction, and ground tactics. The Department of Air Tactics taught courses in the employment of the air force, bombardment, attack, observation, and pursuit. In the air force course, the school stressed the concept of the air force as a combined arms force. Bombardment aviation delivered the real destructive power of the air force and was considered the basic arm.[45] Pursuit, observation, and attack all supported bombardment aviation when not supporting ground operations. Pursuit aviation protected bases and bombers and contested for air superiority. Attack aviation suppressed antiaircraft defenses and attacked enemy air bases and facilities. Observation aviation looked for targets. In the subsequent courses on each of the arms, the faculty addressed support to ground forces. The foundation for airpower's role in American operational art lay in the school's instruction on attack aviation and observation.

Attack aviation's first mission was to conduct air force missions rather than ground missions. The school considered that all counter air missions such as bombing enemy air bases served both the air force objectives as well as other army objectives by shielding the

ground forces. Unlike bombardment aviation, the doctrine and theory for attack aviation did not evolve much in the thirties. At the
beginning of the interwar period, an ACTS text described the goal of
strategic operations against lines of communication as "the disintegration of the whole scheme of movement and flow of supplies to
the troops in the combat zone. This is brought about by directing
operations against the key-points in the enemy's transportation system. Usually it will be found that certain cities, railroad yards or
railway and highway junctions are sensitive points in the enemy's
organization and if they are systematically destroyed the movement
of both troops and supplies to the combat zone will be seriously cur-
tailed or stopped altogether in certain areas."[46] By the end of the interwar period, the Air Corps renamed attack aviation light bombardment, but retained its mission for "the disruption of hostile forces
and their system of supply and replacement, by the destruction of
communication, supply and manufacturing establishments, light
bridges, transportation, equipment, and concentrations of troops."
What did change was the range at which attack aviation might be
used to shape the theater rather than just influence the battle. By
1939, an instructor noted that attack aviation "will be in the theater
of operation and our targets will likely be within three or four hundred miles."[47]

The faculty adamantly insisted that attack aviation should not be
used in close support of troops; nor should its firepower be employed
to replace the firepower of ground weapons. To take full advantage of
its unique characteristics, airpower should be pushed deeper into the
theater of operations, deeper into the enemy's heartland. Students
critiquing a map problem in 1934 complained that the school solution called for an attack on rail lines of communication only fifty
miles behind the front lines. "By destroying the rail lines of communication 150 miles or more behind the front lines, movement of
these particular forces could have been delayed for weeks, and if the
situation warranted the use of attack aviation against the ground
troops, the concentration could have been held up indefinitely."[48]

Observation aviation also pushed deeper into the theater of operations and theater of war. Observation squadrons assigned to corps
performed close or tactical reconnaissance. Operational reconnaissance could be performed by observation squadrons assigned to armies, or more likely to groups working directly for the GHQ Air

Force. Distant reconnaissance would be "accomplished by high flying airplanes carrying cameras and penetrating as far into hostile territory as the capabilities of the airplane and the hostile air resistance
will permit."[49] An illustrative problem in the 1934 observation
course required the students to serve as the air force G-2 intelligence
officer. The problem tasked the students to prioritize the reconnaissance targets for GHQ. Suggestive of the school's philosophy on the
employment of air reconnaissance in the theater, the school solution
suggested this order of targets: hostile air force, sensitive points in rail
lines of communication, interruption of highway transportation,
power sources, and key industries.[50] The Air Corps requirement to
locate and assess targets led invariably to an interest in both military
intelligence and logistics.

Like Leavenworth and the War College, ACTS shared the American military's emphasis on military intelligence and logistics. Obviously, military intelligence was critical in picking and finding the
right targets. The school taught students about intelligence organization from division through theater. In their study of military intelligence, students were tasked with developing intelligence plans that
would then become annexes to operations plans and orders. In 1936,
the commander of the GHQ Air Force distributed a detailed memorandum on intelligence training mandating increased emphasis on
such training and noting that "flying personnel as well as intelligence
personnel should have a clear and definite idea of their intelligence
duties."[51] The air corps was interested in all intelligence: tactical,
operational, and strategic. In the battle for air superiority and in support to ground forces, operational intelligence was critical. In 1940,
the air corps established a system of assistant military attachés
in various embassies to gather strategic intelligence on foreign air
forces and data on economic, industrial, and social analysis to support targeting.[52]

Logistics, both friendly and enemy, was an important part of the
curriculum. The faculty considered it to be "one member of a trinity,
the other two being strategy and tactics. Today all are interdependent
and of equal importance."[53] Military intelligence supported the airmen's interest in targeting and analyzing enemy logistics at both operational and strategic levels. Airmen also recognized that logistics
was also key to their own technology-dependent service. The logistics section taught students the army supply system from division to

theater. This course of instruction included the architecture of the theater of war, which consisted of lines of communication, operations, and the theaters of operations. Specifically regarding air logistics, map problems required students to select air bases and draft administrative orders for their supply. The students recognized the importance of logistics, and in 1934, they recommended even more logistic and operations planning.[54] The faculty drafted their own texts on air logistics and developed a concept for a general system of supply for an air theater of operations.[55] The concept for an air theater of operations was based on airpower's role in coastal defense of the United States.

The requirements for War Plan Orange and expeditionary warfare drove the army to seriously consider and plan for joint operations. The requirement for coastal defense drove the Air Corps to consider joint operations and issues of command and control. Since Brigadier General Mitchell's demonstration of airpower in sinking the German battleship *Ostfriesland* in 1921, the Air Corps staked a claim as the primary force for coastal defense. The confidence of the air corps was summed up in a lecture by Capt. L. S. Kuter: "The Air Corps Tactical School in particular and the Air Corps in general is emphatically convinced and loudly insistent that hostile navies dare not approach within the radius of action of effective land-based air forces."[56] The U.S. Navy did not agree.

The Joint Board directives that established the options for unified command and the principle of paramount interest in coordinating or commanding joint operations aimed largely at resolving the competing claims of the Air Corps and the navy with respect to coastal defense. The Air Corps still maintained that centralized control of air assets by airmen was the key to maximizing the advantages of airpower. However, the requirements for coastal defense and the need to act in support or in lieu of naval forces sparked the corps' interest in maritime matters. Beginning in 1927, the Air Corps Tactical School gave five lectures on naval tactics and strategy. In 1936, a naval aviator joined the faculty and taught a naval operations course in twenty-one periods. Army aviators gathered around demonstrations of fleet tactics laid out on conference room floors. The air corps' claim of responsibility for coastal defense allowed it to argue for long-range bombers, the key air force weapon. The real power of the air force lay in delivering ever bigger bombs over greater ranges, extending the influence and reach of airpower. The bomber was at the heart of air corps theory of airpower in the interwar period.

A few key instructors at ACTS during the thirties shaped their theories of strategic bombing not on experience, but on logic and technology. In 1932, Capt. Harold L. George reported as an instructor at the Air Corps Tactical School. George had plenty of experience to offer the students, particularly in bombardment aviation. He served in the 163rd Bombing Squadron during the Meuse-Argonne campaign in World War I and with Brigadier General Mitchell's Provisional Air Brigade during the battleship bombing tests. Working with 1st Lt. Ken Walker and other instructors, George developed a sophisticated theory for strategic bombing. In his opening lecture George admitted, "We are anxious to develop, logically, the role of airpower in future wars, in the next war. In doing so we realize that airpower has not proven itself under the actual test of war; also we realize that neither landpower nor seapower has proven itself in the face of modern airpower."[57]

It all began with Clausewitz. George accepted Clausewitz's assertion that the purpose of war "is to force the will of one nation upon that of the another—to overcome the hostile will to resist."[58] The official army view, captured in the regulations as early as 1923, explained: "The ultimate objective of all military operations is the destruction of the enemy's armed forces by battle. Decisive defeat in battle breaks the enemy's will to war and forces him to sue for peace."[59] According to George and the advocates for strategic bombing, destruction of the enemy armed forces is only a means to an end: breaking the will of the enemy. Like Douhet, George believed that "modern invention has given to the world a means whereby the heart of a nation can be attacked at once without first having to wage an exhaustible war at a nation's frontiers." Unlike Douhet, the American theorists believed the enemy's will could be broken through precision bombing targeting the key points of the opponent's industrial, economic, and social infrastructure.[60] George informed the students, "It is bombardment aviation which possesses this ability to paralyze a nation, and such paralysis would soon induce the feeling of helplessness and helplessness soon induces hopelessness and it is the loss of hope, not the loss of lives, that decides the issues of war."[61]

George and his fellow instructors studied the economic structure of the United States to determine the kind of key nodes that would paralyze a national war effort. The faculty determined petroleum, the electric power grid, and the main railroad lines as appropriate objectives.[62] This theory depended on cumulative effects rather than the direct battle of annihilation cherished by the army. In discussing how the principles of war might be changed by the advent of airpower, the faculty observed, "This theory of cumulative results constitutes one of the great differences between ground action and air action principally because airpower expresses itself by attack from the air against objectives on the ground."[63] The theory of strategic bombing was not just preached in introductory lectures; it was exercised. The students at ACTS did not plan massive air campaigns, but individually and in groups they participated in map problems that required them to prioritize targets and plan deep bombing raids against industrial targets as bombardment group commanders, with specific scenarios governing how targets should be prioritized. The school solution for the prioritization of bombardment targets in one map problem focused on targets in the United States in 1935 is suggestive. In this problem the school insisted the correct priority was "key industries and rail communications of Sandusky, Ohio, railroad bridges across the Allegheny River, petroleum industry blast furnaces and Bessemer converters at Cleveland, Ohio, steel and aluminum industry at Pittsburg [sic], rail communications at Buffalo, New York, and refineries, tank farms, and gasoline storage facilities at Franklin."[64] Given this list, the target folders for the air corps assault on Germany in the coming war is obvious.

The technological advances favoring bombers in the thirties gave rise to an unwarranted faith in strategic bombing. The ability of bombers to fly farther, almost as fast, and higher than pursuit planes led some to question even the necessity for establishing air superiority. As the primary arm of the air force, airmen compared bombardment aviation personnel to the infantry, insisting that regardless of the challenges "bombardment personnel will reach and destroy their objective."[65]

Airpower's unique advantages in flexibility, speed, and range made it influential at all levels of war. The instruction at the Air Corps Tactical School ensured that future air corps generals and planners absorbed the philosophy as well as the theory in the application of airpower. The school accepted airpower's role in operational art to support and shape the theater—at least with the caveat that air force

operational missions should take priority in the struggle for air superiority. They agreed with the army at large and the senior army
schools on this particular role for airpower. ACTS instruction on
bombardment aviation, however, exceeded official doctrine at least
in spirit, if not in fact. The unofficial doctrines and theories on bombardment aviation taught at the school sowed the seeds for significant arguments over the use of airpower in the coming war.

Not too surprisingly, the students at the Army War College in the
thirties did not plan massive strategic bombing campaigns against
potential or fictional adversaries. War College students planned for
the employment of landpower in an environment in which airpower
was a necessary and ever-present reality in the theater of operations.
They were most concerned with the tactical and operational impact
of airpower in the theater. Typical of missions assigned to the air
corps in exercises was a map maneuver against the fictional Red
coalition in 1938. The Blue commander assigned the following mission: "The GHQ Air Force, operating from airdromes in the Northeast Theater and southern New England is to execute distant reconnaissance, provide close support for ground forces of the theater,
disrupt hostile communications particularly port facilities, and attack hostile overseas expeditionary forces in transport when within
range."[66] In the same exercise, the Red coalition commander assigned
the air force missions that "are those set forth by the theater plan,
namely, to attack Blue air force, including the attack of aircraft factories and munition plants in Connecticut, New Jersey, and Pennsylvania, and secondly, to effect as much delay as possible in Blue concentration."[67]

An ACTS text on air warfare dated February 1, 1938, summarized
the air corps school's view of airpower at the end of the decade. Bombardment was considered the primary arm of the air force with the
mission of executing precision attacks against vital targets in the
enemy's national infrastructure. In order to do this, the air force must
first conduct counter air operations to obtain some degree of air superiority. Support of ground forces was viewed as a necessary but subordinate concern. To succeed, airpower needed to be massed and centrally controlled by the GHQ Air Force. The text did not completely
ignore the subject of supporting ground and maritime forces. When
required, the Air Corps could help ground forces through the direct

destruction of enemy forces or by isolating them by cutting their lines of communication within the theater.[68]

While the text highlighted the importance and potential decisiveness of unilateral air force operations, it also established one of the primary features of joint operational art: how each service's capabilities should be combined to achieve the theater commander's objectives. The text insisted that each service must bring its unique capabilities to bear. In expeditionary warfare, "the success of an overseas invasion depends upon the ability of the naval and air forces to accomplish in an initial phase of action a condition of superiority that will ensure the safe passage of the land force across the sea. Control of the seas is insured only by the defeat or the neutralization of the defending naval and air forces."[69] Likewise, "just as air forces are capable of offering direct or indirect assistance to land and sea forces, so also land and sea forces are capable of assisting the operations of air forces. Land and sea forces may seize territory necessary to the establishment of air bases, they may provide the necessary surface security, and they may contribute to the supply of air forces."[70] Those few words express one of the driving operational concepts for the Pacific campaigns in World War II.[71]

In the 1930s, officers like Henry H. "Hap" Arnold, Ira C. Eaker, and Carl A. "Tooey" Spaatz rose through the ranks of the Army Air Corps embracing Billy Mitchell's vision of American airpower. Committed to the quest for an independent and decisive air force, they looked to the future. Henry Arnold graduated from West Point in 1907 and quickly became interested in flying. In fact, Wilbur Wright taught Arnold how to fly, making him one of the first American military aviators. Arnold spent World War I in Washington, D.C., but his charm, effective leadership, and passion for airpower moved him steadily up the ranks during the interwar period. By 1935, Brigadier General Arnold was the assistant chief of the air corps and in 1938 became the chief. Arnold fully accepted the ACTS theory on the strategic employment of airpower. He would direct the course of American airpower as the chief of the Army Air Corps, later the Army Air Force, throughout World War II.[72]

Ira C. Eaker received a reserve commission in 1917 but missed combat action in the Great War. He distinguished himself in the interwar period, not only by his leadership but also by grabbing headlines through spectacular flight achievements, such as completing the first transcontinental flight purely on instruments in 1936.[73] In 1933, Eaker graduated from the University of Southern California with a degree in journalism. He put this training to good use by co-writing three books with Hap Arnold to popularize and spread the airpower gospel. After taking note of the role airpower played in the German blitzkrieg, Arnold and Eaker published *Winged Warfare* in March 1941. Just as Maj. William Sherman's *Air Warfare* summed up American thought on the employment of airpower in the twenties, Arnold and Eaker's *Winged Warfare* summarized the beliefs of senior U.S. airmen at the end of the interwar period.

The book represented an updated statement of their vision of the employment of airpower. They continued to insist that "the first priority missions are the destruction of opposing air forces, and vital enemy objectives beyond the range or theater of influence of land forces. To take all or part of the air force and remove it from these higher priority missions to missions cooperative in character would be a dangerous error."[74] Independent missions were the primary functions of airpower, but Arnold and Eaker did discuss cooperative missions with the other services, such as observation, working with mechanized forces, paratroops, and pursuit aviation. The authors justly and proudly asserted, "It is now fairly generally agreed that no land or naval battle will be won while the enemy holds superiority in the air and when he is able to bring considerable air pressure to bear on the theater of that battle."[75] Their commitment to airpower's support to ground forces would be tested in World War II. They remained convinced that the best contribution of airpower not only to ground forces but to winning the war lay in destroying the enemy air force and striking at strategic targets. Eaker went on to become a senior American commander with shared responsibility for putting the ACTS theory on strategic bombing into practice as head of the Eighth U.S. Air Force in England.[76]

Tooey Spaatz graduated from West Point in 1914, and unlike Arnold and Eaker, he did not miss combat in World War I. Although assigned to a training command, Spaatz, who loved to fly, went to the front without orders. Within three weeks he shot down three German aircraft. He resented the time he spent at the Command and General Staff School and was by all accounts an indifferent student. Spaatz was a man of action. Along with Eaker, Spaatz participated in a series of spectacular flights, earning the Distinguished Flying Cross for an endurance refueling flight in 1929 in which he stayed aloft for 151 hours. All three men maintained close personal and professional ties and shared a common view of the strategic role for airpower. They were the true apostles of the gospel according to Billy Mitchell.[77]

Arnold, Eaker, and Spaatz all graduated from the Command and General Staff School, but none attended the Army War College. Significantly, the senior air commander who would make the greatest contribution in airpower at the operational level was a graduate of the War College. George C. Kenney enlisted as a flight cadet in World War I. He flew seventy-five missions, shooting down two German aircraft. Kenney had a long involvement with attack aviation, the branch most suited to supporting ground forces. As an ACTS instructor from 1926 to 1929, Captain Kenney wrote the textbooks for the Observation and Attack courses.[78] He attended the Command and General Staff School in 1927 and graduated from the War College in 1933. Nicknamed "Little George" by his friends (he stood a little less than five feet six inches tall), Kenney was confident, energetic, and full of good ideas.[79] In the coming war, Kenney would command both the Fifth U.S. Air Force and the Far Eastern Air Force under General MacArthur. Kenney became the Army Air Force's expert in wielding airpower at the operational level in the Pacific.

Assessment

Airpower truly revolutionized warfare. Its military significance was intuitively evident to the professional soldiers of Europe prior to World War I. Even before the war, European armies organized air units and specified missions. The impact of aerial reconnaissance was immediate as the great armies maneuvered in the opening campaigns. Within eleven years of the invention of powered flight, airpower had become a factor in a land war. By the end of World War I, airpower was influencing military operations on land and sea. Most of the classic missions for airpower at the tactical, operational, and strategic level of war were either demonstrated or considered during the war. This application of airpower gave rise to the debates in the interwar period over who would control air assets and how they would be employed.

During these years, American airmen developed a fairly complete body of air theory and doctrine centered on their firm belief in the primacy of air superiority, the inherent offensive nature of airpower, the need for centralized control of air assets by airmen, and the assignment of targets at the strategic level.[80] The services agreed that gaining air superiority in the initial phase of an operation was critical, if not essential. A good deal of discussion in the thirties about the use of airpower to defend America's coastline led to the agreement on joint and unified action between the services. Still, the importance of airpower in offensive operations was certainly recognized. The debate on centralizing air assets under airmen would not be resolved until the next war. Ground commanders recognized the potential of airpower at the tactical and operational levels both to shape the battlefield and to attrite the enemy. The ability of airpower to affect operations quickly throughout the theater gave it a unique quality that ground commanders were naturally reluctant to relinquish. World War I firmly pointed up this potential at the tactical and operational level. The jury was still out, however concerning the strategic impact of airpower. At bottom, arguments over how to employ airpower came down to the right to chose targets.

Fundamentally, airpower is all about selecting targets—their location and, more important, their purpose. In the 1930s, American military officers recognized that airpower was capable of achieving strategic, operational, and tactical effects. The leadership of both the army and the navy recognized the importance of airpower in modern war, though they might disagree on its decisiveness, and how it should best be employed. Though not yet a full joint partner, it was accepted as a critical component in the prosecution of operations. American theories on the strategic and operational employment of airpower would be tested in World War II; both would be found crucial to victory. In the process airpower would become a full partner in waging war at all levels.

5

SEAPOWER

In January 1865, the commander of the USS *Pontiac* reported to Gen. William T. Sherman in Savannah, Georgia, for duty on the Savannah River. Sherman pointed to the map and in a very few sentences explained his plan of campaign north from Savannah. "You Navy fellows have been hammering away at Charleston for the past three years. But just wait till I get into South Carolina; I will cut her communications and Charleston will fall into your hands like a ripe pear." The naval officer was impressed: "After hearing General Sherman's clear exposition of the military situation the scales seemed to fall from my eyes. Here I said to myself is a soldier who knows his business. It dawned upon me that there were certain fundamental principles underlying military operations which it were well to look into; principles of general application whether the operations were conducted on land or at sea."[1]

Sherman moved north and Charleston fell. The naval officer, Stephen B. Luce, never forgot this lesson in military campaigning and became convinced of the need to study the fundamental military principles and apply them to maritime operations as well. Luce became a chief architect of American naval professionalism and a fierce advocate for naval education. As Luce advanced in rank, he continued to argue for education and training within the U.S. Navy. In November 1882, Commodore Luce wrote a formal letter to Secretary of the Navy William E. Chandler on the subject of a school for the study of naval warfare. In letters to the secretary, he cited the army schools at Leavenworth and the Artillery School at Fort Monroe, Virginia, "as examples worthy of imitation."[2] After a good deal of lobbying, Luce convinced Chandler, and in 1884, the U.S. Navy established the world's first Naval War College at Newport, Rhode Island. The following year, the college opened its doors in what had been the Newport poorhouse.

THEORY AND PRACTICE

In the study of naval warfare, Luce believed in a comparative method and practical application. The methodology used historical study, to compare land and naval operations to discover similarities and differences. Students tackled current naval problems illuminated by the light of naval history. Originally, Luce also wanted part of the Atlantic squadron stationed at Newport to allow for some practical exercises.[3] Stationing fleet units at Newport proved to be problematic, however, and the War College eventually had to find another means to teach practical application in naval warfare.

Having established the college and envisioned its curriculum, Luce's next and perhaps most important step was to gather an appropriate faculty and student body. The college's first class of eight rather unhappy officers, all lieutenants, had been sent from the nearby Torpedo School and "considered they had been shanghaied."[4] Ironically, the only permanent member of the faculty for the first class was an army lieutenant, Tasker Bliss. Luce secured Bliss's assignment in order to insure the comparative approach in the study of land and naval operations. More important, Luce had to fill the critical post of professor of naval warfare. He found the right man in Alfred Thayer Mahan.

Born at West Point, New York, in 1840, Alfred Thayer Mahan was the son of Dennis Hart Mahan, a professor at the U.S. Military Academy. Mahan opted for a career in the navy and pursued a rather undistinguished career until called by Stephen B. Luce to come to the newly established Naval War College. Mahan had served with Luce on the faculty of the Naval Academy and as his executive officer on the USS *Macedonian*. By 1884, Mahan had written a book on naval campaigns of the Civil War and seemed a good fit because of his interest in history and his intellectual outlook. By the time Commander

Mahan got to the college, the navy had ordered Luce back to sea, and Mahan became president of the institution.

Mahan is undoubtedly the most influential naval theorist that America has ever produced. Many of his views became deeply ingrained and exceptionally influential all over the world. Part of the explanation for Mahan's influence was simply timing. He wrote at the end of the nineteenth century, which witnessed the last wave of western expansion fueled by the industrial revolution, growing nationalism, and rampant capitalism. Mahan addressed the role of seapower in establishing and maintaining the great empires. He attempted to answer two questions: what is a navy for, and how should it be used?[5] Interestingly, these are the same questions the United States Army was asking about its own purpose and future at the end of the nineteenth century.

Mahan answered the first question, the purpose of the navy, in a simple but powerful phrase: "command of the sea." Command of the sea, or sea control, allowed the vital commerce on which Mahan claimed national prosperity and greatness depended. Prior to Mahan, American seapower was built for and employed in sea denial. Sea denial is usually the only practical option for an inferior fleet. In the eighteenth and nineteenth century, America enjoyed the Pax Britannica and prospered behind the shield of the Royal Navy. On those occasions where U.S. interests clashed or were threatened by Great Britain, America's only recourse was coastal defense and cruisers. The navy was largely limited to *guerre de course*—making war on the enemy's commerce. By the late nineteenth century, Mahan ably explained the relationship between seapower, prosperity, and national interests. Moving from a strategy of sea denial to sea control mandated a larger navy, from a collection of commerce-raiding cruisers to a fleet capable of projecting seapower. Certainly part of the explanation for Mahan's popularity and influence with naval officers is the fact that his views aligned national interests with the institutional interests of navies.

Mahan used history to illuminate his ideas about grand strategy and naval history. He expressed his ideas in lectures at the Naval War College and then packaged them into widely popular books. His most famous work, *The Influence of Sea Power upon History, 1660–1783*, was published in 1890. In this work Mahan sought to answer the critical question of the purpose of a navy and the role of seapower. The

Spanish-American War in 1898 seemed to validate the new role of the navy for Americans and made creditable many of Mahan's views. Moreover, the United States' acquisition of the Philippines as a result of the war pushed American interests deep into the Pacific with profound and long-lasting results. In 1910, Mahan published his lectures from the Naval War College in *Naval Strategy: Compared and Contrasted with the Principles and Practice of Military Operations on Land*. In this work, Mahan provided his answer to the second question, how should a navy be used. This was the comparison Luce sought to discover concerning the similarities and differences between war at sea and war on land. Mahan intended to find the fundamental principles of warfare at sea.

Just as military strategy in the nineteenth century encompassed operational art, so too did naval strategy embrace the developing notion of operations, and the growing distinction and gulf between strategy and tactics. There were significant similarities between military and maritime theory as discerned by the early naval theorists. Mahan found significant comfort in the works of Jomini as he looked for principles of warfare at sea. Jomini is the most often cited authority in Mahan's *Naval Strategy* and his influence predominates. Jomini's emphasis on the importance of the central position, concentration, lines of communication, interior lines, and choosing bases all are subjects of Mahan's lectures.

Several officers criticized Mahan's illustration of fundamental principles of war at sea by using historical illustrations from earlier centuries.[6] After all, the very rapid pace of technology at the turn of the century had brought the development of the torpedo and the introduction of steam, coal, and the all-armored ship—advances that begged for new answers in the employment of naval force. While recognizing the connection between technology and tactics, Mahan insisted, "From time to time the superstructure of tactics has to be altered or wholly torn down, but the old foundations of strategy so far remain, as though laid upon a rock."[7] The unchanging nature of naval strategy for Mahan lay in determining "the proper function of the navy in war, its true objective; the point or points upon which it should be concentrated, the establishment of depots of coal and supplies; the maintenance of communications between these depots and the home base; the military value of commerce-destroying as a decisive or a secondary operation of war."[8] To the crucial question of how to achieve sea control, Mahan's answer was unequivocal: "the proper main objective of the navy is the enemy's navy."[9] The purpose, role, and employment of navies fascinated Mahan. To him the logic was inescapable: defeat or contain the enemy fleet and sea control is assured. The decisive fleet engagement may be determined by tactics, but the strategy ensures the advantages of concentration and readiness through deployment, maneuver, and basing. Moreover, Mahan insisted that the employment of seapower must serve national policy.[10] This insistence on the subordination of force to policy was important in explaining that navies are true instruments of national power, employed to achieve not just military victory but national strategic objectives. Ultimately, Mahan's greatest contribution lay in his persuading the U.S. Navy to think in terms of a fleet, not individual units. Moving from frigates to fleet meant developing more than just a few heroic captains; it called for the education of a professional class of officers for the tasks of building, sustaining, and employing large-scale units in maritime warfare.

Mahan's influence was global. The German Kaiser, William II, sent copies of *The Influence of Sea Power upon History* to all his ship captains.[11] No one took Mahan more to heart than the Imperial Japanese Navy. More of Mahan's work was translated into Japanese than into any other language. *The Influence of Sea Power upon History* was used as a text for Japan's naval and military staff colleges. Japan's fixation on Mahan's emphasis on the annihilation of the enemy fleet in a decisive battle dominated the Imperial Navy's doctrine "until the Pearl Harbor attack and beyond."[12] Historians have often cited the long-lasting influence of Mahan on the U.S. Navy, but there was another naval theorist at the turn of the century whose influence on the U.S. Navy approached Mahan's in this respect. Sir Julian Corbett was in many ways an unlikely theorist, but his views on seapower better addressed the strategic requirements of the American navy at the turn of the new century.

Sir Julian Corbett was born in London in 1854. Although a lawyer by training, Corbett, like Mahan, came to his theories on seapower through a study of history. He based much of his work, however not on Jomini but on Clausewitz. His two-volume work, *Drake and the Tudor Navy*, appeared in 1898. Subsequent volumes in naval history and articles advocating reform in naval education allowed Corbett to gather a number of important acquaintances in the Royal Navy. In

1902, Corbett was invited to give lectures on naval history at the recently established Royal Naval College. Corbett's lectures took on the nature of applied history—history used to illuminate fundamental principles. By 1905, through the patronage of senior naval officers such as Sir John Fisher, Corbett lectured regularly at the Naval College. His lectures covered not only history, but theoretical subjects such as "the system of Clausewitz," "Limited and Unlimited War," and "The Essentials of a True Naval Defensive."[13]

Corbett published the substance of his lectures and views in *Some Principles of Maritime Strategy* in 1911. This work, used as a text at the U.S. Naval War College, contains his most important observations on the theory of seapower. It begins with a general discussion heavily influenced by Clausewitz on the theory of war in general. Corbett adopted Clausewitz's concept of limited and unlimited war. The nature of any war is determined by its political object, which can be either limited or unlimited. Unlimited war demands the complete overthrow of the enemy and requires the greatest possible exertion. In limited war the effort is proportional to the political objective. The significance in limited or unlimited war lies in the role that seapower may play. Corbett recognized that "since men live upon the land and not upon the sea, great issues between nations at war have always been decided—except in the rarest of cases—either by what your army can do against your enemy's territory and national life, or else by the fear of what the fleet makes it possible for your army to do."[14] In other words, seapower alone is unable or very unlikely to result in the complete overthrow of the enemy. Seapower may, however, play a much greater role in limited warfare. So the first step, as Clausewitz advocates, is to determine the nature of the war. According to Corbett, the senior military leader must ask "what is the political object of the war, what are the political conditions, and how much does the question at issue mean respectively to us and to our adversary."[15]

After discussing the general nature of war, Corbett explores the theory and conduct of naval warfare. Like Mahan, Corbett believed the "object of naval warfare must always be directly or indirectly either to secure the command of the sea or to prevent the enemy from securing it."[16] For Corbett, command of the sea simply meant "the control of maritime communications, whether for commercial or military purposes."[17] Armies are concerned with the conquest of ter-

ritory, and being land based, armies may permanently retain possession of their lines of communication. Navies, however, function in
a hostile medium in which they must share their lines of communication and can never exert permanent or absolute control over sea
routes. Unlike Mahan, sea control for Corbett was a matter of degree,
general or local, and invariably transitory.

Corbett also made a critical distinction between naval and maritime strategy. "By maritime strategy we mean the principles which
govern a war in which the sea is a substantial factor. Naval strategy is
but part of it which determines the movements of the fleet when
maritime strategy has determined what part the fleet must play in
relation to the action of the land forces; for it scarcely needs say-
ing that it is almost impossible that a war can be decided by naval
action alone."[18] If the roots of operational art for landpower lay in
the increased size of armies and the scale of warfare on land, the roots
of maritime operational art lay in Corbett's definition of maritime
strategy.

Corbett divided the conduct of naval warfare into three categories: methods of securing command of the sea, methods of disputing
command, and methods of exercising command. In obtaining sea
control, like Mahan, he believed that seeking out the enemy fleet was
a primary means of obtaining sea control, but the problem was to
bring the enemy into action. Corbett recognized, "If you are in a superiority that justifies a vigorous offensive and prompts you to seek out
your enemy with a view to a decision, the chances are you will find
him in a position where you cannot touch him."[19] The only other
solution was blockade. This accurately forecasted the situation for
the British Royal Navy in World War I.

Corbett differed with Mahan on the notion of concentration and
the important requirement to support the army. A fleet may elect not
to remain concentrated in seeking to either dispute or exercise sea
control; for example, a fleet may dispute sea control through minor
counterattacks rather than by decisive fleet action. According to Corbett, a fleet exercised sea control though the attack and defense of
commerce and the support of military expeditions. It is in support of
landpower that Corbett reached a significant insight in maritime operational art. Developing leverage between seapower and landpower
in a theater of operations is key. Seapower can have significant strategic results through attacks on commerce, but to achieve operational

results that extend its influence ashore and within the theater, sea-power must frequently leverage landpower. As Sir John Fisher, First Lord of the Admiralty from 1904 to 1910 and one of Corbett's chief patrons, famously declared, "The army is a projectile to be fired by the navy."[20] Prior to World War I, seapower contributed to national power by its affect on commerce and military operations ashore. Navies leveraged their own power at sea as well as landpower to implement policy. After World War I, seapower would leverage not only its own inherent capabilities, but landpower and *airpower* as well.

Mahan and Corbett had a few, albeit lesser-known, contemporaries in naval theory. British vice admiral Philip Colomb wrote about naval strategy in a series of books from 1874 to 1891. Colomb was born in Scotland in 1831, and like Mahan, he used history to demonstrate enduring principles of naval warfare. His *Naval War-fare*, published in 1899, emphasized the necessity of obtaining command of the sea. Somewhat later, Admiral Raoul Castex of France wrote a five-volume series on maritime strategic theories from 1927 to 1935. Castex also used a historical method to discuss not only naval warfare but strategy in general. The U.S. Naval War College translated major portions of this prolific author's work. Castex believed that Mahan's prescriptions should be modified based on the specific circumstances of each nation. Although Mahan dominated much of the public and many of the professional minds at the turn of the century, there is, however, much more operational art to be found in Corbett than in Mahan or the other theorists. Mahan defined and made a convincing case for the importance of seapower, but he largely ignored power projection as one of its most important elements. The projection of national power was the most critical strategic requirement for the United States in the early twentieth century, as it would remain. American naval theory became a mix of Mahan and Corbett, but the latter proved more prophetic. Mahan's general theory of sea-power did, however, provide the initial foundation for study at the Naval War College.

Mahan's influence on the early Naval War College is evident. Although he was sent back to sea in 1893, instructors dutifully read his lectures to the students for years after his departure.[21] During the interwar years Corbett's work was widely read and influential at the college. Of course, Mahan's or Corbett's actual influence on the students may be debatable, but the influence of application—the other interest of Stephen B. Luce—is undeniable.

War-gaming became one of the central features of education at the Naval War College. Even as the U.S. Navy moved from a doctrine of sea denial to one of sea control, it had little experience in major fleet actions. Whatever the evolution in naval theory, the question remained of how to apply it. Among the several points on which the developing army and the navy professionalism converged to mutual benefit, one of these was the use of the applicatory system. While serving as an instructor at the Army War College, Cmdr. William L. Rodgers took extensive notes on this system. At about the same time at the Naval War College, Maj. John H. Russell (USMC) came into the possession of an English translation of Gen. Otto von Griepenkerl's *Letters on Applied Tactics*. Another member of the Naval War College faculty, Cmdr. William McCarty Little, formed a committee to adapt Griepenkerl for naval use. Griepenkerl's book, essentially a set of tactical problems, had been in use as a text at the army schools in Leavenworth for some time. Key to the applicatory system was a process for writing orders—developing an estimate and then war-gaming solutions through the use of maps or staff rides. In 1911, Commander Rodgers returned to the Naval War College as president and began implementing the applicatory system as a core element in the curriculum.[22]

At the army schools the officers applied tactics on staff rides and strategy in map maneuvers. At the Naval War College the students applied tactics on a gaming board and strategy on naval charts. The navy's adoption of the applicatory system brought tremendous advantages and some potential problems to officer education. It provided a common problem solving process between the army and the navy based on the estimate and the five-paragraph order. It developed and sharpened judgment and staff skills. Just as at Leavenworth, however, war-gaming could tend to focus students on the process rather than on a broader intellectual inquiry about war. In the American navy, war-gaming was seen as the only practical way to gain peacetime experience in naval operations at the tactical and fleet level. The procedure for the naval war game developed at the War College prior to World War I endured throughout the interwar period.

The image of student officers gathered around the gaming board with small ship models or peering over charts captures the emerging turn-of-the-century professionalism of the U.S. Navy. The procedure for the war game was simple enough and remained essentially unchanged for decades. The instructors divided the students into Red and Blue staff teams. The instructor provided a common general situation to both sides and a special situation known to only one side. Each group of students did an estimate and issued orders. The students handed in their orders to the umpire and plotted their projected move on a chart. Student recorders transferred the moves to the umpire's chart and he judged the contacts between naval forces by using a book of rules.[23] More like the map maneuvers at the Army War College than at the Army Staff School, there was no school solution at Newport. Students and instructors freely critiqued each other in a collaborative attempt to find the best possible solution. With the development of naval war-gaming, Luce finally had all three pieces of his vision for senior education for naval officers: the comparative method, historical study, and application. Naval war-gaming was one of the most celebrated contributions of the college, and in the coming years, application became perhaps the most important part of the curriculum.

As the first institution of its kind, the Naval War College pioneered professional education for naval officers. Luce, Mahan, and their successors worked hard to improve the prestige and influence of the school. The college's early involvement in war planning assisted their efforts. In 1894, the faculty revised the curriculum to include a single war problem, the defense of the East Coast against the British fleet. The growing notoriety of the revolt in Cuba encouraged the faculty to add a special problem the following year to consider war with Spain. The War College submitted its views to the navy, but there was no central war-planning office. The secretary of the navy convened a planning board to consider strategic options. Nothing came of these plans, and Secretary John D. Long of the new McKinley administration convened a second planning board in 1897. Again the Naval War College's views were solicited but not fully accepted. The second board's views became the "official plan" but did not represent the policy of the administration.[24] When the war broke out between the United States and Spain, American strategy was largely ad hoc, but it did include many of the elements of the various prewar plans.

Secretary Long consulted with Mahan and convened a general war board to consider naval strategy.[25] Assistant Secretary of the

Navy Theodore Roosevelt sent orders to Adm. George Dewey, commander of the Asiatic Fleet, to be prepared to engage the Spanish squadron in the Philippines. The ensuing Battle of Manila Bay and the subsequent sinking of the Spanish squadron at Santiago Bay in Cuba made the navy's case as an instrument of national power. This war showcased all the elements of Mahan's and Corbett's theories of seapower. The American navy established local sea control by blockading Cuba, destroyed the opposing naval forces in fleet action, and projected the army ashore to clinch the strategic objective of liberating Cuba. In the postwar estimations of many, the navy outshone the army for its preparation and execution. The problems in the Army's performance led to the Root reforms, which established a general staff and a war college. The Spanish-American War gave the United States new strategic responsibilities and required a two-ocean navy to defend them.

Prior to World War I, the navy made great strides in professionalism and readiness for the challenges ahead. The United States committed to building a large navy, and by 1913, thirty-nine battleships had been commissioned or authorized.[26] There were still problems, particularly in organization. Powerful bureau chiefs under the supervision of Navy Secretary Long ran the service. Although Luce, Mahan, and other reformers advocated for a naval general staff, Long considered this harmful to civilian control and potentially dangerous to American liberties. Persuaded by the success of the Naval Board during the Spanish-American War, Long did authorize a Naval General Board in 1900 to help coordinate war planning and ship construction and generally to advise the secretary on naval issues. This was a long way from the army's progress under Elihu Root in creating a general staff in 1903. Finally, in 1915, with war raging on the European continent and trouble on the high seas, the secretary of the navy authorized establishment of a Chief of Naval Operations (CNO). The CNO had specific responsibilities for planning and coordination but did not control the fleet.

The U.S. Navy made great strides in developing a professional ethos, organization, and fleet by the first decade of the twentieth century. Thanks to Mahan and other reformers, it grew along with America's national aspirations, industrial power, and strategic interests. It was certainly more prepared than the army to contribute immediately to the Allied cause in World War I. It had to be: the navy was the shield behind which the conscript mass American army gained time to mobilize and train for war. If the United States was to make a difference, however, time was short. The U.S. Navy expanded and prepared to play its role in the war—but it would not be Mahan's vision of clashing battle fleets. The Germans had already tested the issue with the British at the Battle of Jutland in 1916. The British got the worst of this massive clash of fleets, but the German High Seas Fleet retreated to port whereupon the Royal Navy reinstituted a distant blockade. The U.S. Navy's participation in the war at sea from 1917 to 18 was much more akin to Corbett's view of exercising general sea control, which the Royal Navy had largely secured before U.S. entry.

The War at Sea

Germany began the war in 1914 with a maritime strategy aimed at damaging the British navy through minor attacks, mines, and submarines. Germany intended to bring its "concentrated fleet to force a battle under favorable conditions," once its "campaign of attrition" had succeeding in achieving "a reduction in the enemy's strength."[27] Initial skirmishes at sea proved inconsequential. The British applied a distant blockade with sufficient effectiveness to prod the Germans into a counterblockade with submarines. The subsequent public outrage after the sinking of the British passenger ship *Lusitania* in 1915 persuaded the German navy to adhere strictly to international law with regard to attacks on commerce. This significantly constrained the effectiveness of submarines and encouraged the Germans to pursue more aggressively their original strategy for obtaining sea control. After the Battle of Jutland, the Germans had to rethink their options. In order to contribute to breaking the deadlock on the western front, the Imperial Navy attempted to dispute allied sea control through a return to unrestricted submarine warfare.

Well before the army could get General Pershing to France, the navy sent Rear Adm. William S. Sims to London. Sims's job was to size up the maritime situation and determine naval requirements. The German submarine offensive was inflicting terrible losses on the Allied merchant fleet. Sir John Jellicoe, British First Sea Lord, pessimistically informed Sims that Germany "will win, unless we can stop these losses—and stop them soon."[28] In the early stages of the submarine offensive the Germans sank 900,000 tons of shipping per month. The total Allied and neutral tonnage at that time was only about 34 million tons, with the rate of new construction at about 177,000 tons per month.[29] If unchanged, the war at sea would be a battle of attrition between submarines and merchant ships, not a battle of annihilation between great fleets.

Sims quickly recommended a convoy system to protect shipping and concluded that the United States must immediately launch a vast construction program to build the necessary destroyers, escorts, and submarine chasers. Contrary to Mahan's theories, Sims believed that the essential U.S. seapower contribution in the war must consist of protecting shipping rather than seeking out and destroying the enemy fleet. This ran directly counter to the wishes of the CNO, Adm. William S. Benson. In 1916, President Woodrow Wilson pushed through an ambitious shipbuilding program emphasizing capital ships. Both Wilson and Benson had their eyes on long-term naval requirements to influence postwar outcomes. Sims insisted that immediate war needs must take precedence. President Wilson decided in favor of Sims, and the navy modified the prehostilities shipbuilding program to forgo battleships in favor of destroyers and antisubmarine craft. Eventually, the U.S. antisubmarine warfare fleet numbered nearly 800 vessels including four hundred wooden sub chasers in the "splinter fleet."[30]

Admiral Sims assumed command of U.S. naval forces in Europe. With headquarters in London, Sims commanded over three hundred ships, five thousand officers, and seventy thousand men on forty-five bases.[31] He assembled a planning staff made up entirely of Naval War College graduates whose job it was to plan and supervise the U.S. contribution to victory at sea. The U.S. Navy had two major tasks in World War I—exercise sea control by defeating the submarine threat, and get the American Expeditionary Force safely to Europe. The western front was three thousand miles from the nation's Atlantic seaboard. Power projection across hostile seas enabled the United States to contribute more than 2 million troops to the fight in France. Landpower proved decisive, but sea control was a necessary condition. It was the transport not only of troops but also of vast quantities of supplies to sustain them and their allies that made victory possible. Secure sea lines of communication allowed the U.S. Navy to escort safely more than 18,640 ships carrying both troops and supplies to Europe.[32]

World War I introduced new technologies that became a permanent part of modern warfare. At sea, submarines and aircraft became
tactical weapons that eventually developed operational significance.
The advantages of the airplane in antisubmarine warfare led to a rapid
expansion in naval aviation. By the end of the war, about 3,120 officers and 45,600 seamen including marines manned and maintained
some 2,100 aircraft. A total of 18,000 seamen and 400 planes served in
Europe.[33] Although the navy employed aircraft primarily in the antisubmarine role, before the war ended it formed the United States
Naval Northern Bombing Group. This group comprised eight squadrons, originally organized to bomb submarine bases on the German-
occupied Belgian coast. By October 27, 1918, the naval and marine
aviators were participating in day-and-night missions against land
targets. Soon the Germans abandoned their submarine bases in their
retreat from Flanders. The navy offered the Northern Bombing Group
to Pershing, but he recommended the group work with the British
army in the north due to proximity. By the time the Armistice was
signed, naval aviation dropped more than one hundred tons of bombs
in the final land campaign.[34]

The U.S. Navy helped the Allied powers to win World War I
through sea control and power projection. The war was not quite the
watershed event for the navy that it was for the army. Necessity
forced the navy to fight not as it imagined by the theoretical lights
of Mahan, but in the more prosaic nature of defensive sea control.
The question now was how would the navy interpret the lessons of
the war, or would it ignore them. How would the navy seek to achieve
sea control in a modern era of submarines, airpower, and expanded
strategic requirements? The General Board, the CNO, and the naval
bureaus were all designed to help chart the course for the navy in
the coming decades, but the Naval War College was the only institution charged not only with helping provide the answers but with
imparting them to the generation of naval officers that would fight
the next war.

The question of how to integrate the limited experience in offensive sea control from World War I and the new technologies into
a naval doctrine and organization that could meet strategic requirements was much discussed at the war college. The Navy leaned
heavily on history, theory, simulation, and fleet exercises. Virtually

all of the senior leadership of the navy in the next war studied these subjects at the Navy War College during the interwar period.

PLANNING AND STAFFS

Admiral Sims had his pick of assignments at the end of the war. He chose to take a reduction in rank to rear admiral in order to become the president of the Naval War College. In his address at the reopening of the college in 1919, Sims announced that the "college aims to supply principles not rules, and by training develop the habit of applying these principles logically, correctly, and rapidly to each situation that may arise."[35] More to the point, Sims reminded the students, "You should never lose sight of the fact that we are all practical fleet officers; that we shall go back to the fleet and be replaced by others from the fleet."[36] In Sims's view, the War College existed to serve the fleet. This was an important distinction that very much affected the college's approach to education. The Army War College's mission statement specifically addressed the need to study joint warfare and provide competent staff officers to the army staff. There was no navy staff.

Successive secretaries of the navy defeated attempts by reformers to establish a general staff. Throughout the interwar period the navy maintained a rather dysfunctional organization that split responsibility among the Chief of Naval Operations (CNO), the powerful bureau chiefs, the General Board, and, beginning in 1922, a commander of the U.S. Fleet. The CNO was "charged with the operations of the fleet, and with the preparation of and readiness of plans for its use in war."[37] The CNO's office included a war plans division, an intelligence division, and a matériel division to coordinate the bureaus from a broad perspective and keep track of readiness. The CNO had no direct control over the bureaus, nor did he have responsibility for supply. The eight bureau chiefs included the Bureau of Navigation (personnel), ordnance, construction, supplies, and others, all worked directly for the secretary. The Commander in Chief of the U.S. fleet, with the unfortunate acronym CINCUS (pronounced "sink us"), actually commanded and directed the fleet. But the effectiveness of this organization depended on the personalities of the senior leaders and their willingness to cooperate.

The ultimate significance of Sims's view of the college and the organization of the navy meant that the War College served the fleet, not the staff. It was more concerned about sharpening judgment and application than about planning. Sims's successor, Rear Adm. William V. Pratt, did much to steer the college toward a broader view of its educational responsibilities, but inevitably the college drifted back to its focus on the application of seapower, not campaign planning. Of the four key functions of operational art—intelligence, command and control, logistics, and maneuver—the Naval War College consistently focused on fleet maneuver, specifically with regard to decisive battle.

In 1925, Admiral Pratt became president of the War College. He brought with him a determination to expand its outlook. By the time Pratt got to Newport, the navy added a Junior War College course intended to expand educational opportunities for its junior officers. The effort followed the army's progressive educational model using the General Staff School at Leavenworth as an example.[38] Pratt wanted to reorganize the college to follow more closely the organization of the Army General Staff and the Office of Naval Operations. He restructured the college into four departments: information, logistics, operations, and policy and command. In the actual conduct of sea warfare, Pratt believed that strategy should merge with tactics into operations. He took a broad and modern view of maritime operations and planning:

> It is during the period of war that the value of a plan of campaign asserts itself. No serious campaign can be undertaken that is not based upon sound strategical premises. Then is the time when the high command, having so trained his fleet to perfection in the tactical field in peace, can, with reasonable assurance, leave the execution of tactical details to his subordinates and devote himself to the study of the strategic situation. For unless the strategic plan of campaign is sound, no brilliancy of tactical execution can entirely overcome a fundamental strategic weakness.[39]

Pratt recognized that tactical brilliance cannot overcome flawed strategy. This was a painful lesson that Germany and Japan would learn in World War II. As important as planning, particularly campaign planning, might be in preparing naval officers, Pratt also recognized that the navy's approach to staffs and planning was fundamentally different than that of the army. In comparing service staffs, Pratt suggested that the army's staff method is bureaucratic and influential in shaping the general's decisions. In contrast, navy staffs are commander-centric and exist simply to implement his decisions.[40] He noted that the "difference in the two outlooks extends even to the training at the two colleges."[41] This distinction and the nature of war at sea goes a long way toward explaining the navy's attitude. Afloat space is at a premium; there is little room for large staffs. Landpower usually affords and requires time for detailed planning. The tactical pace at sea requires more instantaneous decision making by commanders. The platform-centric nature of the navy leads to a natural tactical bias. However, planning is at the very heart of the operational and strategic levels of war that establish the conditions for success and defining victory. As interested as Pratt was in campaign planning and graduating students capable of serving on higher-level staffs, he understood the navy's culture.

The navy's attitude toward staffs, which was shaped by both necessity and tradition, endured throughout World War II and beyond. The modest growth of naval staffs, in response to the requirements of modern warfare, was nothing compared to the growth of army staffs. Jellicoe's staff as commander of the British Grand Fleet in World War I included only sixteen officers. Some thirty years later, Vice Adm. Raymond A. Spruance as commander of the U.S. Fifth Fleet had only twenty officers on his staff at sea, while Vice Adm. William F. Halsey had double that number on his staff as commander of the U.S. Third Fleet.[42] In contrast, a comparable level army operational staff, such as Lt. Gen. Omar Bradley's First Army in June 1944, comprised 361 officers.[43] The Army War College spent much coursework laboriously preparing officers to serve on higher-level staffs. The Army students produced detailed and lengthy campaign plans, sometimes entire volumes taking weeks to compile. The navy essentially split staffs between those afloat and those ashore. The staffs ashore took care of routine procedures and logistic work. The command staffs afloat became separated from the intelligence and supply staffs, whose work was assigned to specialists. As a consequence, the navy took a different approach to staff planning based on its experience, needs, and culture.

The army and navy shared the estimate process and the five-paragraph order, but beyond that the scope and detail of planning

The U.S. Naval War College, established in 1885, was the first naval institution of higher learning of its kind. Located in Newport, Rhode Island, the college helped to pioneer the employment of aircraft and submarines in operational art during the interwar period. Courtesy of the U.S. Naval War College Museum.

War-gaming at the Naval War College was a key part of the curriculum before and after World War I. It played an important role not only in developing the judgment of its students but in working out the operational problems posed by the new era of warfare. Courtesy of the U.S. Naval War College Museum.

differed considerably. Regardless of the differences in service planning, Admiral Pratt recognized the need for an increased focus on
planning, particularly in logistics. There were few logistics experts
available, but Pratt knew where to find them. Pratt recruited four
officers for his new logistics department, though none were navy
line officers. Two civil engineer corps officers, a Marine colonel, and
an army officer—Lt. Col. Walter A. Reed—made up the new department.[44] Pratt put Capt. R. E. Bakenhus, a civil engineer, in charge of
the department. With little time to prepare, Bakenhus simply got the
army doctrine on logistics for overseas expeditions used at Leavenworth, reproduced it, and issued it to the students.[45]

Logistics was soon more prominently featured in the work of the
class during Pratt's tenure. In 1925, the course required each officer to
write a thesis in strategy and logistics. In the fleet problems students
had to address logistics as well as maneuver. In 1927, the class studied
the Pacific area addressing strategic, logistic, and tactical factors. The
problems included the Blue advance across the Pacific: Blue plan for
supply, maintenance, and repair; Orange general plan for war; and
Orange operations to meet Blue attack.[46] Logistics was one of several
areas in which Pratt's initiatives improved education at the college.
He went on to become Chief of Naval Operations, where he continued to exert a positive influence on the navy. Unfortunately for the
service, his initiatives, particularly with regard to logistics, did not
last. In 1927, Rear Adm. Joel Pringle relieved Pratt and promptly reorganized the college. The new president combined the logistics,
strategy, and tactics departments into a single operations department. Logistics was thus subsumed and never got the attention it
received in the army schools.

Throughout the interwar period the Naval War College continued to use history to discover and impart the principles of naval warfare. The 1937 prospectus listed twenty-four of seventy-one lectures
as historical. The study of history taught the principles of naval warfare, exercises taught the art of naval warfare.[47] The prescribed reading contained a good deal of theory, including Clausewitz, Corbett,
and others. There is little evidence the college used official doctrine—
at least not navy doctrine. The 1937 prospectus does list the army's
Staff Officer's Field Manual as required reading.[48] Instead, the college faculty prepared their own texts for instructional use. Teaching
methods changed little in the last decade before World War II, but the

instructional emphasis on how the navy might use modern technology to achieve sea control and meet strategic requirements helped to focus the discussion for the rising class of naval leaders.

All the key navy commanders in World War II passed through the Naval War College. Future admirals Ernest J. King, Chester W. Nimitz, Richmond K. Turner, Raymond A. Spruance, William F. Halsey all came to Newport. How did they conceive of operational art? What did they think? What were they taught? Turner and Spruance not only attended but taught at the Naval War College. After graduating in 1927, having benefited from Pratt's initiatives, Spruance then served two tours as an instructor. He eventually became the head of the operations division from 1935 to 1938. Captain Turner (known by his middle name, Kelly) ran the strategy section for the senior class during these same years. In their lectures on naval strategy and the strategic use of the fleet, both officers demonstrated a sophisticated view of operational art.

Turner was an aggressive and competitive officer who worked the students hard. He was also well known for getting into fiery debates with Capt. Robert A. "Fuzzy" Theobald, another instructor who possessed a sharp intellect.[49] Born in Portland, Oregon, in 1885, Turner served aboard several battleships in World War I. In 1927, he was designated a naval aviator and attempted to emphasize both carrier and amphibious operations. In his lectures on the strategic employment of the fleet Turner outlined his views on the use of seapower, which accepted Corbett's interpretation of the relevance of limited and unlimited wars. Turner insisted that naval force serves national policy and accepted Corbett's categories of naval forces with one exception. Unknowingly foreshadowing his own future contribution to the Pacific war, Turner noted:

> Corbett failed to mention one category of force which, to the Navy, is as essential as either the battle fleet, the control force, or the local naval defense forces. To operate successfully, the Navy requires secure bases. The additional category to which I refer comprises the land forces required for the capture and for the defense of naval bases, at home or abroad. Under the American system the Navy is responsible for the seizure and the defense of advanced bases until the responsibility can be taken over by the Army. At home, a close coop-

eration is required between Army and naval forces; overseas the troops themselves, whether Army or Marine Corps, form an essential category which is an integral part of the fleet itself."[50]

Turner also accepted Corbett's views on concentration of the fleet. He quoted Mahan's stricture on keeping the fleet concentrated at all times, but then paraphrased Corbett's critique "to get results, the fleet must exert its strength at sea, and throughout the entire area that it has taken upon itself to control. Therefore detachments must be sent out in various directions for the performance of the strategic operations that the situation requires."[51] Turner defined the term "operation" as "ordinarily used to describe any arrangement or series of connected military acts."[52] He viewed strategic operations as rather simple and few in number. He listed these as scouting, patrolling, screening, escorting, raiding, supporting, and covering. Though clearly these have tactical application, Turner's discussion describes them more as operational tasks.

Capt. Ray Spruance also had strong views on the conduct of naval warfare. Born in Baltimore in 1886, he became a consummate surface officer, cruising around the world with President Theodore Roosevelt's Great White Fleet in 1907. He also knew something of staff work, having served on Admiral Sims's U.S. Naval Forces Europe staff in World War I. He was a very different personality from the fiery Kelly Turner; his biographer, Thomas Buell, calls Spruance the "quiet warrior." Though quiet, Spruance was competent and thoughtful. His lecture on the nature of naval warfare expressed his convictions as well as much of the conventional wisdom at the War College.

On the whole, Spruance's lectures reflected a good deal of Corbett's influence. The very organization of the curriculum emphasized Corbett's theories, as Spruance noted in his lecture in July 1937: "Corbett classifies naval operations as those undertaken for securing command, those for disputing command, and those for exercising command. In the syllabus of the War College course they are classified as operations for securing command of sea areas, as operations in areas not under command, and as operations in sea areas under command."[53] He further quotes Corbett on the purpose of the fleet: "firstly, to support diplomatic effort; secondly, to protect or destroy commerce; and thirdly, to further or hinder military operations ashore."[54] Lest they miss the essence of seapower, Spruance cautioned the students, "Although you will give much time and thought while at the War College to that climax of naval effort, the fleet action, it is well to remember that, historically, fleet actions on a large scale are rare compared with the number of minor naval actions that are fought."[55]

Spruance touched upon an essential difference between landpower and seapower: "A defeated army may often be rebuilt with fresh men. A defeated fleet is, on the other hand, in large part sunk. Years are required to build new ships and train new crews for them."[56] This again, in part, explains why decisive battle may be possible, even if unlikely, in naval warfare. It also helps explain the navy's tactical bias and attention to technology. Spruance not only understood the differences between landpower and seapower; he understood their relationship. This brought Spruance to an essential operational tenet —leverage:

> We must be prepared to use military operations to assist in the attainment of naval objectives. We must be prepared to use naval operations to help reach military objectives. Only through a sympathetic understanding of the difficulties of each other's problems and a comprehension of the possibilities and limitations of the tools with which the other must work can the Army and Navy achieve the coordination required for success in joint operations.[57]

In 1934, Rear Adm. Edward C. Kalbfus became president of the Naval War College. He was probably responsible for getting Spruance reassigned to the faculty the following year. Kalbfus believed the student text for the estimate and planning process had become outdated. He personally set out to rewrite and expand the text. Kalbfus asked Spruance among others to review the newly revised instructional material entitled *Sound Military Decision*. Spruance didn't think much of Kalbfus's writing style or the fact the president banished the principles of war from the text. Spruance and Turner headed two of the four committees that reviewed the draft. Regardless of Spruance's distaste for Kalbfus's prose and some of the substance (or lack thereof), the college published *Sound Military Decision* in 1936. The text became the bible for planning at the Naval War College. Known as the Green Book (or Green Hornet), the text was slightly revised in 1938, 1939, and again in 1942. It became the primary statement of the Naval War College's views on planning throughout World War II.[58]

Sound Military Decision clearly suggests the navy's grasp of operational art in the last years before World War II. It recognizes three levels of war and indicates that the objectives determine the level. The instructional text discusses campaign plans, operations plans, and operations orders. A campaign plan "indicates what might be called the schedule of strategy which the commander intends to employ to attain his ultimate objective for the campaign. Such a plan usually sets forth the stages into which he proposes to divide the campaign, shows their sequence, and outlines the general plan."[59] The campaign may consist of a series of major operations. The operation plan "covers more complex operations than does an operations order, and projects operations over a greater time and space."[60] An operation is distinguished from a tactical engagement in that the former consists "not of a single act, but of a series of acts"—that is, "of a number of stages or phases of battle, each being a preparation for the one following, until the final stage provides for the attainment of the assigned objective."[61]

In developing the estimate, the text insists that each course of action must be tested for feasibility, acceptability, and suitability. Feasibility ensures the course of action is supportable with the means available. Acceptability relates to cost, and suitability means the course of action will achieve the desired affect. The planners must also prepare alternative or contingent plans "developed from varying sets of assumptions."[62] *Sound Military Decision* called for intelligence and logistic plans but without addressing them in detail. In the matter of command and control, the text insists "that to divide the supreme command in any locality, or to vest it in a body rather than in an individual, is necessarily to diffuse responsibility."[63] *Sound Military Decision* touches upon several key operational functions and mentions several important techniques in operational planning, but essentially, it is about problem solving and decision making. It was referenced, taught, and hammered into the students through lectures and exercises.

The emphasis under Kalbfus on application rather than detailed planning is evident in the student plans and solutions in the war games. Potential war with Japan dominated exercises and planning at the Naval War College. In fact, a good case can be made that the U.S.

Navy in the interwar years was trained, built, and organized for war in the Pacific. Strategic problems on charts set in the Pacific then became tactical problems on the gaming board. The strategic problems routinely war-gamed the War Plan Orange courses of action—either the "through ticket" rush to the Philippines or the step-by-step advance across the Central Pacific. Virtually all of the men who would actually direct the war in the Pacific familiarized themselves with the theater and the operational problems at Newport. The most influential of these, Capt. Ernest J. King, reported to the Naval War College in 1932.

Ernest King was born in Lorain, Ohio, in 1878. He decided on a career in the navy, then graduated from the Naval Academy in 1901. King was an ambitious, hard-driving, hard-drinking, and well-rounded naval officer. He served on destroyers and in submarines, and in 1928 he became an aviator. By the time he got to the Naval War College, he was the senior student in the class and offered the position as Blue fleet commander for Operations Problem IV, the annual war game. The scenario directed the Blue fleet to recapture the Philippines from the Orange (Japanese) forces. The scenario offered three courses of action in the fleet's approach to the Philippines: the Blue fleet could go north of the Mandate Islands (the Marshalls and Caroline Islands) and then through the Marianas, go directly through the Mandate Islands, or sail south of the Mandates in a "through ticket" solution. King argued for the first course, north of the Mandates and through the Marianas, but Rear Adm. Harris Lanning, president of the college, insisted on the southern route.[64] King argued the Japanese could attack the flank of the Blue forces the entire route "and then be caught in a bottleneck between New Guinea, Morotai, and Mindanao."[65]

The head of the Operations Department informed King he could keep his war-game command only if he agreed to play the president's preferred course of action. The ever ambitious King replied "that he would carry out the solution of a berth-deck cook rather than miss the chance to manage a fleet even on the maneuver board."[66] Just as the exercise got under way, King was called away to the funeral of Rear Adm. William A. Moffett, chief of the Aeronautics Bureau, who was killed in the crash of the dirigible *Akron*. In the exercise, the Blue fleet did not fare well. Much as King anticipated, the bottleneck between New Guinea and the Philippines allowed the Orange forces to attrite the fleet. King went on to replace Moffett as the chief of the Bureau of Aeronautics, and eventually became the wartime CNO.

Time and time again, war-gaming at the college demonstrated the risk of the "through ticket" plan to push the fleet rapidly to the Philippines in case of war with Japan. The major problem with this option was logistics. In 1925, War Plan Orange planners calculated the fleet would need 373 support vessels in the fleet train more than 100 of which would be tankers. Although the fleet could transfer fuel at sea, not until 1944 did the navy completely solve the problem of underway replenishment. Strategic requirements demanded that the United States project power across the Pacific; the solution was a measured advance that seized bases along the way.[67] In 1934, the Joint Board adopted the step-by-step advance across the Central Pacific as the only realistic concept of operation for War Plan Orange. This plan always involved some notion of phasing operations, but clearly the Central Pacific advance encouraged a closer look at phasing. Indeed, most all of the student solutions to strategic problems after 1934 included phasing.

Planning at the Naval War College at the end of the 1930s reflected not only the preoccupation with a future war in the Pacific, but the strengths and weaknesses of the curriculum. Most exercises required students to write operations plans, which in format and content would be familiar to any modern military planner. The operations plans included task organization, assumptions, and the familiar five-paragraph format. Brief intelligence and logistic annexes were usually provided. On occasion the strategic problems required students to write campaign plans, but these were brief and typically lacked detail. Operations Problem III in 1935 gave students on the Blue fleet staff the mission "to gain and maintain command of Orange vital sea areas in order to bring about the decisive defeat of Orange."[68] The exercise required the students to write "a campaign plan in general terms to accomplish the CINC's basic mission and an operations plan to carry out the first phase of the contemplated campaign."[69] The staff solution to this problem included a three-page campaign plan consisting of three phases each broken down into a series of tasks. Planning and exercises at the college did not deepen the students' appreciation of campaign planning, logistics, or intelligence commensurate with the students at the Army War College. The exercises did provide a rational problem-solving process that sharpened students' judgment and they certainly familiarized the students with the most likely theater of operations and their most likely enemy. Students were forced to look at operational problems not only from the Blue perspective but also from that of Orange. The exercises provided a vehicle for planning and perhaps, just as important, a way to consider how new technology might be integrated into achieving sea control.

TECHNOLOGY AND SEAPOWER

No discussion of maritime operational art in the interwar period can be complete without a consideration of technology, specifically the advent of the submarine and naval aviation. The battleship too increased in capability during the interwar years, but the submarine and particularly naval airpower would eclipse the battle line as the central means of sea control in the coming war. Mahan asserted that technology may require wholesale changes in tactics, but he insisted that the fundamentals of operational art—strategy as he called it—remained constant. Applied technology is the very essence of the tactical art, but technology can also affect operational art, though to a lesser degree. Just as railroads and the telegraph affected the exercise of operational art on land in the nineteenth century, the new technologies rising from World War I shaped modern operational art both on land and on the sea in the twentieth century. In fleet exercises and in war-gaming at the Naval War College, the U.S. Navy seriously considered how the new technologies should be integrated into the fleet—how they could be used to help achieve sea control.

The exercises at the Naval War College following War Plan Orange used economic blockade as the defeat mechanism for a potential war against Japan. The war scenario directed a return to the Philippines, and securing bases close enough to Japan to establish sufficient sea control to choke Japan into submission. At the tactical level, it was assumed that a climactic battle with the Imperial Japanese Navy was inevitable as the U.S. Navy projected American combat power across the Pacific. For practically all of the interwar period, it was further assumed that the battleship would determine the outcome of such a confrontation. In the 1920s this made perfectly good sense.

Battleships continued to improve in lethality, range, and endurance during the interwar period. The shift from coal to oil gave battleships greater range. The addition of larger-caliber guns and aerial spotting meant they could shoot farther than a torpedo and deliver destructive fire more accurately than aerial bombing.[70] As a Naval War College pamphlet entitled the *Employment of Naval Aviation* pointed out as late as 1937, it took as many as 1,200 planes to carry as much bomb tonnage as contained in the large projectiles of a single battleship.[71] Most important, it had been demonstrated beyond a doubt that a battleship could sink other battleships. Despite the technological improvements and the faith placed in these great ships, the real problem with battleships lay in the fact they could provide only a tactical advantage. The improving capabilities of submarines and aircraft in the thirties suggested that these platforms could contribute at all levels of war—tactical, operational, and strategic.

Submarines first entered the U.S. fleet in the 1880s. Americans continued building a few small submarines suitable for coastal defense all the way up to World War I. Beyond its contribution to coastal defense, the submarine offered another tactical weapon in fleet engagements. The German High Seas Fleet in World War I repeatedly tried to lure the British Grand Fleet onto a screen of submarines. It was with a combination of mines, submarines, and minor surface actions that the Germans intended to attrite the British navy in order to even the odds in a fleet engagement. Early U.S. Navy exercises, however, demonstrated that submarines did not have the speed to keep up with the fleet. As commander of a submarine division, Capt. Ernest J. King led his group of four submarines from New London, Connecticut, to the Caribbean to participate in the first annual fleet exercise in 1923. He soon found his group scattered and plagued with mechanical difficulties. He later admitted, "I had very little to do with the fleet maneuvers that I can remember."[72] Unable to keep up with the fleet, submarines became autonomous weapons of attrition against enemy warships.

The submarine could not become part of the tactical scouting force for the fleet, but it could provide operational reconnaissance. For example, in the Naval War College's staff solution for Operations Problem III in 1935, the submarine force was tasked "to obtain information of enemy movements and dispositions, reduce strength of enemy forces. Observe passages leading into waters of Southern Philippines and passages leading to southward between New Guinea and Borneo."[73] The submarine's tactical and operational missions were codified in doctrine, exercised, and war-gamed. However, its most effective and feared mission was strategic—waging unrestricted warfare against enemy commerce. The German example in World War I was so powerful that attempts were made at the Washington Naval Conference in 1921 to ban the submarine altogether. Failing that, the representative of the United States, Elihu Root, proposed restricting submarines to the surface rules for attacking unarmed merchantmen. This required the submarines to surface, board, and inspect merchantmen.[74] Although the drawbacks of this policy for vulnerable submariners were evident, it became the official American policy.

As early as 1919, Capt. Thomas C. Hart drafted a memorandum to the Chief of Naval Operations pointing out the vulnerability of Japan to unrestricted submarine warfare. The following year he made the same argument in a lecture to the Naval War College.[75] The adoption of diesel engines and steady technological improvements in fire control, torpedoes, and construction techniques made the submarine a real tactical threat with potential strategic significance. After a good deal of debate on the type of submarine needed, in 1936 the General Board recommended a medium-size boat of 1,450 tons with a good cruising range. By the end of the interwar period the navy introduced the *Gato* class submarine of 2,000 tons with an 11,000-mile transoceanic cruising range. Since submarines were still too slow to operate with the fleet, official doctrine called for them to operate alone conducting a campaign of attrition against enemy ships. The same techniques and technology that made them a tactical threat to enemy combatants made lethal against merchant vessels. Unlike tactical combat engagements in landpower, the submarine's random attacks against merchant ships could generate an operational or even strategic effect. The war-ending blockade of Japan as envisioned in War Plan Orange could begin immediately with the deployment of a sufficient submarine force. Ultimately, the advantages outweighed the transitory moral guilt of violating international law. In 1941, the Naval War College recommended that in the event of war, commanders be authorized to declare what amounted to trade exclusion zones in strategic areas in the Far East and near Japan. The joint planners adopted this recommendation in the final prewar war plan, Rainbow 5. The navy abandoned even this modest restraint after the shock of the Japanese attack on Pearl Harbor. Six hours after the attack, the

Navy Department radioed the following concise order: "Execute un-restricted air and submarine warfare against Japan."[76]

Nothing would revolutionize war at sea like the advent of naval aviation. Many historians criticize the navy for a failure to more fully embrace aviation. Some characterize the interwar period as a navy dominated by battleship admirals, "the gun club," who resisted efforts to challenge their Mahanian vision of sea control.[77] In fact, the development of American naval aviation is a great success story. As naval historian George Baer points out, it was not simply a matter of attitude but the allocation of scarce funds.[78] The navy strove to develop a balanced fleet in the interwar period and largely succeeded. An institution fortunate to possess good leadership, the navy did an excellent job integrating airpower into the fleet.

Sufficiently impressed by the experience and promise of naval aviation in World War I, the General Board in 1919 informed the secretary of the navy that "fleet aviation must be developed to the fullest extent. Aircraft have become an essential arm of the fleet. A naval air service must be established, capable of accompanying and operating with the fleet in all waters of the globe."[79] In a lecture at the Naval War College less than a year after the end of World War I, naval innovator Rear Adm. Bradley A. Fiske saw clearly the potential of naval aviation:

> The last war has shown you that our battleships cannot get very near these coasts because of submarines, mines, tor-pedoes, forts, and aeroplanes. What would the fleet do there? Doubtless you answer, "blockade." Doubtless we could blockade; but it may be pointed out that the blockading would have to be done at a considerable distance from the shore, far away from a United States naval base or a drydock and in constant danger from attack by torpedoes, launched from submarines and aeroplanes. If we sent, however, a large number of swift seaplane carriers, like the *Argus* and *Eagle* (British carriers), each carrying twenty torpedo planes and other seaplane carriers equipped with reconnaissance and bombing planes, they could avoid the enemy's battleships, and not only attack all surface craft, but also do very effective work against land defenses; not being deterred by such obsta-

cles as shoals, mines, submarines and forts. It may be that the *Argus* and the *Eagle* are the prototypes of the battleships of the future.[80]

Congress responded to the General Board's recommendations in 1921 by authorizing one aircraft carrier and creating the Bureau of Aeronautics to oversee naval aviation. Rear Adm. William A. Moffett became the first head of the bureau and ably led it until his untimely death in the crash of the airship *Akron* in 1933. Moffett worked closely with the Naval War College to provide the latest data on aircraft and aircraft carriers. In exchange, the college included naval aviation in its war games, gathering valuable insights as a result. Eventually, a relationship developed in which outcomes from Naval War College war-gaming could be tested in fleet exercises and these results then shared with both the college and the bureau. As naval historian Thomas Hone suggests, this created a cycle of ideas providing innovation based on simulation and actual exercises.[81] In 1933, Lt. Cmdr. J. R. Moloney summarized the results of chart and board maneuvers to the students; the war games demonstrated "that a fleet cannot successfully advance into a hostile zone unless it can carry with it an air force that will assure beyond a doubt superiority in the air." He concluded that "aircraft will exert a decisive influence in all forms of naval warfare by the cumulative effect of initial advantages they often gain for a fleet."[82] Surprisingly, in little over a decade naval aviation rose from a novel new technology to an essential part of the fleet and its exercise of sea control.

In the 1920s naval airpower demonstrated its ability to extend the fleet's tactical reach. In the thirties, technical improvements in naval aircraft demonstrated their ability to extend the fleet's operational reach. Billy Mitchell's controversial 1921 demonstration of aerial bombing that sank the German battleship *Ostfriesland* may have sent a chill up the spines of the members of the "gun club," but the admirals knew aircraft had a long way to go before they became true ship killers. Their first inclination was to employ aircraft to support not attack the battle line. Early trials demonstrated that aerial observation could substantially increase the effective range of the battleship's big guns. Eventually, just about every battleship and cruiser had its own spotter plane. By 1927, only five of fourteen naval aircraft squadrons were associated with carriers.[83]

Still, improving technology and a bureaucratic advocacy determined to push the potential of naval aviation as an extension of combat power. In 1922, the Vought VE-7 biplane that made the initial flight from the deck of the USS *Langley*, the first American carrier, had a range of only 291 miles. By 1939, technology extended the range of the Vought SB2U Vindicator carrier aircraft nearly fourfold, to 1,120 miles. Additionally, the Vindicator could carry a 1,000-pound bomb. The development of torpedo planes, dive-bombers, and fighters demonstrated that naval aircraft had become a real element of combat power—ship killers. In 1936, the navy accepted delivery of the PBY (patrol/bomber) Catalina seaplane. Although too vulnerable to be an effective bomber, the Catalina's range of 2,520 miles gave the fleet an operational as well as tactical reconnaissance capability.

Annual fleet exercises and war-gaming at the Naval War College suggested the importance and potential of naval aviation beyond the tactical level of war. Unlike the army, the navy could gather its major units together for annual exercises to test fleet doctrine. From 1923 to 1940, the U.S. Navy held twenty-one major fleet exercises (so-called Fleet Problems). In retrospect, the real value of these exercises was in testing the role of carriers in fleet operations. Specifically, the postaction reports and frank critiques by the participants, particularly flag officers, allowed the navy to benefit from the lessons learned. In the very first post–World War I fleet exercise the Black fleet was assigned the mission to destroy the Panama Canal. Although the Black fleet's two carriers were fictitious, a single very real seaplane flew over the Panama Canal and dropped ten simulated bombs without opposition. The impressive results of the exercise persuaded the commander of the U.S. Fleet, Adm. Hilary P. Jones, to recommend to the Chief of Naval Operations that he accomplish "the rapid completion of aircraft carriers and that all battleships should carry aircraft."[84]

Just four years later, the U.S. Navy commissioned two additional aircraft carriers, the *Lexington* and the *Saratoga*. Fleet exercises in 1929 provided both participating fleets with carriers. Admiral Pratt, former president of the Naval War College and now the Black fleet commander, detached the *Lexington* with an escort and again successfully attacked the Panama Canal, but this time with a sixty-six-plane strike force. By the end of the 1920s, American carriers were operating as a complete tactical division distinct from the battle line. In 1936, Fleet Problem XVIII demonstrated how far naval aviation had come. The navy deployed 152 ships, 496 aircraft and three aircraft carriers. This War Plan Orange exercise ranged from the Aleutian Islands, to Hawaii and Midway. PBY flying boats provided long-range reconnaissance for both fleets. Carrier task forces "attacked" Pearl Harbor and army installations on Hawaii. Navy dive-bombers "attacked" ships of all types. Senior naval air commanders in this exercise included Vice Adm. Ernest King and Capt. William Halsey.[85] It was a convincing demonstration of the potential of aviation in modern naval warfare.

At the Naval War College strategic war games emphasized the operational role for naval aviation. The strategic exercise that began Operations Problem III in 1935 called for the Blue fleet "to gain and maintain command of Orange vital sea areas in order to bring about the decisive defeat of Orange." The Orange forces captured the Philippines, which required the Blue fleet "to cover the landing of troops on these islands, an operation which was to be preceded by air attacks launched simultaneously from three carriers on Saipan-Guam and Peleliu." In the exercise the Blue commander employed his carriers in advance of the main body to achieve air superiority and conduct long-range reconnaissance and air attacks on Saipan, Guam, and Peleliu. The main body covered the 113,000-man expeditionary force assault on the Philippines, and then the game shifted to a tactical battle between opposing fleets. In this confrontation "both fleets launched air attacks as promptly as possible, Blue made Orange carriers and cruisers her bombing objectives. Orange attacked Blue carriers with bombs and Blue battleships with aircraft torpedoes."[86] Clearly, in both simulated and real exercises American naval aviation was used for tactical as well as operational tasks, specifically reconnaissance and fires.

In 1937, Captain Turner lectured on "The Employment of Aviation in Naval Warfare." In contrast to the ambitions of the Army Air Corps, he noted that "aviation produces its maximum effects when employed in supporting the aims of naval or military strategy. It is unable alone to conquer or hold territory, as it must depend upon the other arms for the security of its bases and for consolidation of the gains that it may make possible. We therefore, must reject the idea that aviation alone can achieve unlimited results against well-organized and well trained military or naval forces."[87] Certainly part of the reason for the navy's success in developing naval aviation was its insistence that it serve the fleet. In Turner's assessment, the strategic (operational) impact of aviation lay in the relative increase in the use of attrition and the strategic importance of outlying positions. The true measure of the value placed on naval aviation is the importance the navy assigned to obtaining air superiority. The navy increasingly believed, as the Army Air Corps always insisted, that air superiority is the first mission of airpower because it gives the fleet tremendous advantages in security, information, and attack.

Turner understood that airpower might change the long-held belief in the decisive Mahanian clash of fleets. He suggested that "with this wide deployment of possibly the greater part of the fleet, and the ability of aviation to attack rapidly from long range, we may expect to see an increase in the employment of attrition until it may even become the principal method of inflicting damage upon the enemy."[88] He further understood that shore-based aircraft provided the same advantages and threat as did carrier-based aviation. Turner noted, however, that fleets could not operate near hostile coasts with strong enemy air bases. By the end of the interwar period, the incremental advance across the Central Pacific was mandated by the need not only to obtain logistic bases for the fleet but to secure the fleet from shore-based aircraft. The island-hopping campaign would extend and sustain the projection of American combat power. The strategic and now operational requirement to secure advance bases led to one final ingredient in American maritime operational art of the interwar period—the development of amphibious operations.

AMPHIBIOUS WARFARE

By the end of the interwar period the American armed forces had come a long way from the amphibious descent on Cuba during the Spanish-American War. In 1898, the army rather unceremoniously pitched horses and mules from the transports depending on them to swim ashore. The troops meanwhile were ferried to an unopposed landing in small boats towed by steam launches. The great difference in amphibious capability at the end of the decade was due to operational requirements of the U.S. Navy and institutional requirements of the United States Marine Corps. For the navy, the logic of War Plan Orange was inescapable; fleet exercises and studies at the Naval War College demonstrated the need to seize advance bases. All the same, in an era of scarce funding the navy was more concerned with balanc-

ing and modernizing the fleet than with expanding and fully supporting the requirements of the U.S. Marines.

The Marine Corps wrote another chapter in its storied history in
World War I fighting as the fourth brigade in the U.S. Army's Second
Infantry Division, but this experience also pointed to a certain re-
dundancy in capabilities. To avoid being amalgamated with the army
or becoming permanent "State Department troops" for hard-handed
diplomacy, the marines needed to better define their own corporate
niche.

Prior to 1900, the Marine Corps provided guards and garrisons
ashore and detachments aboard ships. Following the Spanish-
American War the navy's General Board recognized the need for an
advance base force. Apparently, the experience of working with the
army during the war encouraged the board to consider raising its own
naval infantry to defend advance bases that were acquired as a result
of the U.S. victory. Slowly and somewhat reluctantly, the Marine
Corps took on the mission, and between 1910 and 1914 the advance
base force became a reality. The navy authorized a mobile defense
regiment for both the East and West coasts. In 1917, however, anxious
to get into the fight, the marines went into the trenches alongside the
army. Following the war, Commandant John A. Lejeune steered the
Marine Corps back to the advance base mission, fully embracing what
became the corps' most-celebrated expertise—the conduct of amphibious operations.

Born in Louisiana in 1867, John A. Lejeune was short, southern,
and stocky. He graduated from the Naval Academy in 1888, and in
1909 became the second marine to attend the Army War College. His
ambition and sheer competence would propel him to the top of the
corps. His connections, friendship, and reputation among his War
College classmates helped convince General Pershing to give him
command of the Second Infantry Division during World War I. After
becoming commandant of the Marine Corps in 1920, Lejeune quickly
recognized the need for the Marine Corps to fully embrace amphibious operations as its primary mission—and the benefit to the corps of
so doing. Aware that the navy was reviewing War Plan Orange, Major
General Lejeune directed his staff to develop a Marine Corps plan for
the Central Pacific.[89] His director for intelligence, Maj. Earl "Pete"
Ellis, submitted his views in a document entitled "Advanced Base
Operations in Micronesia." This report provided an intelligence esti-

mate along with a sophisticated tactical analysis of the requirements for amphibious operations in the Central Pacific. Ellis stated the case up front: "In order to impose our will upon Japan, it will be necessary for us to project our fleet and land forces across the Pacific and wage war in Japanese waters. To effect this requires that we have sufficient bases to support the fleet both during its projection and afterward."[90] The commandant agreed and embarked on an aggressive effort to convince the navy.

In 1923, Major General Lejeune informed the students at the Naval War College that the primary wartime mission of the Marine Corps "is to provide the Navy with an expeditionary force necessary for the effective prosecution by the fleet of its major mission, which is to gain control of the sea and thereby open the sea lanes for the movement of the Army overseas."[91] Lejeune understood that for the corps to prosper in lean times, it must have a rationale that firmly linked it to national and naval requirements—the marines must be useful to the navy. In the same year that the commandant addressed the Naval War College the advance base force became the Marine Expeditionary Force. To stress the point that the Corps was a fundamental part of American maritime power, it redesignated the Marine Expeditionary Force the Fleet Marine Force in 1933. Significantly, the Fleet Marine Force worked for the commander of the U.S. Fleet. The Marine Corps designed and promoted the Fleet Marine Force as a critical resource in waging a naval campaign, specifically in the Pacific. Not only did the navy's General Board accept Lejeune's interpretation of the corps' new primary wartime mission, but in 1927 it also approved the Marine Corps' responsibility to "provide and maintain forces for land operations in support of the fleet for the initial seizure of advance bases and for such limited auxiliary land operations as are essential to the prosecution of the naval campaign."[92]

The U.S. Marines worked hard to solve the considerable tactical problems involved in modern amphibious operations. Exercises were held in order to gain practical knowledge, and the Marine Corps Schools were put to use. By 1922, the corps ran three schools: the basic course, the company commanders course, and the field officers course. After World War I, the marines used army texts and subject matter. In 1932, Brig. Gen. James C. Breckinridge became commandant of the Marine Corps Schools. Brigadier General Breckinridge brought Lt. Col. Ellis Bell Miller to the school staff. Lt. Col. Miller, a graduate of the Naval War College, the army's General Staff School, as well as the Army War College, dropped all the army courses from the curriculum. Under Miller's supervision the curriculum soon reflected the corps' new mission focus.

The marines studied extensively the failed British Dardanelles-Gallipoli campaign from World War I. The British attempt to open a passage to Russia through the Dardanelles in 1915 pointed up the great difficulty in assaulting a hostile shore. In 1932, each marine student was issued a copy of the official British history of the campaign to familiarize themselves with the problems of major amphibious operations. Students from the field officers course and the Naval War College worked on "the Advance Base Problem." This series of problems, begun in 1931, dealt with defending or seizing a base, usually set in the Pacific in places like Guam, Saipan, or Tinian.[93] In November 1933, the commandant directed the Marine Corps Schools to prepare a landing operations manual. The schools discontinued class to get the job done. The schools' effort and several additional conferences produced the *Tentative Manual for Landing Operations* in 1934. This manual detailed all the necessary tactical pieces for successful amphibious operations, such as command relationships, naval gunfire support, aerial support, ship-to-shore movements, and logistics. A year later, the CNO and commandant approved a revised and renamed *Tentative Landing Operations Manual* as the official doctrine. In 1938, the manual became *Fleet Training Publication 167* (FTP 167).

The marines, to their great credit, devoted themselves to solving the tactical problems in conducting offensive amphibious operations. All the services, however, contributed to the development of amphibious operations. In 1927, the Joint Board not only addressed the amphibious mission of the Marine Corps; it essentially gave the army the same mission to "conduct land operations in support of the navy for the establishment and defense of naval bases."[94] In 1929, the Joint Board published *Joint Overseas Expeditions—Tentative* to provide more-detailed guidance to the army and navy on landings against opposition. Three years later, the army and navy held a joint exercise involving six hundred soldiers and seven hundred marines simulating a two-division expeditionary force to test procedures. By all accounts the exercise was a poor performance that served to highlight the need for adequate landing craft and better command and control.

Based in part on this exercise, the Joint Board issued a revised *Joint Overseas Expeditions* in 1933. This manual contained a good deal of technical information, recommended task organization tailored to amphibious operations, and placed an emphasis on detailed planning. It was in an effort to build on these ideas that the Marines published the *Tentative Manual for Landing Operations* the next year.

Army interest in amphibious operations continued into the late thirties. In 1935 and 1936, the army sent observers to fleet exercises designed to test amphibious doctrine. To encourage army participation in fleet exercises, Adm. William H. Standley, the Chief of Naval Operations, sent a letter to the Army Chief of Staff, Douglas MacArthur, reminding him that under War Plan Orange and Joint Board agreements, the army was expected to make amphibious landings.[95] MacArthur's successor, Gen. Malin Craig, enthusiastically supported army participation in 1937 and 1938. The army formed the First Expeditionary Brigade around the 30th Infantry Regiment and support units for the 1937 fleet exercise. Despite this effort at cooperation, CNO William D. Leahy declined further army participation at the end of the decade claiming the conduct of amphibious operations was a maritime responsibility. By 1940, the threat of war served to force an end to interservice rivalry, and joint amphibious training resumed. Within another year division-size landing exercises were planned and executed. These exercises revealed many problems, but the tactical doctrine developed in years before proved invaluable.

Even though the fundamental tactical doctrine was in place, much of the needed equipment and organization would have to wait until the actual prospect of war shook loose the purse strings. At the operational level, amphibious operations and joint problems were little studied or exercised at the Naval War College. The navy recognized the need for amphibious operations but remained solidly focused on fighting the decisive sea battle. Students at the Army War College did plan major amphibious operations in connection with War Plan Orange—specifically, the retaking of the Philippines. The navy raised six marine divisions in World War II, all of which were committed to the Pacific. For its part, the army committed twenty divisions to the Pacific, and even in this maritime theater it conducted more and larger amphibious operations than did the Marine Corps. Still, it was the Marine Corps that pioneered the tactical doctrine and the Army recognized this fact. Army Chief of Staff George

Marshall was sufficiently impressed with FTP 167 that he authorized its publication in 1941 as U.S. Army *Field Manual 31-5, Landing Operations on Hostile Shores*—little changed but the cover.[96]

The professionals in all the service colleges understood the essential truths of modern operational art. Addressing the Naval War College, U.S. Marine colonel Ellis B. Miller insisted: "War will not be conducted by the Navy. War will not be conducted by the Army. War will be conducted by a combined Army and Navy Team and we must train, organize, learn our signals and be ready to play ball as one fighting unit."[97] Modern operational art even in a maritime theater requires sea, air, and land power. By the end of the interwar period, the U.S. Marine Corps was not just another tactical weapon available to the navy, army, or State Department but a tactical force with an operational mission available to a theater commander. Amphibious operations had become both a recognized form of naval warfare and a necessary element of modern American operational art.

ASSESSMENT

Historically, the role of seapower consists in achieving and maintaining sea control. Sea control allows the navy to defend the sea lines of communication, isolate adversaries, and to project and support landpower to win wars. During the interwar period advances in submarine and aviation technology extended the range and lethality of seapower. The role of the navy in modern operational art continued to center on establishing sea control, but in the modern era this role is fulfilled by supporting airpower as well as landpower.

The achievement of the Naval War College lay in instilling in naval officers a consistent and rational approach to problem solving. The War College did not emphasize planning; rather, it focused on execution—the application of seapower. While it is true the college studied the Battle of Jutland and spent a good deal of time on the climactic tactical battle between fleets, it is also true that in their strategic problems, the faculty and students dealt with operational challenges. Moreover, the college taught about the three levels of war and understood their relationship. Though he only mentions the classical two levels, Capt. G. B. Wright expressed the essential relationship to students in 1934: "Strategy should be the master, tactics the servant, as strategy is the servant of policy."[98] At the strategic level the college accepted that military force is used in a political context. Naval objectives must be nested in national policy. At the operational level, the navy recognized the importance of the essential functions of modern operational art—command and control, logistics, and intelligence. The study of these functions were driven largely by the requirements of War Plan Orange, but the Naval War College focused on maneuver.

The senior educational institution of the U.S. Navy was not about doctrine or planning, but about command judgment and decision making. All this was a product of the navy's service culture, its traditions, as well as operational requirements. The American navy developed a maritime operational art during the interwar period that recognized the importance of geography and technology in achieving sea control. Beyond a simple clash of fleets to achieve sea control as suggested by Mahan, the U.S. Navy understood that sea control in the future would require the combination of seapower, landpower, and airpower to achieve operational and strategic objectives in a theater of war. All this proved very necessary, as the United States faced its greatest military challenge in the coming world war.

6

The European Theater of War

The Japanese attack on Pearl Harbor in the early hours of December 7, 1941, swept away the isolationism and pacifism that gripped many Americans throughout much of the interwar period. The Japanese attack galvanized American public opinion and thrust the United States into another world war. The United States entered this war much better prepared than in 1917. After France's defeat in 1940, increased military budgets, and military production as well as the expansion of the armed forces, together quickly shifted the nation into mobilization.[1] The tasks of developing and applying American military power were now in the hands of officers who had spent the last twenty years studying modern war so that they would be prepared to fight this one.

Although the events that drew the United States into World War II occurred in the Pacific, the first strategic agreements America made were with its new ally Great Britain, to give the war in Europe first priority. Even before U.S. entry into the war, the Joint Board began to set aside the old color plans and draft new ones considering coalition warfare against the Axis powers. One of these plans, Rainbow 5, called for the disproportionate commitment of U.S. forces to Europe or North Africa while assuming a defensive posture in the Pacific. In November 1940, a separate study of strategic options by Adm. Harold R. Stark, Chief of Naval Operations, which became known as Plan Dog, confirmed this strategic approach. Admiral Stark also advocated talks between the American and British military staffs. President Franklin D. Roosevelt accepted Stark's recommendation, and the first staff talks took place on March 29, 1941. This conference and the subsequent staff meetings following America's entry into the war confirmed the Europe-first strategy. Once that was accepted as a key principle of Allied strategy, the question became how to apply Allied force in Europe against the Axis.

OPERATION TORCH

Just two weeks after Pearl Harbor, the British and American chiefs of staff met at the Arcadia conference to discuss Allied strategy. The situation was grim. In Europe, the Soviet Union seemed on the verge of collapse, and in the Middle East, the Germans challenged British control everywhere in the Mediterranean. Japanese victories in the Pacific would continue for the next six months. The Allies agreed to form a Combined Chiefs of Staff to provide strategic direction for the war. The British chiefs also provided their American counterparts with their strategic proposal for "closing and tightening the ring round Germany."[2] This involved limited offensives on the Continent while taking the offensive in the Mediterranean. These talks inaugurated a running debate between American and British planners, reflecting their respective strategic cultures and relative states of preparedness.

To facilitate working within the framework of the Combined Chiefs of Staff, the Americans formed their own Joint Chiefs of Staff (JCS) consisting of Gen. George C. Marshall (ASC 1908), Gen. H. H. "Hap" Arnold (CGSS 1929), and Adm. Ernest J. King (NWC 1933), the new Chief of Naval Operations, with Adm. William Leahy, a former CNO, as its presiding officer.[3] For Marshall and the War Department, the War Plans Division (WPD) did the strategic planning. Early in 1942, General Marshall decided to reorganize the War Department. The WPD became the Operations Division (OPD), and on March 9, Maj. Gen. Dwight D. Eisenhower (CGSS 1926, AWC 1928) became its first chief.[4] The OPD proved critical in shaping not only strategic planning, but logistics and, in the early years, even operational planning. Eisenhower noted that the officers who staffed the OPD consisted of a carefully "selected body of officers, which had, between the two World Wars, truly absorbed the teachings of our unexcelled system of service schools."[5] This select group included Thomas Handy, Lyman Lemnitzer, and Mathew B. Ridgway—all graduates of the Command and General Staff School and the Army War College—plus Albert C. Wedemeyer, who graduated from Leavenworth but attended the German Kriegsakademie instead of the War College. These officers made significant contributions to the strategic and operational direction of the war, during which they rose to prominence.[6]

In February 1942, the OPD expressed its strategic views in a memorandum to Marshall insisting that an attack through western Europe was the proper course of action. The OPD based its recommendation on the overriding need to keep the USSR in the war, the superior lines of communication offered by England's proximity to the Continent, and the ability to mass Allied power and "attack our principal enemy while he is engaged on several fronts."[7] The OPD's perspective underscored the U.S. preoccupation with logistics, mass, and concentration. Theoretically, at least to those familiar with Clausewitz, this made perfect sense. According to the Prussian theorist, "not by taking things the easy way—using superior strength to filch some province, preferring the security of this minor conquest to great success—but by constantly seeking out the center of his power, by daring all to win all, will one really defeat the enemy."[8] The American planners saw Germany as the center of gravity and believed it should be struck directly and as soon as possible. This was the shortest route to victory. The OPD began work on Bolero, a plan for an attack across the English Channel.

Bolero provided for the buildup of U.S. strength in Britain to establish sufficient force for a cross-channel attack. Two variants of the plan emerged: Sledgehammer and Roundup. Sledgehammer called for a diversionary strike of up to two divisions in 1942 that would force the Germans to undertake significant operations in the west. The Operations Division conceived Sledgehammer as a desperate contingency plan in case the Soviet Union appeared on the verge of defeat. The planners viewed Roundup as the main event projected for 1943. By July, the British concluded that Sledgehammer was not feasible, but Prime Minister Winston Churchill remained anxious to keep pressure on the Axis and seize the initiative. In a telegram to President Roosevelt, he again urged an operation to liberate French North Africa as "by far the best chance for effecting relief to the Russian front in 1942."[9]

The British had long considered operations in French North Africa. The British chiefs raised the project during the Arcadia conference under the code name Gymnast. Eisenhower and his planners in the OPD convinced Marshall that any such operation constituted a dispersion of Allied effort and would delay the decisive cross-channel attack. Like that of the Americans, the British perspective strongly reflected their strategic culture. The British Empire had a long history and predilection for peripheral operations.[10] Moreover, no British leader was in a rush to get into a bloodbath on the Continent like in World War I. The British wanted to ensure that peripheral operations or the war in Russia significantly weakened German power before getting into a death match with the Wehrmacht in France. Although sympathetic to Stalin's need for a second front, Churchill would not sacrifice British interests or increasingly slender British resources in a premature cross-channel attack. American planners suspected the British might use American military power to sustain the interests of the British Empire. Of all the many good reasons to begin offensive operations in North Africa, in the end it may have been politics that decided the issue—American politics.

Churchill became an extremely persuasive and persistent advocate for a North African operation. Added to the British prime minister's arguments, President Roosevelt desired to boost American morale and get U.S. troops into the fight against the Germans before the upcoming congressional elections.[11] Marshall and his planners felt so strongly that any major effort in the Mediterranean was a diversion of strength that they proposed shifting to offensive operations in the Pacific and adopting a defensive posture in the European theater.[12] Roosevelt sent Harry Hopkins, Admiral King, and General Marshall to London in July to hammer out an agreement with the British. His instructions left them with few choices: "It is of the highest importance that U.S. ground troops be brought into action against the enemy in 1942."[13]

The British believed Sledgehammer was a recipe for disaster. Since Bolero could not be ready before 1943, the conclusion was inescapable. At the end of the meeting, Marshall called Eisenhower to his London hotel room. A month earlier, on June 24, 1942, Marshall had appointed Eisenhower as commander of the European Theater of Operations U.S. Army (ETOUSA). Now Marshall informed Eisenhower that the Allies would invade North Africa and that Eisenhower would command the Allied operation. Eisenhower immediately began planning for the first major Anglo-American offensive operation of the war, Operation Torch.

There was little in Eisenhower's career that prepared him for this immense responsibility beyond his military education at the Command and General Staff School and the Army War College. He missed combat action in World War I; indeed, he had never commanded above battalion level. Just a year before his assignment as commander of ETOUSA, Eisenhower was a colonel on an army staff. His first combat experience would be as a theater commander. It is little wonder that such decorated British veterans from World War I as Field Marshal Sir Alanbrooke, Gen. Bernard Montgomery, or Gen. Sir Harold Alexander already with two years experience in fighting the Germans might look down their aristocratic noses at this mechanic's son from Abilene, Kansas.[14] Eisenhower's character would serve him well in the constant give-and-take of coalition warfare. But the years of professional study and education in the decades between the wars provided his only real preparation to command and plan major operations. His first responsibility as the designated commander for Torch was to offer the Combined Chiefs a workable plan for establishing the first Allied theater of operations in North Africa.

Politically, both Churchill and Roosevelt wanted to launch Torch as soon as possible. Militarily, the operation had to be executed before winter weather further complicated amphibious landings. Conscious of both requirements, Eisenhower immediately utilized the U.S. Army officers from his ETOUSA staff. He placed Brig. Gen. Alfred M. Guenther (CGSS 1937, AWC 1939) in charge of the combined planning team. The combined planners moved into Norfolk House on St. James Square in London on August 4. Fully aware of the complex joint and combined issues involved in the planning, Eisenhower decided to pull Maj. Gen. Mark W. Clark (CGSS 1935, AWC 1937) from command of the American II Corps then in England and name him deputy commander in charge of the combined planning staff.[15] This planning staff became the nucleus of the Allied Force Headquarters.

In World War I, only a small group of exclusively French staff officers served Marshal Foch as Supreme Allied Commander. Eisenhower determined that his staff must reflect both the military expertise and the coalition participation necessary to win. From the British, he asked for "two officers each from the Navy, the Army, and the Air Force. In each service, one of these officers should be especially qualified in operational planning and one in intelligence work."[16] The American officers selected for Eisenhower's staff duplicated this arrangement. Eisenhower adopted the American G staff system. In the final staff organization, the British provided the chiefs of the naval staff, air staff, and the G-2. Americans served as deputy commander, chief of staff, and most critically, the G-3 and G-4. This was the very first joint and combined staff in history.

Eisenhower insisted on his authority as commander. When Marshall appointed him to head the European theater of operations (ETO), U.S. Army, Eisenhower made sure that all U.S. forces conformed to the principle of unity of command. He personally went to see Marshall's counterpart, Admiral King, to ensure the latter would support him in unified command over army and navy forces. King assured Eisenhower "that he wanted no foolish talk about my authority depending upon cooperation and paramount interest."[17] Likewise, as commander of the Allied force, Eisenhower insisted on a new level of authority for Allied theater commanders. During the North African landings, the three task force commanders reported directly to Eisenhower. Following the landings, he exercised command through a limited circle of key subordinates consisting of a U.S. army commander, a British army commander, an Allied naval commander, and separate British and American air commanders. (See figure 4.) Later in the campaign, the Allies eventually consolidated the land forces and air forces under single subordinate commanders. This command arrangement employing Allied land, air, and naval commanders under a single theater commander was a thoroughly modern and original command structure.

As Lt. Gen. Kenneth Anderson prepared the British First Army for Operation Torch, the British government drafted a set of instructions to govern his relations with his American allies. In reality, Anderson's political masters simply intended to send him the same guidelines issued to Field Marshal Sir Douglas Haig, who commanded the British Expeditionary Force in World War I. Haig's instructions in 1918 allowed him to refer to the British government any order issued by Foch that he felt imperiled his force. The British allowed Eisenhower to review the document, and he promptly made some changes. Anderson retained the right to communicate directly

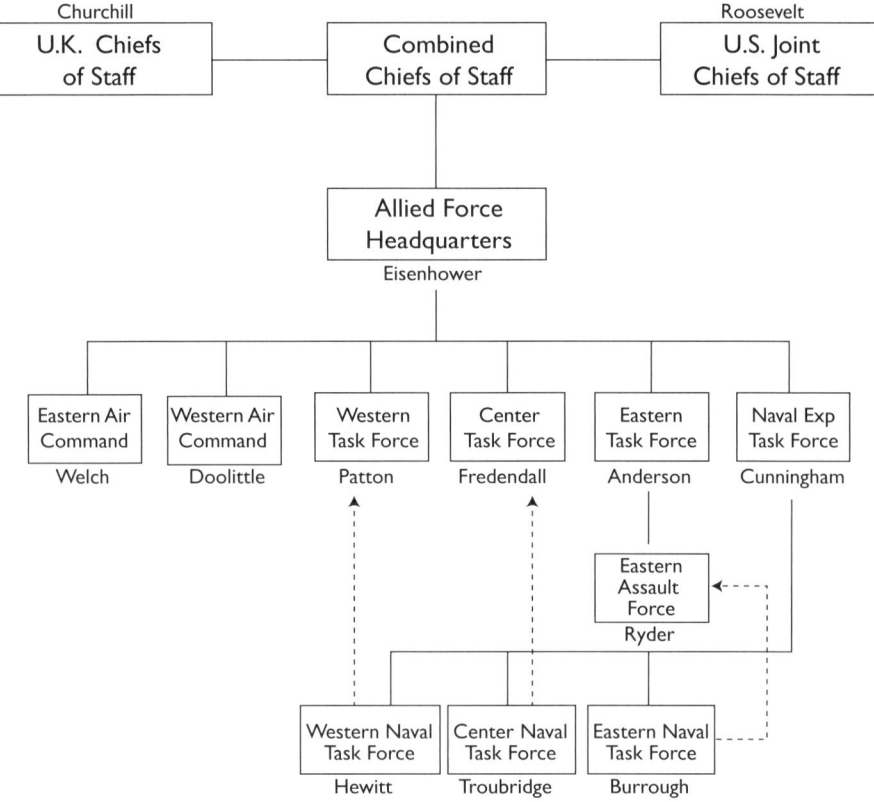

Figure 4. Allied command and control for Operation Torch.

with London, but Eisenhower insisted he be informed first. Most important, the revised document instructed Anderson: "You will carry out any orders issued by [Eisenhower]."[18] This complete subordination of a combined, joint, and integrated command and staff remained the Anglo-American model for the rest of the war in Europe.

Politically and militarily, the assault on French North Africa as the first combined major offensive operation was ambitious and challenging. The Allied force would have to stage from both England and the United States to assault an ostensibly neutral country. Tunisia, Algeria, and Morocco made up the larger portion of the French colonial empire on the northern and northwestern coasts of Africa. The armistice that marked the French defeat in 1940 pledged military forces loyal to the Vichy government to defend the French Empire from any intrusions by Allied forces.

Even if the Allied invaders survived enemy submarines and air-power and reached North Africa, they still faced the challenges of operating in a hostile and largely spartan environment. The poor roads and few sizable ports made French North Africa a relatively underdeveloped theater. Casablanca, Oran, and Algiers were the most important ports capable of supporting major operations. More than 1,200 miles separate Casablanca on the Atlantic coast from Tunis on the Mediterranean. A main coastal road and a parallel interior road provided the primary hard-surface transport system. A "long rickety railway line" ran from Casablanca through Oran, Algiers, and into Tunis.[19] All this meant that the lines of communication and operation available to the Allies lay along the coast.

The French in North Africa possessed a significant capability to resist. Eisenhower estimated the French had "fourteen divisions rather poorly equipped but presumably with a fair degree of training and with the benefit of professional leadership."[20] They had some five hundred aircraft available. The Allies considered the bombers obsolete but believed France's fighters superior to Allied carrier aircraft. By comparison, the Allies could initially muster only three hundred carrier aircraft to cover the landings.[21] French naval power was also considerable. The Allied G-2 intelligence staff estimated the French navy maintained two battleships, four cruisers, fifteen destroyers, and thirty-six submarines in North African waters.[22] Obviously, the success of Operation Torch depended greatly on whether the French would resist. Roosevelt insisted that U.S. troops lead the invasion in hopes of French cooperation rather than resistance.[23] Churchill's order to the British navy to sink the French fleet at Oran, Algeria, after the fall of France had soured Anglo-Vichy relations. The neutrality of Francisco Franco's Spain was also critical in preserving Allied access to the Mediterranean. French military power in the theater required the Allies to project significant combat power. This meant that from an operational point of view, logistics and airpower drove Allied planning.

The Combined Chiefs' directive for Operation Torch specified three objectives: establish lodgments in the Oran-Algiers-Tunis and Casablanca areas; conduct a rapid exploitation to control the entire area to include Tunis; and finally, annihilate Axis forces opposing British forces in Egypt and Libya.[24] This made for a tall order. Logistics, particularly shipping, determined how much Allied power could be gathered and projected ashore. Along with airpower, logistics also helped determine where that power should be put ashore. The crux of the operational problem was where to make the landings. The major ports of Oran and Algiers were obvious objectives, but British planners insisted that every attempt should be made to land as far east as possible to facilitate the rapid occupation of Tunis. Only by occupying that port and Bizerte might the Allies forestall rapid German reinforcements into the theater that would upset Allied plans. The American planners, particularly in the OPD, insisted on a landing at Casablanca on the Atlantic coast, outside the Mediterranean, due to concerns about the Spanish.

The American planners focused on logistical and airpower considerations. The closer to Tunis the Allied forces moved, the more vulnerable they became to German airpower based in Sicily and Sardinia. Due to an insufficient number of carrier aircraft, the Allies had to depend on land-based air support.[25] This meant that airfields in North Africa had to be seized as soon as possible so that aircraft transiting from Gibraltar could rapidly enter the fight. Without land air bases in the theater, amphibious operations would have to depend on whatever carrier aircraft the Allies could muster. From the U.S. perspective, the Allies could not project enough airpower into North Africa to cover landings east of Algiers.[26] Regardless of how the French air force reacted, German airpower in the Mediterranean threatened Allied forces. In August, the Axis sank twelve of thirteen British ships in a convoy bound for Malta. They sank the remaining ship after it reached harbor.

In addition to airpower, American planners remained very conscious of the lines of communication to the Mediterranean. Eisenhower worried that only a landing at Casablanca could secure the Allied lines of communication.[27] Spanish Morocco lay adjacent to the Strait of Gibraltar. If Franco joined the Axis or even permitted German use of Spanish airfields, the Allies might lose Gibraltar, which would effectively cut them off inside the Mediterranean. Casablanca secured the lines of communication through the Atlantic to and from the United States. Rickety as it might be, the rail line running from Casablanca to Tunis provided a logistic lifeline. Because of his concerns about Spain, Eisenhower insisted on a contingency plan that called for additional forces if the Spanish intervened.[28]

The critical operational decisions centered on where to land and timing. The initial plan called for simultaneous landings at Casablanca, Oran, Algiers, and Bône. In his explanatory cable to Marshall, Eisenhower indicated the landings at Casablanca might be deferred five to ten days due to a lack of air cover.[29] The British chiefs balked at the plan. British planners rescrubbed the available lift and air support and determined that simultaneous landings were not possible. They insisted the Allies accept the risk of canceling the Casablanca landings. To address the British concerns, Eisenhower proposed to limit the assault to two American-led landings inside the Mediterranean at Oran and Algiers. The force at Algiers "would push rapidly eastward." A later second convoy would land additional armored forces at Oran and "from there strike toward the rear to open up communications and seize Casablanca."[30] Eisenhower published the new plan on September 21 and provided it to the Combined Chiefs. Eisenhower's attempts to reconcile the British and American concepts of Torch initiated what he called the "transatlantic essay contest."[31]

The resulting exchange between the British and American chiefs included several proposals and counterproposals. OPD planners countered that if the Allies could provide naval support for only two landings, these should be made at Casablanca and Algiers. The British again objected that this approach would lose the opportunity to get to Tunis before the Germans. The senior Allied brass had reached an impasse. Churchill appealed directly to Roosevelt, making his case for landings at Algiers and Oran. Eventually, Roosevelt offered to reduce the size of the U.S. landing force at Casablanca to provide troops and resources for a third landing.[32] The final plan essentially reverted to Eisenhower's proposal of September 9 for simultaneous landings at Casablanca, Oran, and Algiers.

The planning for Torch continued throughout the debate over the number and location of landings. Gen. Mark Clark assembled thirty-seven British and American planners and announced, "Some of you men are less confused than others about Torch. Let's all get equally confused."[33] The planners developed Plan A and Plan B to anticipate the final decision on landings. They also developed deception and contingency plans. By the middle of September, the supply situation fell into such disarray that the Allies postponed the attack to

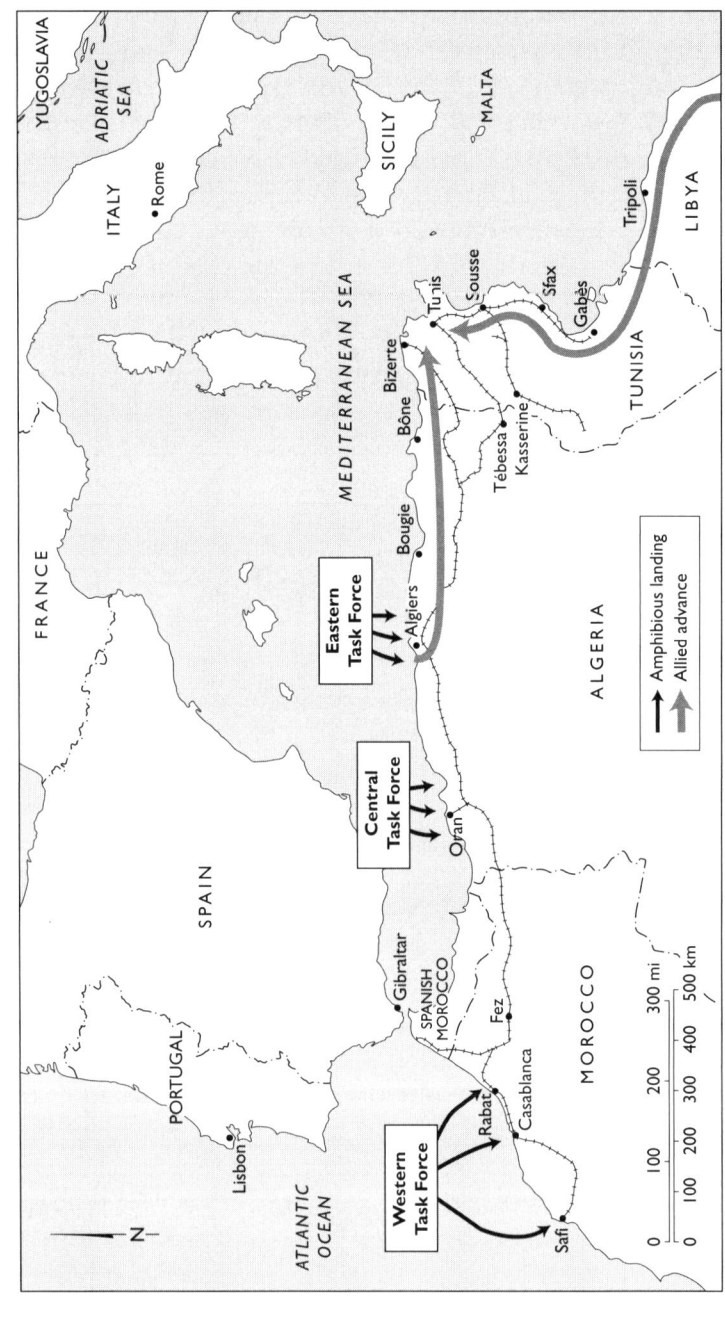

Map 3. Operation Torch, November 8, 1942, to May 13, 1943.

November. The planners, now under the supervision of the G-3, Brig. Gen. Lyman Lemnitzer (CGSS 1936, AWC 1940), kept refining concepts and working the details. Clark handpicked Lemnitzer based on his experience in planning and his status as a graduate of the Army War College. Crediting his experience at both Leavenworth and the War College, Lemnitzer recalled, "I could not have asked for any better preparation for all of this in that period of service which included school, teaching, and practical experience."[34]

The outline plan for Torch coordinated three major operations—the landings at Casablanca, Oran, and Algiers. Maj. Gen. George S. Patton (CGSS 1924, AWC 1932) led the Western Task Force with 35,000 men to assault Casablanca. This task force sailed from Norfolk, Virginia, in thirty-nine ships. The Central Task Force under Maj. Gen. Lloyd R. Fredendall (CGSS 1923, AWC 1925) sailed from England with thirty-nine thousand men in forty-seven ships to seize Oran. Maj. Gen. Charles W. Ryder (CGSS 1926, AWC 1934) commanded the Eastern Task Force with the British ground force contingent. It sailed from England with thirty-three thousand men in thirty-four ships with Algiers as its objective. The planners sequenced Torch in two phases, the assault landings and the buildup. (See map 3.)

The plan did not look much beyond the landings and the initial race to Tunis. The final plan listed the objective, political considerations, and the missions of each task force, as well as aviation and naval support. The base plan included outline plans for each task force, with the naval and aviation support as annexes. In the initial phase, the plan stressed the importance of the early seizure of airfields and consolidation of the ports to support a rapid buildup of combat power.[35] The plan called for a bold use of airborne troops and Army Rangers. The Second Battalion, 503rd Parachute Infantry Regiment, was given the mission of seizing the airfields at Tafaraqui and La Senia near Oran. The British opposed this use of airborne troops flying directly from England, noting "the paratroopers would have to be transported approximately 2,000 miles in darkness and hit a pinpoint target at a scheduled minute, involving a rather remarkable feat of navigation and timing."[36] Eisenhower and Clark favored the operation and retained the airborne assault. The plan tasked the First Ranger Battalion to seize the coastal defense batteries at Arzeu. Similarly, other forces specifically targeted coastal defense batteries, airfields, and ports. The planners hoped the small direct assaults on the harbors at Oran and Algiers would forestall French sabotage of vital port facilities.[37]

These tactical features of the plan represented bold, if not innovative, thinking. Operationally, the plan charged the Eastern Task Force with getting ashore and pushing on to Tunis as quickly as possible. Although U.S. troops with a small British contingent under General Ryder would make the initial landing, Eisenhower tasked Lt. Gen. Kenneth Anderson with quickly organizing the British First Army ashore and striking for Tunis. The plan directed the Western Task Force to build up forces to secure Morocco and prevent any Spanish intervention. Eisenhower planned to create the U.S. Fifth Army specifically for this purpose.[38] Beyond getting ashore, enlisting French cooperation, and building combat power in theater, there was little the Allies could do to prevent or resist a rapid German reaction to the landings. The Allies possessed insufficient airpower to isolate North Africa operationally beyond the initial landing areas. The naval outline plan recognized "it is essential for the success of the Army plan that no substantial enemy reinforcements should reach Tunisia." Eisenhower asked the navy "to use every endeavor to prevent seaborne traffic between Italy or Sicily and Tunisia."[39] Despite the many changes, Eisenhower's headquarters published the final plan by mid-October. The invasion date was set for November 8. Success now depended on Allied execution and German reaction.

The Invasion and the Race to Tunis

On November 5, 1942, Eisenhower and some of his staff flew to Gibraltar to supervise the operation. His concern for Axis airpower was well founded. The following day, two German JU 88 aircraft attacked a B-17 carrying Eisenhower's G-3 and the U.S. air commander, Maj. Gen. James Doolittle, en route to Gibraltar. When the pilot was wounded, Doolittle seized the controls while Brigadier General Lemnitzer manned a 50-caliber machine gun.[40] The assault convoys had better luck and arrived off the coast of North Africa with little trouble.

Despite Allied efforts to win French cooperation, the Americans met significant resistance at Casablanca. The French navy sortied, but the Allied naval covering force quickly crushed its resistance. The Western Task Force plan to seize Casablanca called for landings at Fedala, Mehdia, and Port Lyautey. Gaining the shore, American troops overcame resistance and concentrated for an attack on Casablanca itself. The French surrendered before Patton could order the assault on November 11. The Central Task Force also met determined resistance. Its direct assault on the harbor at Oran with a small force in two Coast Guard cutters failed. Despite the resistance, the landings succeeded and a coordinated attack on Oran overcame French defenses on November 10. At Algiers, the direct naval assault on the port also failed, but Allied forces soon surrounded the city, which capitulated on the same day as Oran. The day after Algiers surrendered, Eisenhower directed General Anderson to push toward Tunisia.

In a letter to Maj. Gen. William D. Connor (AWC 1909), a former commandant of the Army War College, Eisenhower explained his decision to rush Anderson's force eastward before the Allied buildup was complete. Some of his staff argued that the Allied force should amass before racing to Tunis poorly provisioned and with inadequate numbers. Eisenhower commented:

> When that argument was going on, I recalled the particular War College problem that made such an impression on me. We had been working on a problem of resisting invasion in Connecticut, and all the statistical technicians had worked out in detail the most advanced line that they could defend consistent with getting the logistics properly arranged and the necessary forces on the field. Your criticism of the problem was that it was one that obviously called for instant and continuous attack. I remember you said: "Attack with whatever you've got at any point where you get it up, and attack and keep on attacking until this invader realizes that he has got to stop and reorganize, and thus give to us a chance to deliver a finishing blow.[41]

The Allies achieved operational, if not strategic, surprise with their landings, but the Germans reacted quickly. Within a day of the invasion, the Germans established a bridgehead in Tunisia and began flying in ground troops and fighter aircraft. Despite repeated efforts by the Allied navies to cut the enemy's lifeline to Europe, Axis ships also carried men and equipment into Tunis and Bizerte. By the end of November, the Germans shipped 159 tanks and armored cars, 1,097 other vehicles, and 127 artillery pieces to Tunisia.[42] Eisenhower urged British forces under General Anderson eastward, but it was a case of too little too late. Anderson tried to get east using all possible means by land, sea, and air. The Eastern Task Force reserve, the British Seventy-eighth Infantry Division, landed at the small port of Bougie, 100 miles east of Algiers, on November 11. The next day, airborne and seaborne forces secured Bône, another 125 miles east of Bougie. From this point, a small British mobile group, "the Blade force," drove down the coastal road into Tunisia while airborne forces secured the railroad center at Souk el-Arba and a southern airfield at Youks-les-Bains near Tébessa. Behind this thin screen of Allied troops, the bulk of the British Seventy-eighth Division moved east into Tunisia.

On November 17, the Allies clashed with German forces at Mateur, twenty miles south of Bizerte. Anderson continued his advance with two prongs, one aimed at Bizerte and the other at Tunis. Eisenhower rushed U.S. forces eastward to support the drive. On November 27, the British took Tebourba, only twenty miles from Tunis. By the end of the month, Anderson still only had two brigade groups on line supported by a regiment of armor. On December 1, the Germans counterattacked and threw the Allies out of Tebourba, capturing over one thousand prisoners. For the next three weeks, the two foes battled over the approaches to Bizerte and Tunis. Constant rains hampered Allied efforts at supply and support. Allied aircraft operated at the end of their range from airfields more than 125 miles distant. Axis aircraft gained local air superiority by massing airpower launched from nearby all-weather airfields.

The Allies planned one more major attempt to break through on Christmas Eve, but after visiting the front and personally observing battlefield conditions, Eisenhower realized his forces had reached their culmination point, beyond which any further attacks risked defeat by overextending his resources. He reluctantly called off the attack. The race to Tunis was over. Resigned to an operational pause, Eisenhower authorized a withdrawal to a more defensible position and now raced to build up combat power to finish the job. The Germans also rushed reinforcements into the theater. In December, Hitler designated Gen. Hans-Jürgen von Arnim to command the newly established Fifth Panzer Army in northern Tunisia.

The British army's old nemesis, Field Marshal Erwin Rommel, eclipsed Arnim by engineering the most serious Allied setback in

Tunisia. Pressed by Montgomery's Eighth Army, Rommel's forces withdrew from Egypt and crossed Libya to southern Tunisia. To avoid an attack in his rear from Eisenhower's forces in Tunisia, Rommel struck hard at the U.S. II Corps in February 1943. The defeat of American forces in the Battle of Kasserine Pass embarrassed Eisenhower. He weathered the storm by directing changes in command in II Corps, streamlining the command organization, and again rushing Allied reinforcements to the front. Montgomery's continuing pressure from the south together with the Allied forces in the west forced the Axis troops into a shrinking perimeter in the northeastern corner of Tunisia. The Allies launched the final assault the first week of May. After hard fighting, U.S. forces entered Bizerte on May 7, and British forces captured Tunis the same day. Six months after the invasion, the Allies finally secured North Africa, capturing more than 240,000 Axis prisoners, including 125,000 Germans.[43]

Assessment

Strategically, Operation Torch allowed the Allies to seize the initiative and clear North Africa of Axis forces. It eventually achieved all its strategic and operational goals, but not on the original time line and not as soon as hoped. At the operational level, the Allies learned important lessons in joint and combined operations. Since the combined planning staff included both British and U.S. officers, it is difficult to differentiate between operational perspectives beyond a few key decisions. Indeed, there was apparently a good deal of common professional judgment on many topics. Key operational decisions revealed differences in the Allies' approach to modern warfare. The American commanders and planners emphasized the importance of logistics, airpower, and unity of command consistent with the interwar instruction they received at the Command and General Staff School and the Army War College.

The American planners' insistence on a cross-channel attack in 1943 rather than peripheral operations in the Mediterranean reflected their strategic and operational culture. Massing forces in Britain allowed for an adequate logistics base, a suitable platform for airpower, and a chance to strike directly at the center of gravity, Germany. Once President Roosevelt decided to postpone the invasion of France in order to get American troops quickly into combat with the Axis in 1942, the American planners in Washington and England began to tackle the enormous operational challenges involved in invading North Africa. The combined planners' operational design synchronized three major operations in two phases. The OPD's insistence on the Casablanca landing at the risk of early seizure of Tunis fully reflected the American belief in the importance of logistics in modern war. Eisenhower also made some key operational decisions that affected the campaign and the later course of the war.

The Allied failure quickly to seize Tunis proved frustrating. Eisenhower cited four reasons for this failure: the weakness of General Anderson's force due to a lack of shipping, a shortage of motor equipment, poor weather, and the proximity of Tunis and Bizerte to Axis bases in Sicily and Italy.[44] The lack of shipping was a limitation of means. The weather and the close proximity of Axis bases were simply operational realities. The lack of motor transport and the proximity of enemy bases required operational decisions related to risk.

In organizing the invasion force, Eisenhower noted, "The Allied force was initially loaded and dispatched with a principal purpose of getting ashore and seizing three main ports. To accomplish this mission, it came woefully short in motor transport and other auxiliaries normally making up the tail of an army."[45] Concerned about French resistance, he insisted on more combat power and less tail.[46] He willingly accepted the risk of not being able to move quickly to Tunis in order to ensure capture of the key ports. Only the ports could provide the logistic infrastructure needed to sustain large operational forces. Eisenhower undoubtedly believed he was justified in taking this risk because the British intelligence estimate provided to him indicated that it would take two weeks for the Germans to get significant forces into Tunisia.[47] Eisenhower stripped American units in Casablanca and Oran of their trucks, in order to provide additional transportation to units moving east. By January, Eisenhower was pleading for more trucks. He personally asked Admiral King for additional escorts for a special convoy.[48] The War Department rushed 5,400 trucks to North Africa, which eased conditions considerably, but they did not arrive until late February.[49]

Eisenhower believed that one of the most important contributions Torch made to the Allied war effort was in the development of an effective combined command and staff organization.[50] Eisenhower insisted on the principle of unity of command and created the first combined and integrated staff in history. Allied command arrangements continued to evolve throughout the North African campaign. The campaign began with Eisenhower supervising six subordinate commanders. In January 1943, he consolidated Allied air forces under Maj. Gen. Carl A. Spaatz (CGSS 1936). A problem remained with the ground forces. The defeated French refused to serve under British command, which required U.S. forces to split between Morocco and Tunisia. At the Casablanca conference in January, the British chiefs recommended changes in command arrangements to help coordinate Montgomery's Eighth Army and Eisenhower's forces. In the new organization, a single ground commander coordinated all Allied land forces in North Africa, including the British Eighth Army. (See figure 5.)

This arrangement provided the theater commander with single component commanders for land, sea, and air. The British succeeded in appointing their own officers to each of these subordinate commands.[51] British motivation for the change in command arrangements went beyond a rationalization of Allied organization. Field Marshal Lord Alanbrooke, chief of the British Imperial Staff, noted in his diary, "We were pushing Eisenhower up into the stratosphere and rarefied atmosphere of a Supreme Commander, where he would be free to devote his time to the political and inter-Allied problems, whilst we inserted under him one of our own commanders to deal with the military situations and to restore the necessary drive and coordination which had been so seriously lacking."[52]

Eisenhower, while welcoming the new organization, again insisted on an American understanding of unified command. He rejected what he called the British committee system, in which the component air, land, and sea commanders simply cooperated with each other in the planning and conduct of operations. He insisted on his right to organize, coordinate, and control the forces under his command.[53] Eisenhower's concept of operational control became a model for Allied command in Europe for the rest of the war and later for the North Atlantic Treaty Organization (NATO). This integrated combined command and staff organization provided the Allies with a significant operational advantage. By contrast, the Axis powers failed to establish combined, integrated, or even joint command arrangements. (See figure 6.)

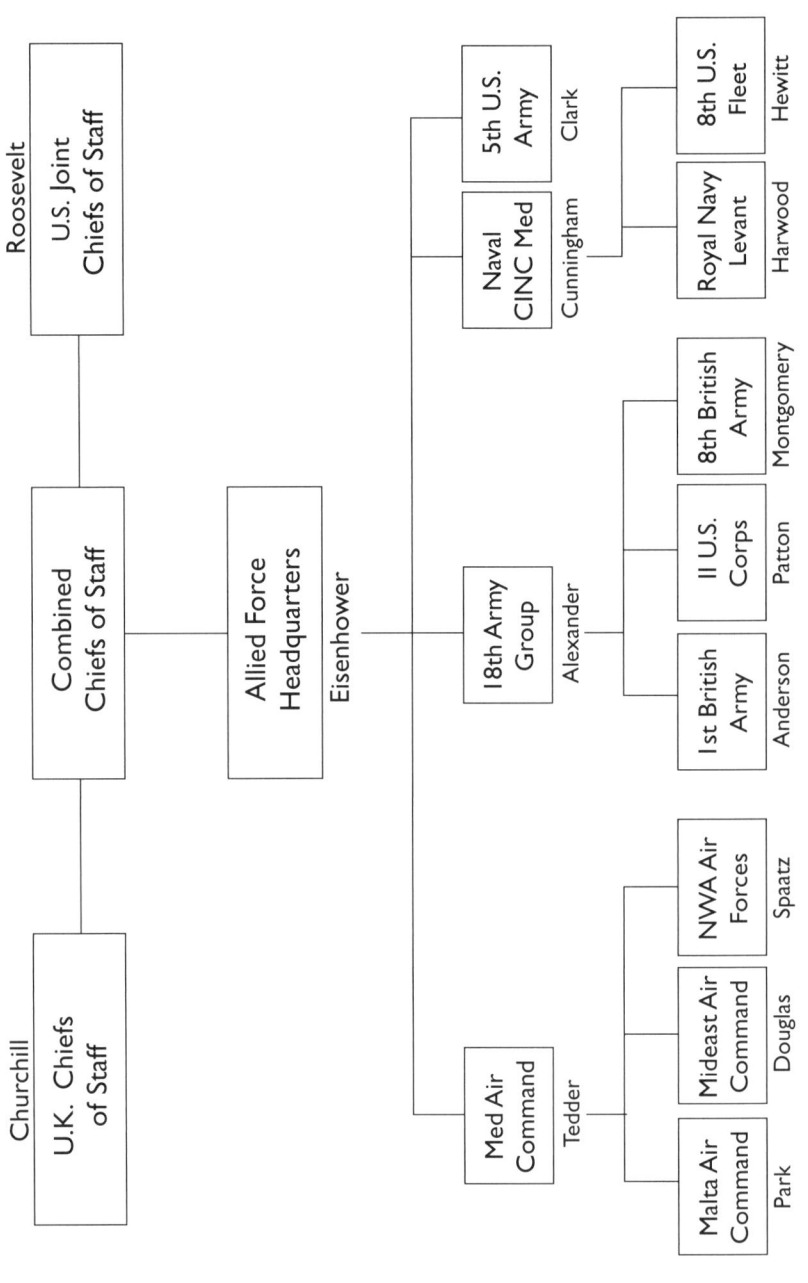

Figure 5. Allied command and control in North Africa, March 1943.

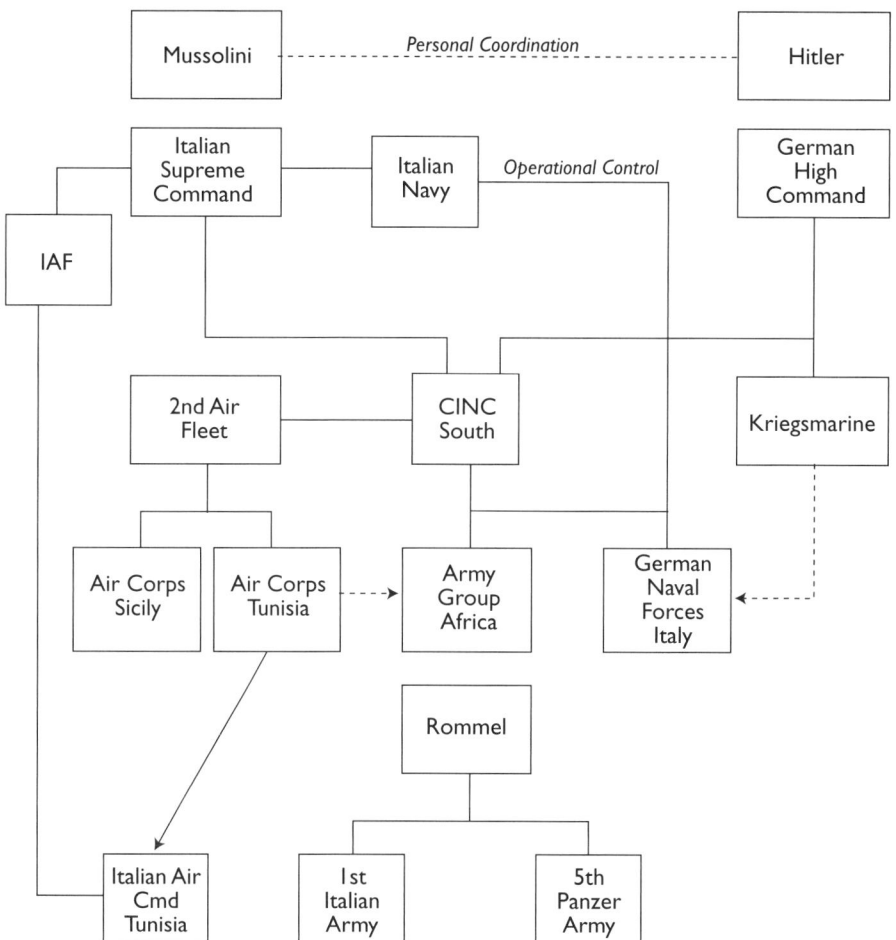

Figure 6. Axis command and control in North Africa April 1943.

There was no unified Axis Mediterranean theater. The Italian commando supremo provided operational direction to the Italian North African Command in collaboration with the Germans. Although nominally subordinate to the German-Italian panzer army, Rommel frequently received direct guidance from Hitler. Field Marshal Albert Kesselring headed the Luftwaffe in the Mediterranean and as senior officer directed von Arnim's panzer army, but initially not Rommel's force. Not until April 15, 1943, did the Axis create its Army Group Africa, with Rommel as commander of both the First

Italian Army and the Fifth Panzer Army. At no time did the German or Italian navies come under the command of a single theater commander.

Eisenhower's employment of airpower reflected much of the prewar military instruction in American military schools. The strategic and operational circumstances combined with the ambitious objectives for Torch to preclude the achievement of air superiority or even parity prior to the campaign. Once ashore and in theater, Allied airpower consistently targeted enemy airfields in an effort to develop air superiority. Eisenhower's operational employment of airpower to hammer away at the enemy lines of communication and to isolate the enemy in theater was very much in keeping with instruction at the Command and General Staff School and the War College. To protect the Allied invasion force, Eisenhower made the German submarine bases and operations in the Bay of Biscay the priority target for Carl Spaatz's Eighth U.S. Air Force.[54] This control of long-range, land-based aircraft for operational rather than strategic missions set a precedent for future operations.

The Allies organized airpower for Torch into two commands. Brigadier General Doolittle commanded the newly organized Twelfth U.S. Air Force charged with supporting American forces in the west. The British Eastern Air Command supported the push eastward for Tunisia. Inevitably, Doolittle's outfit became involved in the battle for Tunisia. The Twelfth Air Force consisted of bomber, fighter, and air support commands. XII Bomber command handled the operational missions of attacking the enemy lines of communication and isolating Axis forces in North Africa. B-17s pounded the Axis ports in Tunisia and Sicily. B-24s attacked Axis shipping. Eisenhower was convinced that success of the Tunisia campaign hinged on interrupting enemy communications.[55] To this end, the theater commander harnessed long-range, land-based aircraft to operational missions.

Tactically, American air-ground coordination was poor for much of the campaign. Eisenhower adhered to the prewar arrangement of letting ground commanders set priorities for the XII Air Support Command. Not until February did Spaatz convince Eisenhower that air commanders tasked with ground support should have more control over their units.[56] U.S. air power benefited greatly from the experience of the British in tactical air support gained by the Royal Air Force (RAF) and Montgomery's Eighth Army. By the end of the campaign, tactical air commands such as Spaatz's Northwest African Air Force had been established and placed on a status equal to that of army commanders. Allied airpower became much more effective at all levels once the leadership sorted out the command organization.

Command arrangements for Allied airpower evolved during the campaign, demonstrating the need for centralized control and unified command under the theater commander. Eisenhower eventually appointed General Spaatz to command all Allied airpower. The Allied command reorganization that provided Eisenhower with a single subordinate ground commander also led to a consolidation of Allied airpower throughout the Mediterranean. General Spaatz was left in command of what became the North African Air Command under the overall Allied Mediterranean air commander, Air Chief Marshal Arthur Tedder. This completed the Allied command model that endured for the rest of the war. Centralized control and direction of airpower was critical to its development as a true instrument of operational art.[57]

From theater commander to squad leader, the Americans learned a great deal from the North African campaign. After the failure to win the race to Tunis, a frustrated Eisenhower reflected, "I think the best way to describe our operations to date is that they have violated every recognized principle of war, are in conflict with all operational and logistic methods laid down in textbooks, and will be condemned, in their entirety, by all Leavenworth and War College classes for the next twenty-five years."[58] The textbooks and exercises of the interwar years all stressed mass, concentration, the need for air superiority, and the necessity of adequate logistic support.

In the race to Tunis, Eisenhower hastily forwarded U.S. units piecemeal as they became available. Without regard to the Allies' ability to sustain forward forces and in the face of Axis air superiority, the theater commander rushed to accomplish the ambitious objective most cherished by the British—seizure of Tunis before the Germans. Seizing Tunis would shut the back door on Rommel's Afrika Korps and finally trap the Desert Fox between two Allied forces. Well before the invasion, Eisenhower, Patton, and Clark believed the chances of beating the Germans to Tunis at less than 50 percent. The only real chance the Allies had of denying Tunis to the Axis lay not with Allied forces but with the French. By denying or resisting

General Eisenhower confers with his senior American ground commanders (in helmets, left to right), Twelfth Army Group commander Lt. Gen. Omar Bradley, Third Army commander Lt. Gen. George Patton, and First Army commander Lt. Gen. Courtney Hodges. All were graduates of the Army War College. Courtesy of USAMHI.

During the interwar years the U.S. military recognized the critical need to understand and plan for the immense logistic challenges involved in projecting, employing, and sustaining large armies across great distances. This key feature of American operational art is vividly demonstrated in this photograph of Omaha Beach shortly after the invasion of France in June 1944. Courtesy of USAMHI.

Strategic bombers were frequently employed to support operational objectives. These B-17s are flying over the Mediterranean en route to bomb southern France. Courtesy of USAMHI.

Leveraging airpower was critical in the operational maneuver of land forces. In this photograph air and ground commanders meet in Nancy, France, to discuss operations for the advance into Germany in 1944. From left to right: Lt. Gen. Carl Spaatz, Lt. Gen. George Patton, Lt. Gen. James Doolittle, Maj. Gen. Hoyt Vandenberg, and Brig Gen. Otto Weyland. Courtesy of USAMHI.

German access to French airfields and ports, the French could have bought time for the Allies to move in. Unfortunately, the French in Tunisia cooperated with the Axis.

Ironically, the extended campaign for North Africa probably worked to the Allies' advantage. On the strategic level, the eventual victory demonstrated the Allied ability to persevere, helping to sustain national will in a long and demanding war. One of the objectives of Operation Torch included engaging and drawing additional German forces into North Africa to relieve the Soviets. The Germans obligingly decided to contest the Allied offensive pouring even more troops into the bag. Operationally, the experiences of conducting major amphibious operations and of sorting out the Allied command and staff arrangements proved invaluable. Finally, the benefit of combat experience for American commanders and troops provided many tactical lessons.

American involvement in planning Torch reflected the best instructional practices of the interwar period at Leavenworth and the War College. The plan was adequately phased and included both contingency and deception plans. It provided for the political context and clearly linked strategic objectives to military operations in the theater of operations. The American emphasis on airpower and logistics, which contributed to an unwillingness to accept more risk in the operational design and execution, was characteristic of the study of war in the interwar years. And the insistence on unity of command and a rationalized joint and combined staff organization represented real progress in operational art. All the lessons from Operation Torch would soon be put to good use in the most important Anglo-American major operation of World War II—Operation Overlord.

OPERATION OVERLORD

American planners always believed a cross-channel attack into France would be the decisive operation of World War II. Within four months of the U.S. entry into the war, Eisenhower sent a memorandum to Marshall arguing "that the principal target for our first major offensive should be Germany, to be attacked through Western Europe."[59] Marshall agreed and very quickly Eisenhower's Operations Division drafted Bolero, an outline plan for an invasion of France in 1943. This early American vision of a cross-channel attack served to drive indus-

trial production decisions to ensure that the equipment and personnel would be ready when needed, and it also reflected, conceptually,
American military thinking from the interwar period. Moreover,
much of substance of these early ruminations on a cross-channel
attack survived into the final plan.

Early Planning for the European Theater of Operations

General Marshall briefed President Roosevelt on Bolero on April 2,
1942. The president approved the plan and directed Marshall to go to
London to secure agreement with the British on the main outline of
the proposed Allied effort. The key operational features of this early
plan centered on where to land, in what strength, to what purpose,
and how to employ airpower. The final version of Bolero called for a
three-phase operation. The first phase included not only the necessary buildup of forces but also continuous raiding both to gain combat
experience and to deceive the Germans. The second phase, the cross-
channel attack, called for six divisions to strike between Le Havre
and Boulogne. The final phase described the consolidation and expansion of the beachhead and the beginning of a general advance. (See
map 4.)

The Americans estimated the total force required as forty-eight
divisions, 5,800 combat aircraft, and 7,000 landing craft. Airborne
troops would help secure the beachhead, and airpower would "prevent rapid movement of German reinforcements toward the coast."[60]
Once the Allies established the beachhead, "strong armored forces
would be rushed in to break German resistance and seize the line of
the Oise–St. Quentin. A movement towards Antwerp will then follow to widen the salient and permit movement of additional forces
across the channel between Boulogne and Antwerp."[61]

According to U.S. military thinking during the interwar period,
direct assault on the main force of one's strongest enemy such as
contained in Bolero was the surest and shortest route to victory. The
plan reflected an American optimism and faith in the tremendous
industrial potential of the United States to match the scale of logistic effort necessary. The American planners proposed employing airpower in keeping with interwar concepts, certainly at Leavenworth
and the Army War College. Airpower would operationally isolate
the battlefield, interdict or prevent enemy reinforcements and buy
time for the Allied buildup. The leveraging of airpower to shape the

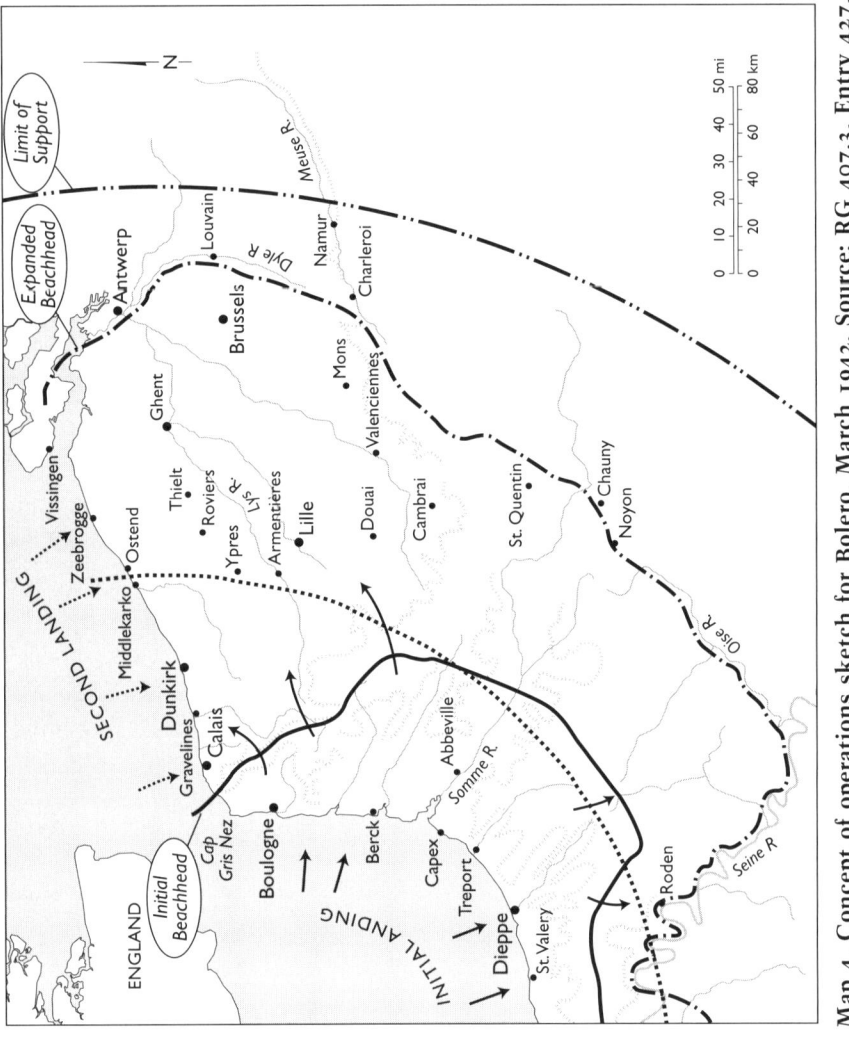

Map 4. Concept of operations sketch for Bolero, March 1942. Source: RG 407.3, Entry 427, Box 24325, File 308, NARA.

battlefield would enable the gathering of forces and logistics necessary to defeat the German army.

Hap Arnold shared his thinking on Bolero with General Marshall in a memorandum dated March 27, 1942. Arnold proposed a four-phase operation: preparation, air offensive, surface invasion, and exploitation. Arnold foresaw that any invasion of France required air superiority over the landing area. He believed that airborne operations would probably play a decisive role. He noted, "Bombardment of all classes will be employed in close support of the invading force. It will be sound military procedure to return the major elements of the heavy bombers to their normal strategic role against vital industrial and or civil objectives as soon as the ground situation will permit."[62] Two years later, in the final planning for Overlord, the employment of strategic airpower became a major point of contention between the Allies. Arnold rather presciently recognized the need to focus all elements of joint power to achieve the immediate operational goal before returning to the cherished concepts of strategic bombing.

A month after Arnold's memorandum, Maj. Gen. Carl Spaatz met with Marshall to present his concept of the operation. Spaatz believed an attritional air battle would be necessary first to defeat the Luftwaffe in order to gain air superiority for the invasion. To force the Luftwaffe to fight, Spaatz proposed striking targets of such economic importance that the Germans would have to defend them or lose the war.[63] This concept of the operation supported both the air force's interwar belief in strategic bombing and the War College's insistence on supporting the ground campaign. This plan conveniently matched strategic air theory with the operational requirement for air superiority over the invasion area.

In April 1942, General Marshall secured British approval for the plan. After a year of sobering defeats, the British welcomed the promise of deploying over 1 million U.S. soldiers to Great Britain for operations in 1943. More bad news, however, soon followed. On June 21 in North Africa, Tobruk fell to Rommel's Afrika Korps. Churchill now became an advocate of using Allied forces to redress the deteriorating situation. The British prime minister's eloquence combined with Roosevelt's desire to get U.S. troops into action on the ground against the Germans in 1942 led directly to Operation Torch. The North African invasion refocused Allied resources and priorities on the North African (later Mediterranean) theater of operations.

The plan for an invasion of France recovered some momentum after the Casablanca conference in January 1943. The British continued to push for a peripheral strategy in which the Allies would pursue operations in the Mediterranean. Marshall and the U.S. planners continued to argue that priority should be given to the invasion of northwest Europe and the establishment of the European theater of operations. In the end, the Allies agreed to continue operations in the Mediterranean, push the combined bomber offensive, and establish an Anglo-American planning staff for the cross-channel attack. The British selected Lt. Gen. Frederick E. Morgan to become Chief of Staff to Supreme Allied Commander (COSSAC) for planning the invasion of Europe.

The COSSAC Plan   While the Combined Chiefs deferred the decision on who would be the Supreme Commander for the ETO, General Morgan went ahead with the planning effort anticipating a British commander. He therefore asked for a British staff organization that amalgamated American staff officers. Morgan wanted a small staff modeled after the one that served Marshal Foch during World War I.[64] This initial staff was not integrated but parallel. Each branch had a British and American principal staff officer with separate staff sections. Complete integration occurred in the fall when Maj. Gen. Ray Barker (CGSS 1928, AWC 1940), the deputy chief of staff, reorganized the complete staff along functional rather than national lines.[65] By July, this staff produced an outline plan for Operation Overlord.

The COSSAC plan called for four phases: preliminary, preparatory, assault, and buildup. The preparatory phase included air operations to "reduce the effectiveness of the German Air Force in that area and will be extended to include attacks against communications more directly associated with the movement of German reserves which might affect the Caen area." [66] The area around Caen, a major road hub, was fairly open. One of the largest towns in Normandy, it sat across the Orne River and the Caen Canal, significant water obstacles. Three divisions would carry out an assault in the Caen area assisted by an airborne seizure of the town itself. This would secure the lodgment and establish a base for future operations. The planners envisioned a series of battles to gain the initial foothold, concentrate a sufficient force, and then proceed "by bounds cracking the enemy lines with separate, massed, and carefully prepared attacks for each new objective."[67] The task of the COSSAC plan was not to annihilate the German army but to secure a lodgment.

At the Quebec conference in August 1943, the Americans finally got the British to agree to make Operation Overlord the priority effort for 1944. The Combined Chiefs of Staff approved the COSSAC plan. They also approved a request by the Overlord planners for a diversion against southern France and directed General Eisenhower, then the Allied commander in the North Africa theater of operations, to draw up the plans. The COSSAC plan resolved the issue about where to land and to what purpose, but the matter of strength levels was contested. Churchill suggested at Quebec that the assault should be strengthened by at least 25 percent.[68] The availability of landing craft remained the key factor. As always, logistics would determine the art of the possible.

Landing craft had been the key planning assumption in putting together a feasible plan to invade France. Initially, the British planners asked for 8,500 landing craft to lift ten divisions. Reviewing potential requirements for the Mediterranean theater, American planners believed that only 4,500 landing craft would be available. At a conference in Washington, D.C., in May 1943, the Allied planners agreed to the lower figure. This necessarily constrained the COSSAC planners to a five-division assault, with three divisions in the first wave followed by two more. While still commanding in North Africa, well before he knew he would command the European invasion, Eisenhower saw the COSSAC outline plan. He shared with Montgomery his view that the amphibious assault with three divisions was inadequate and that the plan failed to provide for the quick capture of Cherbourg.[69] Montgomery came to the same conclusion and upon his return to England made his criticism known.

The Revised Plan   In December 1943, while visiting the Mediterranean theater, President Roosevelt informed General Eisenhower that he would command Allied forces for Overlord.[70] Eisenhower arrived in London the next month and began reviewing the plan with his new staff and making changes. The COSSAC staff formed the nucleus of a combined joint staff along the lines of Eisenhower's earlier experience in the Mediterranean. As in North Africa, the British retained command of the component Allied forces: Montgomery would command the Twenty-first Army Group, the land force; Adm. Sir Bertram

Ramsey, the Naval Expeditionary Force, and Air Chief Marshal Sir Trafford Leigh-Mallory, the Air Expeditionary Force. (See figure 7.) Gen. Omar Bradley (CGSS 1929, AWC 1934) led the U.S. First Army under Montgomery's supervision during the initial invasion. When the subsequent flow of more numerous U.S. forces to the Continent required the creation of another army, Eisenhower planned to directly supervise both the British Twenty-first Army Group (which would become an all-British affair) and a newly organized American army group without a separate land component commander. This arrangement reflected not only the Supreme Commander's belief that another level of command was unnecessary, but also his growing confidence as a theater commander. With command arrangements settled, the control issues surfaced again as Eisenhower made some key operational decisions.

It is sometimes difficult to distinguish between the operational approaches of allies who plan and conduct joint and combined operations. American and British planners comprised both the COSSAC and Overlord combined staffs. Several key operational decisions in the planning for Overlord illustrate both similarities and differences between the American and British approach to operational art. The first major operational decision focused on the need to expand the initial amphibious assault to five divisions and enlarge the landings in the Cotentin Peninsula closer to Cherbourg. The COSSAC planners had long advocated a larger assault force, but they were constrained by the limitation on landing craft. Both Montgomery and Eisenhower insisted on the changes to ensure the success of the lodgment and to secure a port.[71] Subsequent decisions by Eisenhower on important operational matters generated more friction with his British colleagues.

The Combined Chiefs of Staff allocated General Morgan's COSSAC planners two airborne divisions. Morgan planned to use the airborne forces to seize Caen and critical river crossings. Eisenhower decided against an airborne seizure of Caen. He wanted to increase the mass of the airborne drops and use them to help seal and then expand the lodgment. General Marshall endorsed an even bolder plan by Hap Arnold to drop airborne forces to threaten crossings over the Seine River and Paris itself.[72] Eisenhower thought this use of airborne forces too bold. He agreed that distant vertical envelopment was an operational advantage but argued that the airborne forces would be

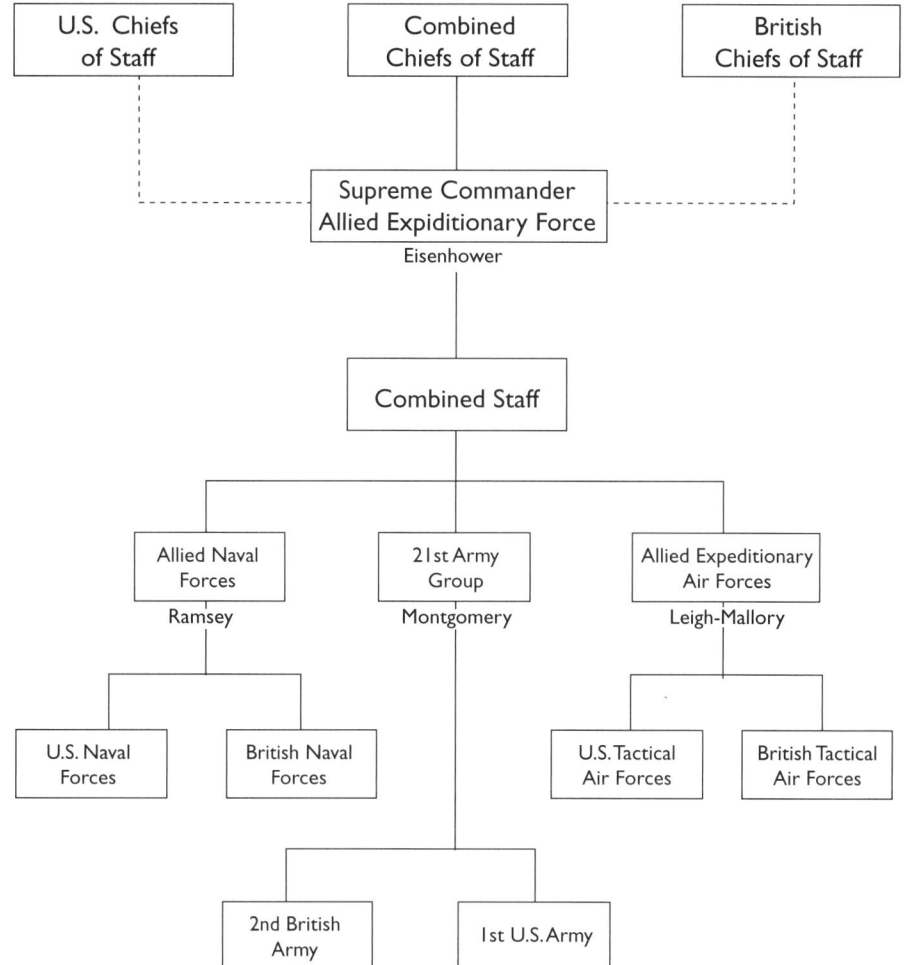

Figure 7. Allied chain of command for Overlord.

immobile once on the ground and that the initial assault force would lack the operational mobility to reach them. Eisenhower believed the Germans would ignore or contain the airborne units in order to strike at the more critical amphibious forces.[73] Recognizing the potential operational value of airborne forces, Eisenhower later authorized the creation of the First Allied Airborne Army commanded by U.S. Lt. Gen. Lewis H. Brereton (CGSS 1926). Eisenhower used this joint and combined headquarters as a command-and-control and planning staff for the employment of his theater reserve. Constituting a permanent joint and combined headquarters to employ the theater reserve represents a unique operational innovation in World War II.[74]

With regard to Overlord, Air Chief Marshall Leigh-Mallory, Eisenhower's air commander, objected to even the tactical drops in the American sector. He predicted losses as high as 75 to 80 percent. Eisenhower agreed with the risks but insisted that "a strong airborne attack in the region" was "essential to the whole operation and it must go on."[75] Eisenhower was convinced the U.S. airborne drops would hasten the capture of the Cotentin Peninsula and secure the port of Cherbourg, the prize so necessary to sustain the lodgment. He understood that sometimes the tactician must pay a premium price in order to assure operational results.

Another key operational decision revolved around the use of airpower. In the early Bolero planning, Hap Arnold underscored the need to employ airpower to isolate the battlefield. Indeed, this had been a key feature of interwar instruction at both Leavenworth and the War College. Likewise, Eisenhower insisted that all airpower, including the strategic air assets, should be utilized to ensure the success of Overlord. From his perspective, the Allies agreed that Overlord would be the main effort in 1944; it would be the decisive operation of the war. The Allied Air Expeditionary Force, Eisenhower's air component, developed a plan for a three-month bombing attack on railroad and transportation centers in France and Belgium. The plan called for the Fifteenth U.S. Air Force to attack targets in southern France from its bases in the Mediterranean. Additional targets would be attacked by the British Bomber Command.

General Spaatz, now commanding the U.S. Strategic Air Forces (which included the Fifteenth Air Force), objected to any diversion of strategic airpower from the Combined Bomber Offensive. True to his roots in strategic bombing theory at the Air Corps Tactical School during the interwar years, Spaatz offered to attack German oil resources instead of diverting his forces to tactical targets. Eisenhower realized that the potential operational benefits of the rail-target plan far outweighed any tactical or strategic outcomes in the near term. Operation Overlord was a race to see whether the Allies could irrevocably secure a foothold on the Continent or the Germans could build up sufficient force to defeat the invasion. The Supreme Commander's early thinking on the matter was revealed to Spaatz just as they both assumed command early in January. "In establishing your headquarters," Eisenhower wrote, "please bear in mind that your command and [Air Marshal Sir Arthur] Harris' organization [British Bomber Command] are to be the two big guns in the preparatory phase."[76] The British Bomber Command opposed the rail plan as adamantly as Spaatz did, but their objections centered more on submitting to Eisenhower's control. Not too surprisingly, Air Marshal Charles Portal, chief of the RAF, opposed surrendering virtually all British airpower to Eisenhower. Churchill backed his air chief.

Eisenhower did not oppose the continuation of strategic bombing. He understood that the major operational advantage that would accrue from strategic bombing was the destruction of the Luftwaffe, which would also be crucial to the success of Overlord. Foremost in Eisenhower's mind, however, was the need to concentrate all Allied power on the immediate task at hand, getting Allied landpower back onto the European continent. The Supreme Allied Commander remained determined: "If a satisfactory answer is not reached I am going to take drastic action and inform the Combined Chiefs of Staff that unless the matter is settled at once I will request relief from this command."[77] Now Churchill backed down and the subsequent compromise left Allied heavy bombers under the direction, but not the command, of the Supreme Commander for the duration of Overlord. Still, the British war cabinet debated for two weeks over authorizing an operation that would result inevitably in significant French and Belgian civilian casualties. With the insistence of the British chiefs of staff and the prime minister's support, the war cabinet gave its approval. On April 14, Eisenhower took over direction of the heavy bombers and within days issued his directive for attacking the French rail system.

Eisenhower's determination to mass Allied airpower was no less evident in his desire to concentrate Allied landpower to ensure the success of Overlord. As early as August 1943, Overlord planners requested an additional assault into southern France as an important diversion to assist the main attack. Eisenhower became an insistent advocate for this operation, code-named Anvil. The Supreme Commander viewed Anvil not only as a diversion for German forces in France but also as a key avenue of approach that would get as many as ten more Allied divisions into the fight for France, open up additional lines of communication, and secure the port of Marseille. He insisted to the Combined Chiefs as early as January that "Overlord and Anvil must be viewed as one whole."[78] He even suggested delaying the invasion for thirty days in order to obtain the necessary lift for a two-division assault. The British saw it differently. Montgomery favored canceling Anvil in order to strengthen the Overlord assault. The British chiefs believed that the five hundred miles of rugged terrain that would separate the two operations made impossible any real support for forces engaged in Overlord.[79] Likewise, Churchill, with an eye on the Mediterranean theater of operations, did not want to launch Anvil if it meant sacrificing prospects for the Italian campaign, which had begun in 1943.[80]

In an effort to overcome the stalemate in Italy, the Allies had landed two divisions at Anzio in order to bypass stiff German resistance and threaten Rome. By February, this effort stalled and the requirement for amphibious lift to continue support for Allied forces in this operation temporarily killed the prospects for a landing in southern France. By March, Eisenhower conceded Anvil was no longer possible given the current state of Allied resources. He remained interested in Anvil, not only from a desire to concentrate Allied forces but also because he suspected the Normandy ports might be slow to open even if Overlord met most other expectations.[81] As Eisenhower crafted the operational design for the revised plan for the Normandy invasion, the importance of logistics, concentration of forces, and airpower dominated his thinking.

The Campaign

The Combined Chiefs directed Eisenhower to "enter the continent of Europe and, in conjunction with other United Nations, undertake operations aimed at the heart of Germany and the destruction of her armed forces."[82] The Allies designed Operation Overlord to secure the lodgment from which further operations would carry their forces into the German heartland. The cross-channel attack was the first and the most critical of the series of major operations that would achieve the strategic objective of defeating the German armed forces in central Europe. Planning focused on this immensely complex and detailed undertaking of just getting and staying ashore.

The Overlord planners envisioned a secure lodgment area eventually expanding to the Seine River within ninety days of the initial assault. Eisenhower believed that the Ruhr, Germany's great industrial center, constituted the heart of Germany. To get there, according to Eisenhower's concept, Overlord operations following Overlord would advance on a broad front with two army groups; the aim of the main effort in the north would be to secure the Belgian ports in order to sustain the drive. He hoped to complete the destruction of enemy forces west of the Rhine while looking for any opportunities to seize bridges over that river. To take the Ruhr, Eisenhower envisioned a double envelopment in the north and in the south by way of Frankfurt.[83] This plan of campaign was thought out well before the first infantrymen waded ashore in northwest France.

By the end of May 1944, the Allies massed thirty-seven divisions, 3,134 aircraft, and 3,601 serviceable landing ships and craft in England. The naval force included six battleships, twenty-two cruisers, and ninety-three destroyers.[84] To oppose this armada, the German forces in France had sixty divisions of all types and four hundred fighter aircraft, only half of which were available for support in Normandy.[85] The German navy was in no condition to offer serious resistance; naval forces amounted to two destroyers, thirty-three operational torpedo boats, and a few smaller patrol craft and minesweepers.[86] German submarines could not operate in the close confines of the channel. Germany's hopes for success depended on its army, but it was weak in precisely the areas the Allies excelled: command and control, logistics, intelligence, and airpower.

In 1942, Hitler appointed Field Marshal Gerd von Rundstedt as the commander in chief in the West. Despite the grand title, von Rundstedt had direct control over army forces, and only limited control over German SS and Luftwaffe parachute units. The German forces had no joint command or unity of command. As von Rundstedt's chief of staff noted after the war, "The chain of command was very complicated and muddled; there was no absolute responsibility as was given to Field Marshal Montgomery or General Eisenhower."[87] (See figure 8.) German naval and air forces did not report to Rundstedt and he could only request their cooperation in the country's defense. In any case, German airpower and seapower had little to contribute; the successful defense of the Atlantic Wall depended on the German army. The German Army in the West was organized into two army groups. Rommel's Army Group B consisting of the Seventh and Fifteenth armies defended the channel coast.

Since Rommel's appointment as the commander of Army Group B in December 1943, Rommel did his utmost to strengthen German

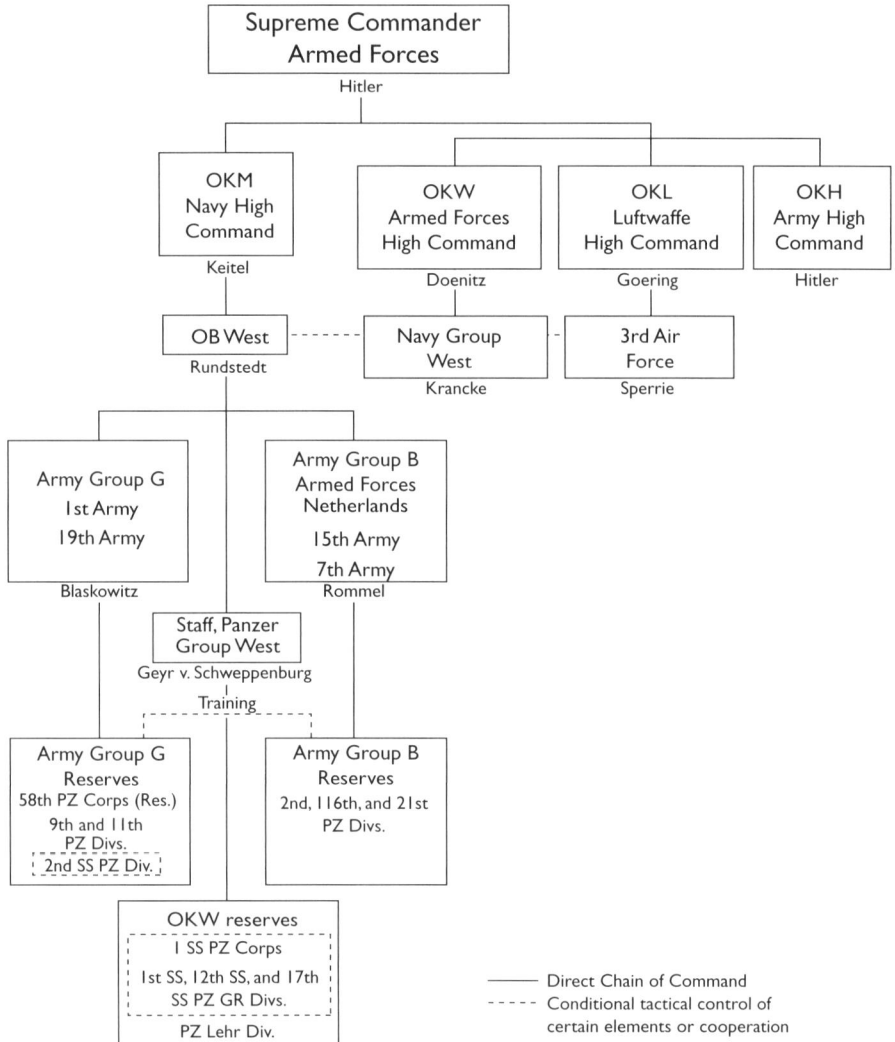

Figure 8. German chain of command in the West, May 1944. Source: Gordon A. Harrison, *Cross-Channel Attack* (Washington, D.C.: U.S. Army Center of Military History, 1951), 244.

defenses, but the German High Command could not agree on the best way to defeat the coming invasion. The debate centered on whether the Germans should conduct a linear or mobile defense. Rommel's experience convinced him that German tank reserves must be positioned close enough to the invasion beaches to counterattack Allied

forces immediately, while they were most vulnerable. It was his belief that "elements which are not in contact with the enemy the moment of invasion will never get into action, because of the enormous
air superiority of the enemy." In Rommel's view, "If we do not succeed in carrying out our combat mission of warding off the Allies or
hurling them from the mainland in the first 48 hours, the invasion
has succeeded and the war is lost."[88] Von Rundstedt and Gen. Geyr
von Schweppenburg commanding Panzer Group West favored a more
conventional mobile defense: static coastal divisions would attrite
and funnel the Allied attack while local reserves immediately counterattacked to fix the enemy until a large-scale counterattack with
theater reserves could be mounted to throw the Allies back into the
sea. In the end, the Germans adopted a compromise of sorts, driven
more by the limited means available than by operational theory.

German intelligence was abysmal. Spies in England could gather
little information and photo reconnaissance was limited. Logic alone
convinced the German High Command that the Allies would invade
through the Pas de Calais. The Pas de Calais was the shortest route to
France and to the Ruhr.[89] The Allied deception plan Fortitude reinforced this German misassumption by using Lt. Gen. George S. Patton's fictional First U.S. Army Group as a decoy. Accordingly, Rommel gave his Fifteenth Army covering this area priority in building
fortifications and strengthened it with twenty-five divisions, six of
which were panzer divisions. In the Seventh Army covering Normandy, Rommel concentrated nine divisions and one panzer division. In keeping with his goal of defeating the Allies at the water's
edge, Rommel moved his local reserves close to the coast. Rundstedt
held only three panzer divisions and one panzer grenadier division as
a theater reserve. Even to employ these, however, he needed Hitler's
express approval. Unhindered and with effective command and control, the Germans could still mass considerable force against the vulnerable assault divisions wherever they might come ashore.

On June 6, 1944, the Allies launched Overlord. The Germans
fought tenaciously with their usual tactical skill. The Allies completely outfought the Germans, however, at the operational level.
Thanks to poor German intelligence, the Allies achieved both tactical and operational surprise. The complicated and muddled German
command and control inhibited anything other than a tactical response. German leaders remained so convinced that the main Allied

attack would come in the Pas de Calais, that the first substantial reinforcements did not arrive until well past Rommel's estimate of the crucial forty-eight hours. Aided by the German fixation on the Pas de Calais, the Allies successfully isolated the battlefield with air-power. Rundstedt's chief of staff noted, "The crippled rail net forced us to unload troops and supplies far behind the front and resulted in an extraordinarily long supply line."[90] Another German general asserted more pointedly, "This decisive role [of Allied airpower] is not so much to be seen in the support of the Allied landing units, but rather in the fact that all movements of the German forces and their supply troops, were made almost impossible during the day by the Allied air force."[91] Airpower's greatest impact was at the operational rather than tactical level; Eisenhower's insistence on the transportation (rail-target) plan paid handsome dividends.

Schweppenburg agreed that "the supply system depended too much on railroads and on centralized supply depots."[92] Germany's inadequate supply system also fatally handicapped its defense efforts not only because of Allied attacks on German lines of communication, but also because the German system in general was flawed. Gen. Gunther Blumentritt complained that "the unusual command channels in the service of supply made strategic leadership more difficult."[93] Much of this sounds very much like the criticisms that American Army War College students noted in their study of German supply methods in World War I.

Regardless of difficulties in supply or in rushing reinforcements to the front, the tenacity of the German tactical defense upset the Allies' timetable and general plan for expanding the lodgment. The Germans quickly contained the British efforts to break out in the Caen area. Although the Americans took Cherbourg, German defenders there destroyed the port facilities. The Allies quickly secured the lodgment, but again the Germans succeeded in containing Allied forces and preventing a breakout. Eisenhower again urged Anvil as part of the operational solution to prevent a stalemate. Two weeks after D-Day, Eisenhower cabled the Combined Chiefs arguing that "it is imperative that we concentrate our forces in direct support of the decisive area of northern France. Anvil provides the most direct route to northern France where the battles for the Ruhr will be fought."[94]

Eisenhower solicited support from Marshall and British general Maitland Wilson, the Supreme Allied Commander, Mediterranean theater of operations. Wilson would command Anvil forces until they physically linked up with Eisenhower's. The slow expansion of the Normandy lodgment helped overcome British objections. By the time the Allies got around to launching Anvil, however, the breakout was well under way. The Allies finally invaded southern France on August 15. Two weeks earlier, Operation Cobra provided the long-awaited breakout from the Normandy beachhead, signaling a general collapse of German defenses in western France. Anvil forces quickly drove up the Rhone Valley, and in fourteen days they effectively destroyed a German army, capturing close to eighty thousand prisoners, and more important, the ports of Toulon and Marseille. The Sixth Army Group, commanded by Gen. Jacob Devers (CGSS 1925, AWC 1933), linked up with Bradley's Twelfth Army Group on September 11, 1944. This bold and well-executed operation was characteristic of the U.S. insistence on concentration and concentric attack.

In an ironic stroke, following the breakout in the American sector, the Allies reached the Seine River just slightly beyond the ninety-day window stipulated in the original plan. Once past the Seine, the campaign followed the course outlined by Eisenhower prior to the invasion. The Supreme Commander insisted on a broad-front strategy that made the main effort in the north the securing of ports. After arriving at the German border, the Allies secured crossings over the Rhine, enveloped the Ruhr, and systematically eliminated German resistance.

Assessment

The American official history of the cross-channel attack noted a major difference between Americans and British in strategy: the British took a more opportunistic approach; the Americans took the longer view. The debate over when to return to the Continent dominated the strategic discussions between the two Allies for the first three years of the war. "The British said in effect, 'how can we tell what we should do six months or a year hence until we know how we come out of the next month's action?' The Americans retorted, 'how do we know whether next month's action is wise unless we know where we want to be a year from now?'"[95] The military planning system that was established and taught in the advanced American military schools in the interwar years may help to explain this difference, in part. The planning system as taught at the War College

for twenty years insisted that once the national authorities established strategic objectives, a joint plan, followed by a theater plan,
must follow. This hierarchical planning system ensured that all
plans, from the tactical to the operational and the strategic would be
nested in and harnessed to national objectives.

At the operational level, Overlord demonstrated differences in
approach and capabilities not only between the Allies but between
the Anglo-American perspective, on the one hand, and the German
view, on the other. Marshall and Eisenhower consistently advocated
focusing combat power at the decisive point in the decisive theater of
operations. Both British and American planners recognized the importance of logistics and airpower, but when decisions had to be made
regarding competing priorities or military views, Eisenhower relied
on his military education and hard-won experience. His insistence on
his prerogatives as a commander, such as in expanding the invasion
area in order to ensure the quick capture of Cherbourg, in favoring the
rail-target plan, and in promoting Operation Anvil, was consistent
with principles taught at both Leavenworth and the War College.
Eisenhower fundamentally grounded his understanding of operational art in his firm convictions with regard to the decisiveness of
concentration of forces, the leveraging of joint power, the importance
of logistics, and the necessity of unity of command.

The Allies, of course, made mistakes as well. Martin Van Creveld
criticizes the Allied High Command for allowing the logisticians to
dominate planning.[96] Russell Weigley and many other historians are
quick to point out the failure of the Allies to close the Falaise Gap and
thereby miss an opportunity to destroy the German Seventh Army.[97]
The Allies did not have to be perfect, however, only better than the
Germans. The Germans did not emphasize logistics or intelligence to
the degree the Allies did. Eisenhower concluded that "throughout the
struggle, it was in his logistical inability to maintain his armies in
the field that the enemy's fatal weakness lay."[98] The Supreme Commander credited Allied airpower for this circumstance, but the German army did not hold logistics in the same high regard as did the
Americans.

The German General Staff system was dominated by the operators. Unlike the American staff system, in which the intelligence,
logistics, and operations sections were coequal, the operations officer
was always the senior and dominant staff officer the German sys-

tem.[99] This bias did not serve the German army well when it came to intelligence. As one German general noted, "A weakness of the German Army was the lack of instinct and knowledge in practical intelligence work. In the unwritten tradition of the *Heer*, intelligence work had a slight odor of not being respectable—at any rate, not as important as the work of operational personnel who controlled the fighting."[100] Germany's military leadership simply failed to recognize that operational art is more than drawing arrows on a map. It is the ability to project, sustain, and employ force in theaters of war to achieve strategic objectives.

Overlord's success and, in sum, Allied operational success in the entire European theater lay in the simple fact that the Allies proved more capable than their opponents at projecting, conducting, and sustaining operations in theaters of war. It was not a matter of which side possessed the greater means, but how the Allies used those means. In contrast to the Allies, the Germans proved tactically superb but operationally inept. Particularly in intelligence, logistics, and command and control, German operational art was flawed. The Allies matched their strengths against German weaknesses, effectively leveraging landpower, seapower, and airpower. This is the essence of modern operational art. Yet, for all its success, American operational art reached its peak, not in Europe, but in the Pacific.

7

THE PACIFIC THEATER
OF WAR

For the United States of America, World War II began and ended
in the Pacific. Unlike the war in Europe, the war in the Pacific
was long anticipated, in terms of American involvement—in both
fact and fiction. Homer Lea's *The Valor of Ignorance*, published in
1909, a work on geopolitics, predicted an inevitable war in the Pacific
with the Japanese capturing the Philippines, Hawaii, and even portions of the United States' western coast.[1] Hector C. Bywater pub-
licly suggested the probable strategics for a Pacific war as early as
1922 in *Sea-Power in the Pacific: A Study of the American-Japanese
Naval Problem*.[2] Three years later, he described the conflict in a
novel, *The Great Pacific War: A History of the American-Japanese
Campaign of 1931–1933*. Bywater was not so prophetic in his vision
of the war, but he expressed forward-looking views on the role of
naval aviation and the continued reliance on decisive main-fleet surface engagements.[3]

The American military had been preparing intellectually for the
possibility of war with Japan since as early as 1906. As I have discussed in earlier chapters, War Plan Orange, the plan for hostilities
with Japan, was the most realistic of the war plans developed in the
interwar period. More important, an entire generation of U.S. officers
studied and exercised the probable course and requirements for such
a conflict with Japan in the senior service colleges. For most of War

Plan Orange's existence, it called for a powerful drive across the Central Pacific to relieve the American garrison in the Philippines (presumably captured by the Japanese) and progressively obtain bases from which to blockade and defeat Japan. In the smoking ruins of the Pacific Fleet after December 7, 1941, however, the United States found itself without much combat power even to defend its Pacific possessions, let alone drive across that vast ocean to threaten Japan. The Japanese attack at Pearl Harbor initiated a series of near simultaneous offensives that within six months led to the capture of Malaya (now Malaysia), the Netherlands Indies (present-day Indonesia), and the Philippines. By July 1942, the Japanese pushed into the Bismarck and Solomon islands. At the beginning of the Pacific war, the United States assumed a defensive strategy not only out of necessity, but out of choice. Strategic choice, almost unilateral American direction, and the nature of the Pacific theater ensured that the war there would be fought differently than the war in Europe.

The Arcadia conference, held in Washington in December 1941, officially confirmed the Europe-first strategy, making the Pacific a secondary theater of war. By the following March, the Allies agreed on the general strategic responsibilities for the global war. Together, Britain and the United States would oversee the main effort—the European theater of war—but the British had primary strategic responsibility for India and Burma while the United States assumed strategic responsibility for the Pacific, including China.[4] These basic decisions fundamentally affected how the war in the Pacific would be fought not only strategically, but operationally. In Europe, both Eisenhower and the senior commander in the Mediterranean theater of operations reported to the Combined Chiefs of Staff, but in the Pacific, the senior American commanders reported to the U.S. Joint Chiefs of Staff. The Pacific would largely be an American show. Although in both Europe and the Pacific senior American leaders and planners shared a common view of staff organization, process, and large-unit operations, thanks to their staff and war college experience, the Pacific provided a much better opportunity for the exercise of American operational art. The expeditionary nature of the Pacific theater, with its vast distances and almost unilateral American direction, gave full scope to the development of American operational art as studied during the interwar years.

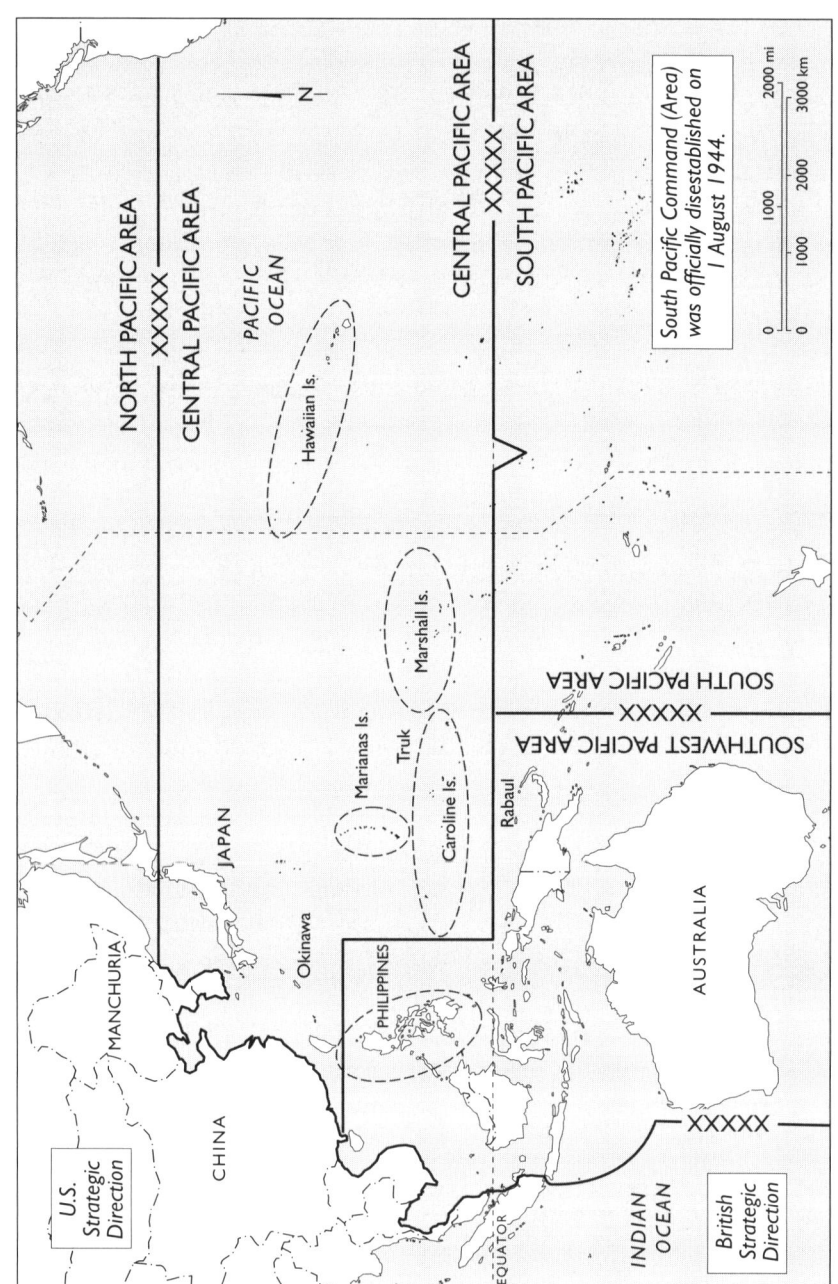

Map 5. Pacific theater of war—operational boundaries.

204

Unity of Command and Theater Organization

In the Army War College exercises dealing with the Pacific theater during the interwar period, the students expressed a strong preference for joint staffs with a single theater commander. Joint staffs became a reality, but the effort to develop a single unified command for the Pacific theater ran aground on the rocks of interservice rivalry as well as senior officer personalities. As Marshall wrote to Douglas S. Freeman, "To one in my position the matter of personalities of higher commanders will always be a major consideration, having far more importance than the blue-print solutions of Leavenworth and the War College would lead the student officers to anticipate."[5] Virtually all senior American military leaders, as well as President Roosevelt, understood the advantages of having a single theater commander.[6] The problem was, who would it be? By the summer of 1942, General MacArthur was a well-known public figure cast in a heroic light by his doomed defense of the Philippines and by virtue of his leading the fight against Japan from Australia. Adm. Chester Nimitz (NWC 1923), the new commander of the Pacific Fleet, was relatively unknown to the public and quite junior to General MacArthur. The army could not ignore MacArthur, but Admiral King, Chief of Naval Operations, would not entrust the Pacific Fleet to an Army officer. In the end, the Joint Chiefs settled on the creation of two theaters of operations in addition to the China-Burma-India theater: the Pacific Ocean Areas (POA), with Admiral Nimitz as commander; and the South West Pacific Area (SWPA), headed by General MacArthur. (See map 5.)

At the end of March 1942, the Joint Chiefs of Staff appointed MacArthur as the Supreme Commander of the South West Pacific Area. His area of responsibility included Australia, the Bismarck Archipelago, and all of the Netherlands Indies except Sumatra. MacArthur's designation as Supreme Commander stemmed from the inclusion of Australian and New Zealand forces into his command. As an Allied commander, he quickly organized his force into component air, naval, and land commands; Vice Adm. Herbert F. Leary (NWC 1932) commanded the SWPA naval forces; Lt. Gen. George Brett (CGSS 1930, AWC 1936), the Allied air forces; and Australian general Sir Thomas A. Blamey, the Allied land forces. After becoming unhappy with his air and naval commanders, MacArthur replaced them within a year with Maj. Gen. George C. Kenney (CGSS 1927, AWC 1933) and Rear Adm. Thomas C. Kinkaid (NWC 1930), respectively. Throughout the war, MacArthur depended heavily on Kenney and Kinkaid, but generally ignored General Blamey. As soon as American army units began arriving in large numbers, MacArthur began relying directly on American task forces and units outside Blamey's control.

In January 1943, MacArthur asked Marshall to transfer Lt. Gen. Walter Krueger (ASC 1907, AWC 1921) to the SWPA, along with a field army headquarters. MacArthur's command eventually included two American field armies, the Sixth commanded by Krueger and the Eighth under Lt. Gen. Robert L. Eichelberger (CGSS 1926, AWC 1930). Krueger became MacArthur's primary land force subordinate. Despite being an Allied commander, MacArthur reported directly to the U.S. Joint Chiefs of Staff rather than the Combined Chiefs of Staff. Also uniquely, the establishing directive for SWPA insisted that "the Joint U.S. Chiefs of Staff will exercise jurisdiction over all matters pertaining to operational strategy."[7]

At the same time the JCS created MacArthur's command, they appointed Admiral Nimitz as the commander in chief of the Pacific Ocean Areas (POA). Nimitz's command included the rest of the Pacific except for the band of ocean off the coast of Central and South America.[8] His area of responsibility was subdivided into North, Central, and South Pacific areas. Unlike MacArthur, Nimitz did not function as an Allied commander, and he directly controlled all U.S. forces in the Central and North Pacific areas. He exercised control over the South Pacific area through a subordinate naval commander. Even though there was no singular unity of command in the overall theater of war, both Nimitz and MacArthur exercised unity of command in their respective areas of responsibility.[9]

Many of the commanders in the Pacific, and certainly most historians of the Pacific campaigns, criticized the failure to unify the command under a single theater commander.[10] Clearly, interservice rivalry and personality conflicts provided root causes for this failure. If national rivalry between British and American officers plagued Eisenhower, U.S. interservice rivalry plagued the Pacific theater. MacArthur was deeply suspicious of the navy.[11] Admiral King, Chief of Naval Operations, rightly considered the Pacific a maritime theater and firmly believed the navy should run the show. With regard to the army, Marshall tried to suppress interservice rivalry by edict and personal example.[12] If the theater had to be separated into naval and army areas of responsibility, the practical matter was how to make it work. Much as anticipated in the interwar college exercises, the solution lay not just in the ultimate goodwill of senior commanders but in the development of joint staff organizations and planning.

Development of Joint Staff Planning

In Europe, Eisenhower as Supreme Commander could make major operational decisions and shift resources—albeit carefully, under the eyes of the Combined Chiefs of Staff. Even though he was a Supreme Commander, Eisenhower was shaped and limited by national and political considerations in his exercise of theater command in coalition warfare. In the Pacific Nimitz and, to a lesser degree, MacArthur were less constrained than was Eisenhower in this regard. For the Pacific theater, the Joint Chiefs of Staff balanced the needs of the global war and reserved the right to make not only theater strategic decisions but even major operational decisions. The JCS left the conduct and detailed planning of operations to the American theater commanders, but the Joint Chiefs participated in the campaign planning for the Pacific in both a directive and collaborative fashion. Historians often overlook the role of the JCS staff planners in shaping the Pacific campaigns. Strategic and operational direction of the Pacific war was a collective enterprise, as MacArthur and Nimitz put forward recommendations and the JCS deliberated and decided after considering their own operational concepts. The Joint Chiefs, and Marshall in particular, quickly realized that modern war on a global expeditionary scale required joint staffing for realistic planning.

Curiously, the initial impetus for improved joint staff planning did not come about from an interest in better prosecuting the war against the Axis, but from a desire to compete more effectively in the councils of strategy with America's Allies. Within a month of the U.S. entry into the war, the Allies agreed to form the Combined Chiefs of Staff to provide for military cooperation between the British and the Americans. The U.S. Joint Chiefs of Staff appeared soon thereafter, though without any formal charter or directive detailing the group's functions or responsibilities. The army and the navy agreed to organize a small Joint Staff to support the Joint Chiefs and to

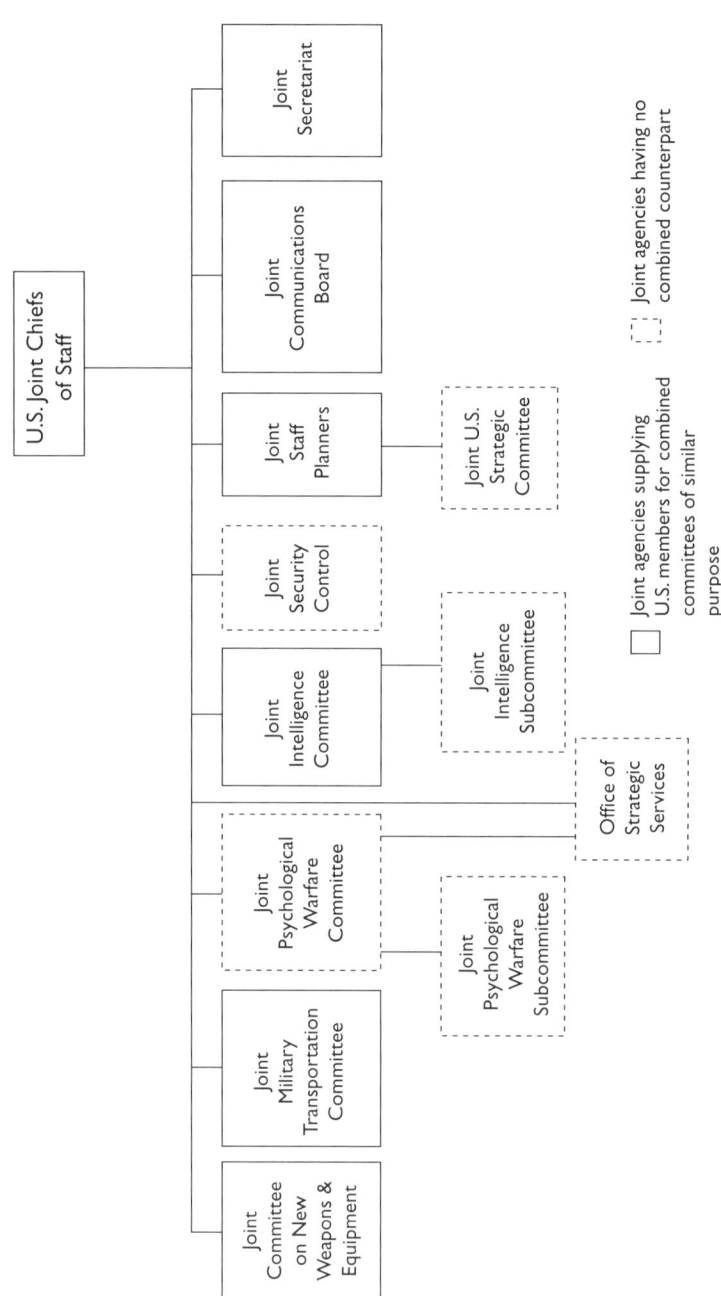

Figure 9. Joint Staff organization, November 1942. Source: *A Concise History of the Organization of the Joint Chiefs of Staff, 1942–1978.*

provide representation on the Combined Planning Staff. This small staff included a Joint Intelligence Committee and a Joint Planning Staff (JPS). (See figure 9.) The JPS consisted of an army, a navy, and an air planner. A Joint U.S. Strategic Committee integrated from the old Joint Board supported the JPS for detailed studies. In comparison to the longer-established British planning system, the initial American attempt at joint planning demonstrated many shortcomings.

In April 1942, Brig. Gen. Albert C. Wedemeyer became the army planner on the Joint Planning Staff. Wedemeyer accompanied General Marshall to the Casablanca conference in early 1943. Marshall believed this conference was crucial to moving the Allied focus from the Mediterranean to the invasion of France. The British were much better prepared to present their case. Wedemeyer noted the presence of "swarms of British officers of all ranks, representing the three services." He identified several "weaknesses in the planning work of the American staff: we lacked pre-prepared studies, and were forced to rely on memory."[13] In a letter to Maj. Gen. Tom Handy, Chief of the OPD, Wedemeyer paraphrased Julius Caesar to complain: "We came, we listened, and were conquered."[14] Gen. Joseph T. McNarney (CGSS 1926, AWC 1930), Marshall's deputy chief of staff, agreed and recommended a committee to provide a thorough study of the Joint Chiefs and all its subordinate agencies.[15]

The special committee reported in March 1943, and the JCS adopted its recommendations. The reforms included the creation of a Joint War Plans Committee (JWPC), a Joint Logistics Committee (JLC), and a reorganized Joint Intelligence Committee. (See figure 10.) The JWPC was at the heart of the Joint Chiefs' strategic and operational planning. The JCS directed the JWPC to make sure "all studies of combined action and joint war planning should be undertaken by joint action from the time the studies or war plans are initiated."[16] In other words, war plans should be born jointly. From the time it was established until the end of the war, this committee produced more than one thousand studies and plans. The JWPC studies covered every major strategic decision and every major operation. Not surprisingly, these studies took the form of staff studies as taught at Leavenworth.[17] The Staff School taught the basics of the military problem-solving process, and this process provided a systematic way to analyze and solve the complex problems of modern global war. The operational art and campaign planning taught at the war colleges, likewise,

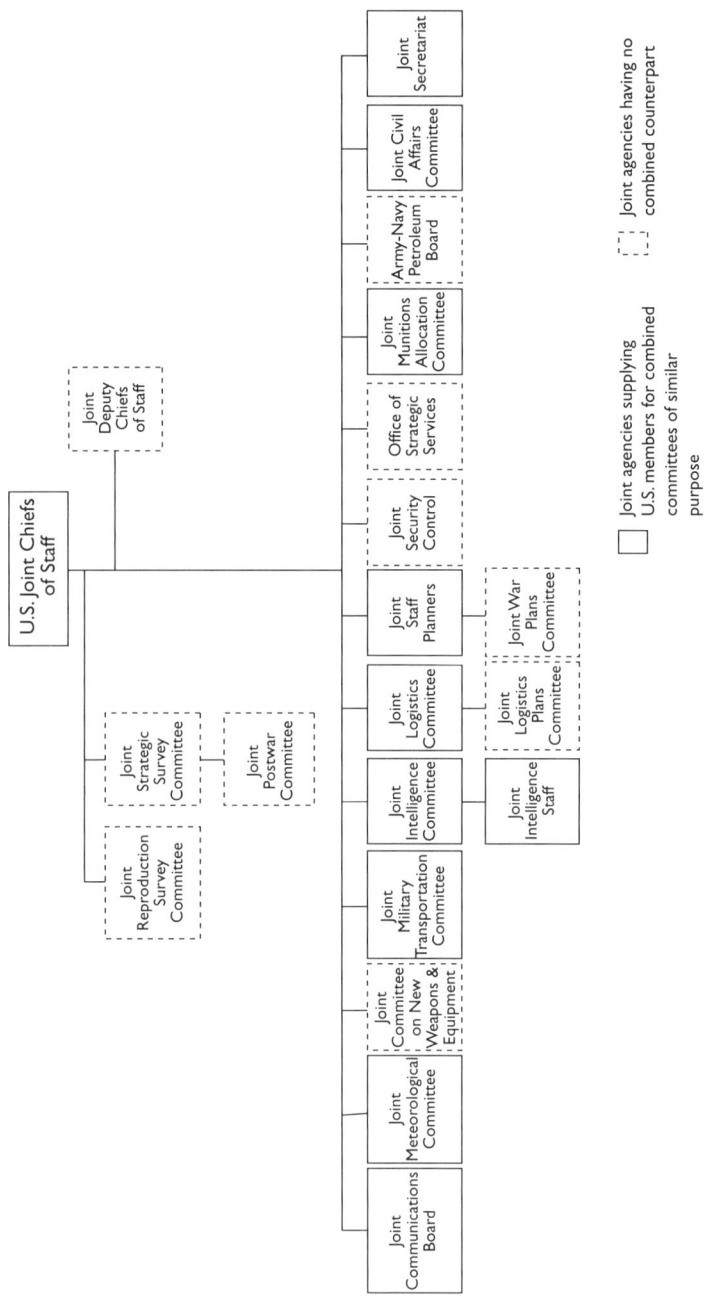

Figure 10. Joint Staff after reorganization. Source: *A Concise History of the Organization of the Joint Chiefs of Staff, 1942–1978.*

provided method, format, and insight into developing the operational plans drafted by the Joint Chiefs.

Supported by the Joint Intelligence and Logistics committees, the JWPC developed operational outline plans for all the major operations in the Pacific theater. Inasmuch as the theater commanders worked directly for the Joint Chiefs, this joint planning played a crucial role in the development of the final operational plans. The JWPC developed extremely detailed plans. For example, the JWPC outline plan to recapture the Philippines ran to more than one hundred pages of analysis and planning. The campaign plan was phased, and major operations were sequenced and timed. The plan included a political and economic estimate as well as detailed relief maps.[18] The outline plans routinely included detailed logistic and intelligence estimates provided by the respective committees. In summary, these plans in form and format followed the interwar methodology as taught in the service schools and colleges.

The Joint Logistics Committee provided expertise to both the Joint Planning Staff and the Joint Chiefs. The JLC advised "the joint staff planners in the consideration and preparation of joint war plans as to the logistic aspects of such plans in order that the Joint Staff Planners may insure the integration of logistics with strategy in joint war plans."[19] The planners were keenly aware of the overriding importance of logistics in waging global war. The JLC kept the Joint Chiefs informed on the "logistic implications of proposed U.S. commitments relating to joint and combined operations."[20] Utilizing a parallel structure to the Joint War Plans Committee, the JLC created the Joint Logistics Planning Committee to prepare detailed logistic plans and studies. The structure of the JPS in Washington highlighted the American recognition of the importance of jointness, logistics, and intelligence at the highest levels.

The joint planning in Washington did not duplicate planning efforts in the theater. The JPS developed outline plans for future operations. They provided the Joint Chiefs with an informed way to adjudicate resources, evaluate proposals from the theater, and direct operations in the Pacific. For Nimitz and MacArthur's staffs, the detailed work of the joint planners provided an excellent starting point and wealth of information. Most important, the joint work accomplished in Washington provided for U.S. service consensus on the major decisions facing the president, the Allies, and the theater of operations commanders. The mechanism of joint planning developed in Washington shaped and directed the campaigns in the Pacific beginning in 1943.

Joint planning in Washington alone could not overcome the potential friction and failure that might be generated by interservice rivalry in theater. Final planning and conduct of operations required a good deal of cooperation both between and within the separate theaters of operations in the Pacific. The very nature of the overall Pacific theater demanded more intimate cooperation between land, sea, and air forces than in Europe. Nimitz's Pacific Ocean Areas required army garrisons, engineers, and logistics for captured islands in the fleet's drive across the Central Pacific. MacArthur, obviously, could go nowhere in the Southwest Pacific without the navy, and everybody needed airpower, both land-based and carrier aviation. The development of joint staffs and joint operations in the Pacific theater varied according to the service and the personalities of the commanders.

In reluctant compliance with a presidential order, General MacArthur left the Philippines before their capture in March 1942 in order to organize Allied defenses in Australia. He took with him fifteen members of his staff, including Maj. Gen. Richard K. Sutherland (CGSS 1928, AWC 1933), his chief of staff, and Col. Charles A. Willoughby (CGSS 1931, AWC 1936), his G-2. Brig. Gen. Stephen J. Chamberlin (CGSS 1925, AWC 1933) and Col. Lester J. Whitlock (CGSS 1928), both already in Australia, served as MacArthur's G-3 and G-4, respectively. The president suggested to Marshall that a few senior Australian or Dutch staff officers might be welcome. Marshall passed the suggestion along, but MacArthur replied that there were no "qualified Dutch officers present in Australia and that the Australians with a rapidly expanding army did not have nearly enough staff officers to meet their own needs, let alone to serve on his staff."[21] Marshall also pushed for a more joint staff in the SWPA, but MacArthur stuck to a traditional army staff only modestly integrated with his Allies and the navy.

Admiral King and General Marshall recognized the need for joint staffs and a clearer definition of unity of command in the theater. In April 1943, the JCS published a directive on "Unified Command for U.S. Joint Operations." This directive set the pattern for U.S. joint operations for the rest of the war and beyond. The directive unequivocally stated that joint force commanders' "responsibilities are the same as if the forces involved were all Army or all Navy." In practice, this meant joint force commanders assigned missions to subordinate forces but kept their involvement in administration to a minimum and left military discipline up to the respective service. Most important, the JCS directed that "a joint staff of appropriate size will be organized to assist the Joint Force Commander. It will comprise representatives of each of the several component parts of his force in such a manner as to insure an understanding of their several capabilities, needs and limitations, together with the knowledge essential to maximum efficiency in integration of their efforts."[22] Progress in providing more joint representation was slow on MacArthur's staff. By April 1944, MacArthur reported he had twelve naval and two marine officers on his staff with two more on the way.[23] Jointness in MacArthur's command was achieved through staff coordination, and primarily through the interaction of his subordinate commanders.

Admiral Nimitz's path to joint staffing was more complete, but indirect. Until 1943, as commander of the Pacific Fleet, Nimitz had a well-established naval staff, but no separate joint staff as the unified commander of the POA.[24] General Marshall assigned army officers to navy staffs, in both the Central and South Pacific areas, to look after army interests and assist in operations. Marshall also designated Lt. Gen. D. C. Emmons (CGSS 1934), and later Lt. Gen. Robert C. Richardson, Jr. (CGSS 1924, AWC 1934), to be Nimitz's army component commander, responsible for providing and administering all army assets operating under Nimitz.[25]

Army concerns over logistics renewed Marshall's interest in providing Nimitz with a joint staff. Brig. Gen. Edmond H. Leavey (CGSS 1938), an army logistician, who was assigned to the Pacific Fleet's supply service, wrote a blistering letter to his former boss, Gen. Brehon Somervell (CGSS 1925, AWC 1926), commanding general of the Army Services Forces. Soundly criticizing the lack of a true theater staff, Leavey noted, "There was no section or officer in Nimitz' headquarters or elsewhere either designated for or capable of, coordinating and controlling the Service of Supply activities in the theater."[26] Leavey's reports and those of others on the same subject reached Marshall, who then pressed King. Naval planners in Washington had already begun to study the problem, even requesting information on Eisenhower's joint and combined organization in Europe.[27] On September 6, 1943, Nimitz preempted further discussion by announcing

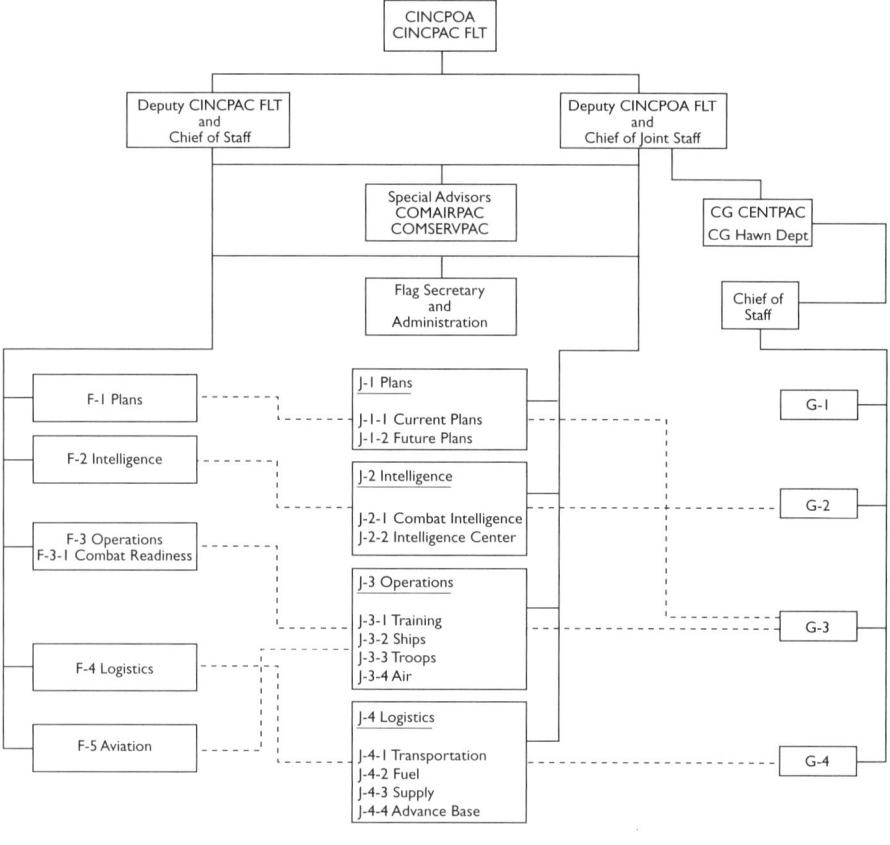

Figure 11. Admiral Nimtz's Pacific Ocean Areas Joint Staff. Source: Louis Morton, *Strategy and Command: The First Two Years*, 497.

the formation of a joint staff. Nimitz appointed navy officers for the operations and plans staff sections, while Army officers filled in as chiefs of the intelligence and logistics sections, the latter officer being Brigadier General Leavey. (See figure 11.)

Nimitz's joint staff represented a significant step forward in American operational art, yet like with MacArthur, jointness in the conduct of operations largely depended on his component commanders. Both MacArthur's and Nimitz's theaters of operations provide excellent examples of the development of American operational art in the planning and conduct of operations in World War II. For MacArthur, the defining moment of his service in the war was his triumphant return to the Philippines. In fact, the Philippines campaign was the largest and most decisive campaign in the Pacific war. In Nimitz's main area of responsibility, the Central Pacific, Operation Iceberg, the battle for Okinawa, provided the final expression of American operational art in World War II.

RETURN TO THE PHILIPPINES

The Japanese pressed their advantage for most of 1942. In July, Japanese imperial forces expanded into the Solomons and New Guinea. Following the Battle of Coral Sea in May and the clear Japanese defeat in the Battle of Midway in June, American forces began to assume limited offensive operations. Nimitz's forces in the South Pacific contested Japanese advances at Guadalcanal in the Solomon Islands chain, and MacArthur attacked the Japanese in New Guinea. Tactical opportunism within a general defensive theater strategy characterized early American efforts. By early 1943, the Americans were no longer interested merely in defending in the Pacific. In May, the Joint Chiefs approved a general strategy to defeat Japan.

That JCS "Strategic Plan for the Defeat of Japan" proposed recapturing Burma and the Philippines, followed by "an overwhelming air offensive against Japan from bases in China."[28] In keeping with the formula of the old War Plan Orange, the planners favored progressively developing bases from which U.S. airpower and seapower could isolate and destroy Japan's warmaking capacity, followed by an invasion only if necessary. Gen. Hap Arnold especially wanted to obtain bases for the new long-range B-29 bombers. He remained convinced these new instruments of airpower could apply the kind of bombing necessary to win. The debate centered on which line or lines of operations best suited Allied strategic and operational needs. By 1944, three options became apparent: bomb Japan into submission from air bases along the China coast, advance from the Southwest Pacific through the Philippines, or advance through the Central Pacific in line with the old War Plan Orange.

Both MacArthur and Nimitz agreed in their recommendations for 1944 that the Philippines should be retaken, at least in part. MacArthur's Plan Reno III, submitted to Marshall in October 1943, provided for a five-phase campaign plan leading to the invasion of Mindanao, the southernmost of the Philippine Islands.[29] The Joint Staff planners studied these recommendations and agreed that Mindanao was necessary as a base to neutralize Japanese airpower on Luzon, the major and most important island in the archipelago. The planners' eyes, however, were fixed on Formosa as the prize that would yield the greatest operational advantage due to the island's proximity to both China and Japan.[30] The Joint Chiefs made their decision and issued instructions to their Pacific theater commanders on March 12, 1944. This directive informed Nimitz and MacArthur "that the most feasible approach to the Formosa-Luzon-China area is by way of Marianas-Carolines-Palau-Mindanao area."[31] The JCS ordered MacArthur to occupy Mindanao with a target date of November 15, 1944, "preparatory to a further advance to Formosa either directly or via Luzon."[32] Furthermore, planning responsibilities for Formosa fell to Nimitz while MacArthur planned for an invasion of Luzon, if required.

At last the JCS ordered MacArthur to return to the Philippines. MacArthur argued passionately to both President Roosevelt in person and to the Chiefs of Staff through his representatives for an invasion of Luzon.[33] The Joint Chiefs debated the wisdom of bypassing Luzon for Formosa. Formosa may have been strategically the best option, but after further study it appeared operationally impractical. For Nimitz to take Formosa, he would need U.S. Army troops—lots of them. The Joint Logistics Committee and army planners estimated that Nimitz would need between 77,000 and 200,000 service troops to build and maintain the air bases and infrastructure. The planners calculated this could not be obtained until troops could be transferred from the European theater. MacArthur, however, could invade Luzon before the end of 1944 with the troops on hand.[34] On October 3, 1944, just weeks before the return to the Philippines, the JCS authorized MacArthur to invade Luzon. MacArthur's staff completed the final planning for the largest operation in the Philippines while the campaign was already under way.

Campaign Planning

The campaign plan to liberate the Philippines reflected some of the best practices studied in the interwar years, as well as the hard-earned experience gained from two years of war in the Pacific. MacArthur had long planned to return to the Philippines. He submitted his first Reno plan for that purpose in February 1943. The SWPA planners constantly updated the Reno plan to reflect the evolution of the Joint

Chiefs' decisions on Pacific strategy. MacArthur submitted the last of the series, Reno V, to the Joint Chiefs in June 1944. The campaign plan provided for a series of major operations leading to the reoccupation of the southern and northern Philippines. The plan adhered closely to the JCS directive of March 1944. The overall strategic objective was to penetrate into the Formosa-Luzon-China area to "establish bases for a final assault upon Japan."[35] The JCS assigned the operational objective to MacArthur's forces to occupy Mindanao in order to establish "air forces to reduce and contain Japanese forces in the Philippines preparatory to a further advance to Formosa, either directly or via Luzon."[36]

The campaign mutually leveraged airpower, seapower, and landpower. The key element in the scheme of maneuver depended on airpower—extending the reach of land-based bombers through the occupation of successive bases. Airpower also protected the flanks of the SWPA drive and, in combination with seapower, delivered the ground forces. Ground forces would "displace forward by water and air, covered by naval and air forces, to seize and establish air bases in each successive objective."[37] Reno V called for bypassing or neutralizing enemy strength "by air, land, and sea action."[38] The planners organized the campaign into four phases, all sequenced and timed, ending in the invasion of Luzon by January or February 1945. (See map 6.)

Once the JCS approved Reno V, MacArthur quickly produced a more detailed campaign plan, Musketeer, that dealt solely with operations in the Philippines. Like the Reno plans, Musketeer went through several iterations. The Musketeer plans directed a series of major and minor operations to complete the liberation of the Philippines. The planners code-named the major operations King, Love, Mike, and Victor. In King I, Allied forces would secure Sarangani Bay in southern Mindanao by November 15, 1944, and in King II secure Leyte on December 20. Operation Love was designed to seize bases in Central Luzon in January and February. Operation Mike would capture key positions and secure Luzon in April and May. Finally, Operation Victor would destroy bypassed enemy garrisons and free the rest of the archipelago.[39]

While MacArthur planned his return to the Philippines, his forces and Nimitz's Central Pacific forces both maintained continuous pressure on the Japanese. MacArthur advanced on the island of Morotai just north of New Guinea and ever closer to the Philippines. This

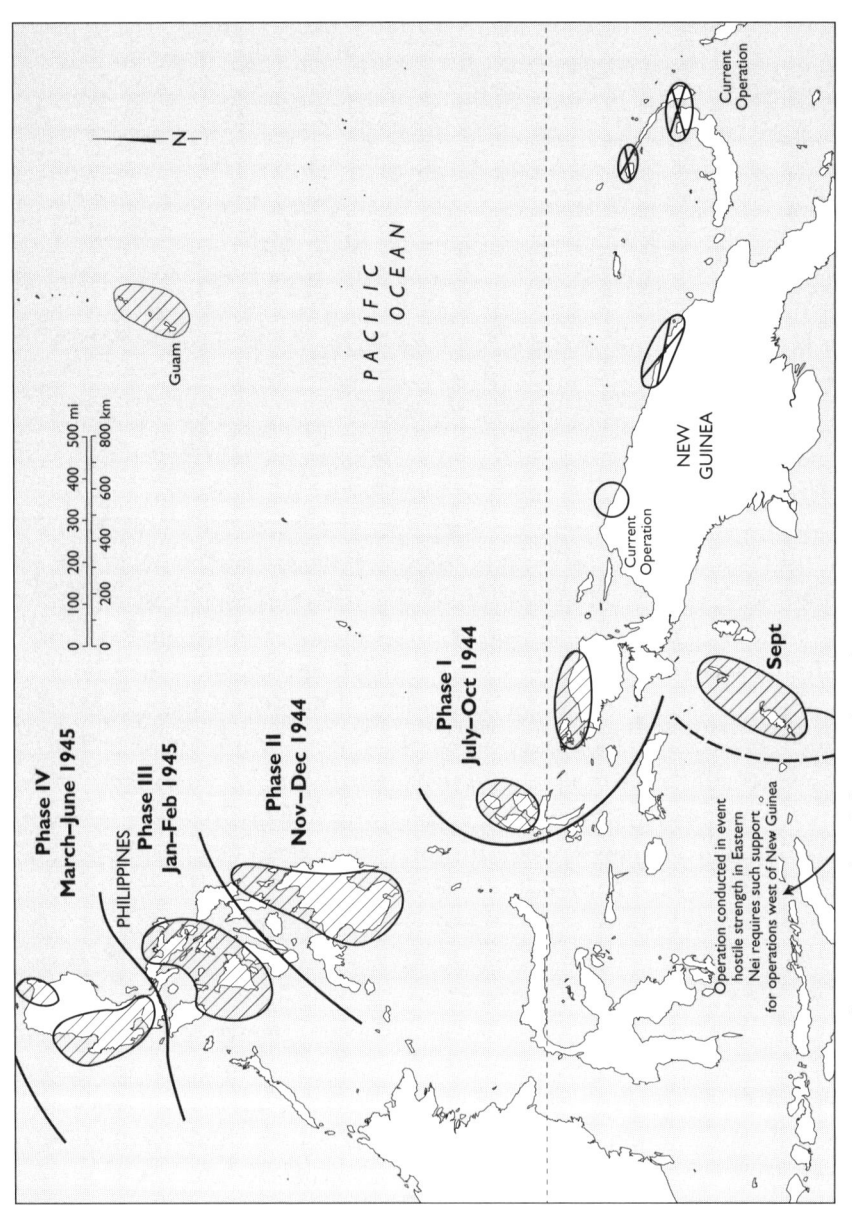

Map 6. Concept of operations sketch for Outline Plan Reno V, June 1, 1944. Source: USAMHI.

shaping operation was designed to keep extending the operational reach of MacArthur's air component, the Fifth Air Force. Nimitz's Central Pacific forces took Saipan, Guam, and Tinian in the Marianas between June and August. In a coordinated move with MacArthur's assault on Morotai, Nimitz's forces struck at Pelelieu in the Palau islands. In support of these operations, fast carriers from Halsey's Third Fleet attacked Yap and swept the Philippine coast from September 7 to 14. Halsey encountered little resistance, and immediately he recommended to Nimitz that his planned attack on Yap should be canceled and MacArthur should bypass Mindanao and advance directly on Leyte.

Nimitz agreed and relayed these recommendations to the Joint Chiefs of Staff then meeting with the Combined Chiefs at the Quebec conference. Nimitz offered not only to cancel the planned operation against Yap but to loan MacArthur the army's XXIV Corps, then loading at Pearl Harbor for that operation. Marshall fired off a message to MacArthur requesting his views. MacArthur was unavailable, aboard ship and en route to observe the Morotai landing under radio listening silence. No doubt with some trepidation, his chief of staff, Major General Sutherland, accepted the proposed change in operations in MacArthur's name. Sutherland's message reached the Chiefs while they were attending a formal dinner for the Canadian prime minister. They excused themselves from dinner and deliberated in another room. Within an hour and a half the Chiefs ordered Nimitz and MacArthur to cancel the intermediate operations and invade Leyte on October 20.[40] This example of extraordinary strategic and operational flexibility advanced the war in the Pacific by thirty days and changed the direction of the campaign.

In keeping with the new decision to bypass Mindanao, MacArthur updated his plans. His staff published Musketeer III on September 28, 1944. Like previous versions, this plan took the form of the five-paragraph field order, but it included assumptions on both friendly and enemy forces. This plan reflected not only the decision to bypass Mindanao and go straight to Leyte but also the JCS decision to invade Luzon. The plan listed two operational objectives: occupation of the Manila–Central Plain area of Luzon and the establishment of bases as directed by the Joint Chiefs in support of further operations against Japan. The plan also listed the ultimate or strategic objective to "re-establish and defend the constituted government of the Philippine Islands."[41]

The Musketeer campaign plan retained most of the major operations identified in the earlier versions, including a preliminary move to seize Mindoro (Love III) and several contingent operations. The main effort (Mike I) called for an amphibious assault at Lingayen Gulf and overland operations through the Central Plain to Manila by Krueger's Sixth Army. One of the contingency operations (Mike II) directed a landing at Dingalen on the eastern coast of Luzon by Lieutenant General Eichelberger's Eighth Army, to be executed "if required to turn the eastern flank of hostile defense force in the northern Central Plains or exploit southward."[42] An excellent example of operational art, the final campaign plan included a series of timed, phased, even contingent major operations to secure operational and strategic objectives in the theater.

The Reno and Musketeer campaign plans looked beyond immediate operations to forecast future operations. They served as "a guide covering the larger phases of allocation of means and of coordination between projected operations of Southwest Pacific forces."[43] MacArthur's headquarters also published more-detailed operational instructions. These operations orders provided the necessary details to allow MacArthur's subordinate commanders to complete their own tactical plans. These major operations orders also took the form of the five-paragraph field order developed during the interwar period. The orders included tasks for each of the components—land, air, and sea. In the instructions for both Leyte and Luzon, the theater Services of Supply (USASOS), the organization responsible for logistics, was treated as a major separate component. Ultimately, the Services of Supply would take responsibility for base infrastructure, but the operations plans specifically tasked Krueger's Sixth Army "to initiate the establishment of naval, air and logistic facilities for the support of subsequent operations."[44] Landpower would be used to build air and sea bases, so that jointly all elements of the theater commander's combat power could be brought to bear. (See figure 12.)

Just as planning was a collective enterprise between the Joint Planning Staff and the theater staffs, it was also a collective effort at the theater level. After publication of the Reno plan in June, MacArthur's headquarters held a series of meetings with the component planners in Brisbane, Australia, from July 20 to August 6, 1944. For

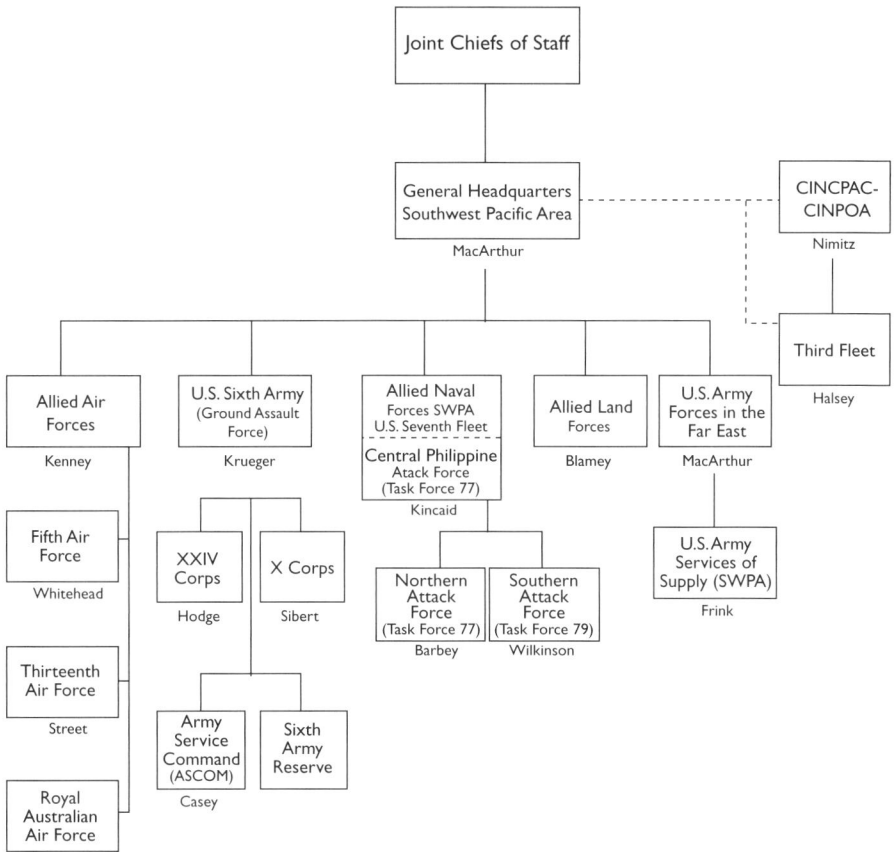

Figure 12. Organization for Leyte operation. Source: M. Hamlin Cannon, *Leyte: The Return to the Philippines*, **25.**

two weeks, Allied air, naval, and ground planners met to hammer out requirements and coordinate their efforts. Following the conference, each component headquarters made its own detailed plans and issued its tactical plans.[45] By the end of September, the campaign plan, the major operations plan for Leyte, and the component tactical plans were finalized. All that remained was execution.

The Campaign

The Philippine Islands are centrally located between China, Japan, and what were then Malaya and the Netherlands Indies. Luzon in the north, the Visayan Islands including Leyte and Samar in the central portion, and Mindanao in the south make up the Philippine archipelago. The Philippines lay directly across Japan's lines of communication to its resource area in the Netherlands Indies, Malaya, and Sumatra. Japan's need to protect this line of communication prompted the Japanese decision to capture the Philippines early in the war.[46] To remain in the war with access to oil and other critical resources, the Japanese needed to take and hold the Philippines at all costs.

In the summer of 1944, the Japanese began strengthening their positions in the Philippines and developing plans for the archipelago's defense. The steady Allied advance along New Guinea and the invasion of the Marianas threatened the empire's inner defense area. The Japanese High Command developed the Sho-Go (Victory) plans for the defense of the Philippines, Formosa, and the Japanese home islands.[47] Once the main enemy effort could be determined, the Imperial General Headquarters (IGHQ) intended to wage a decisive battle against any enemy penetration into the inner defense area. The Sho-1 plan dealt with the defense of the Philippines. Although the Japanese had suffered considerable losses through the attrition of their naval and air assets as a result of the Allied offensives in the Central and Southwest Pacific, they still possessed the capability to mass significant combat power to oppose U.S. forces in the battle for the Philippines. Tactically, the Japanese had often proved superior to their opponents, but there were significant weaknesses in the Japanese military machine at the strategic and operational levels.

Strategic direction for Japan's Pacific war effort resided with the Imperial General Headquarters (IGHQ). If army-navy rivalry occasionally complicated American prosecution of the war in the Pacific, it was nothing compared to the rivalry and lack of cooperation that plagued the Japanese armed forces at virtually every level. Separate army and navy sections made up the Imperial Headquarters. The IGHQ met twice a week with the chiefs of staff of both services presiding. It was not a joint command and possessed none of the staff or joint mechanisms of the U.S. Joint Chiefs of Staff. Instead, as one historian of the war observed, "it was a façade to cover two separate organizations with strong competing interests and rivalries."[48] Both services developed their plans and orders separately, issuing them through their own chains of command. (See figure 13.)

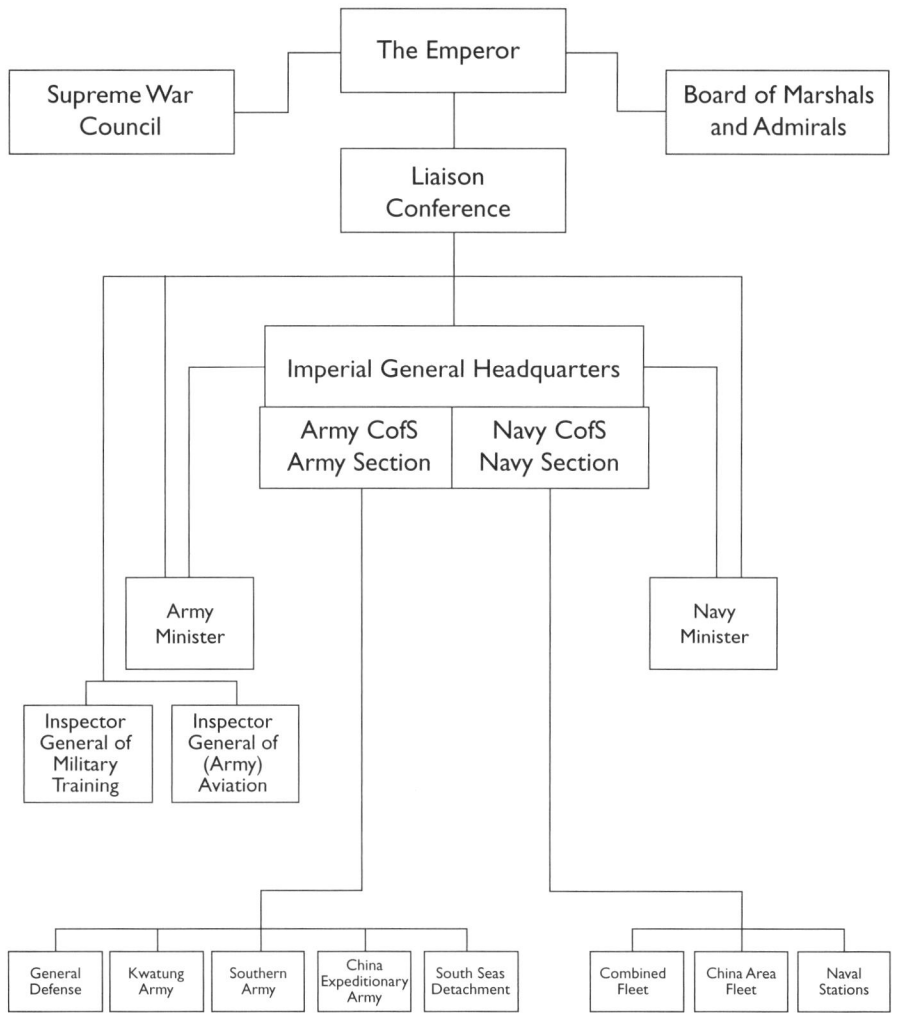

Figure 13. Organization of the Japanese High Command. Source: Louis Morton, *Strategy and Command: The First Two Years,* **238.**

At the operational level, the Japanese did not organize their land, sea, and air forces under a single joint commander. Nor did they establish separate geographic areas under a theater commander; rather, they simply maintained separate army or fleet headquarters each commanded through its own service channels. Coordination between the army and navy was based on the principle of cooperation,

This photograph shows a flight of U.S. Army Air Force B-25 Mitchell bombers. These bombers were used extensively by Maj. Gen. George Kenney's Fifth Air Force in the South West Pacific Theater against targets on both land and sea. Land-based airpower was critically important in projecting U.S. military power in both European and Pacific theaters. The operational and strategic direction of the Pacific campaign was largely determined by the need to obtain sea and air bases to extend American operational reach. Courtesy of USAMHI.

The development and employment of the naval air arm through fleet exercises, as planned at the Naval War College during the interwar years, was essential in the success of American operational art in the Pacific during World War II. Courtesy of the USAMHI.

The chief architects of Pacific victory at sea during World War II, from left to right: Admirals Chester Nimitz, Ernest King, and Raymond Spruance. All were graduates of the U.S. Naval War College. Courtesy of the U.S. Naval War College Museum.

which did not always extend to the routine practice of each service keeping the other informed of its activities.[49] The Japanese
emphasized fighting spirit, offensive operations, and maneuver. The
Japanese military gave considerably less emphasis to supply and
intelligence.

There were no logisticians on the general staffs of either Japan's
army or its navy. Responsibility for supply was vested in the minis-
tries of war and the navy.[50] The Japanese devoted few resources to
developing a system of sophisticated intelligence gathering or analysis. Despite the lack of an effective intelligence service, it was easy to
predict many of the U.S. offensives. For example, it was obvious to
the Japanese that the carrier raids on the Philippines, Palau, and Formosa in September 1944 portended American operations aimed at
the Philippines. The question was when and where in the Philippines
the Americans would strike.

The Sho-1 plan called for the Japanese army to delay in the southern and central Philippines while preparing to conduct the decisive
battle on Luzon. A late change in plans and disagreement among senior Japanese army leaders complicated efforts for an effective defense. Field Marshal Count Hisaichi Terauchi, who commanded the
Southern Army, had the Fourteenth Area Army and the Fourth Air
Army to defend the Philippines. The Japanese navy also stationed the
First Air Fleet in the Philippines, but this unit reported directly to the
commander of the combined fleet in Tokyo. The Imperial General
Headquarters and Field Marshal Terauchi decided to place their hopes
on airpower and abandoned the plan to make the main effort on Luzon.
They believed that land-based airpower could win at least temporary
air superiority from American carrier-based aircraft, enough to allow the Japanese to reinforce garrisons on the other islands. Lt. Gen.
Shigenori Kuroda, commander of the Fourteenth Area Army, had little faith that Japanese airpower was up to the task of defeating the
invasion and stressed that the decisive battle would be fought on
land.[51] Less than two weeks before the invasion, the Japanese High
Command replaced Kuroda with Lt. Gen. Tomoyuki Yamashita.

Yamashita was convinced that a decisive battle on Leyte or any
island other than Luzon would waste Japanese resources and fail.
Terauchi overruled Yamashita and insisted that every effort be made
to reinforce Leyte.[52] On October 19, Terauchi issued this order:
"The Southern Army, assembling all its fighting power, will seek

decisive battle with the main strength of the enemy forces landing in the Philippines."[53] Dutifully, Yamashita ordered his subordinate responsible for the southern and central Philippines, Lt. Gen. Sosaku Suzuki, commander of the Thirty-fifth Army (equivalent to a U.S. Army corps), to make the maximum effort to defend Leyte. The Southern Army committed ten divisions and five brigades, almost 180,000 men, to the defense of the Philippines. An additional division and one brigade in China and Formosa stood by in reserve. At the time of the American landing, 16,000 men of the Japanese Sixteenth Division occupied Leyte. Throughout the battle, Yamashita shuttled as many reinforcements as possible to Leyte, while the Southern Army labored to provide air and ground reinforcements to the Philippines.

On October 17, troops from the U.S. Sixth Ranger Battalion landed on Sulaun Island and Dinagat to secure the approaches to Leyte Gulf. Two days later, a vast armada of seven hundred ships carrying 200,000 men of Krueger's Sixth Army arrived off the gulf. On the morning of October 20, after a heavy naval bombardment, the Sixth Army went ashore with two corps abreast. The X Corps and XXIV Corps made rapid progress against light resistance. By midnight, more than 132,000 men and 200,000 tons of supplies and equipment were put ashore. Two days earlier, as soon as the Japanese became aware of the preliminary operations by the Rangers, the Imperial Navy issued its order for Sho-1, setting in motion the largest naval battle of the war.

Japan's combined fleet moved from widely scattered bases to the long-awaited decisive battle with the U.S. Pacific Fleet. The Japanese planned to lure away Halsey's powerful carrier groups with Vice Adm. Jisaburo Ozawa's Northern Force, which consisted of four aircraft carriers, two battleships, three cruisers, and eight destroyers. With Halsey's covering force gone from the immediate area, three other Japanese naval strike forces converged on Leyte to destroy MacArthur's landing and supporting forces. Vice Adm. Takeo Kurita's Center Force with five battleships, twelve cruisers, and fifteen destroyers approached from Malaya and Borneo toward the San Bernardino Strait between Samar and Luzon. The Southern Force, commanded by Vice Adm. Shoji Nishimura, consisting of two battleships, one heavy cruiser, and four destroyers, was supported by yet another naval force of two heavy and one light cruiser with four destroyers commanded by Vice Adm. Kiyohide Shima. The Southern Force was to pass through the Surigao Strait between Leyte and Mindanao.

The Battle of Leyte Gulf, the greatest naval battle the world has ever seen, exposed one of the chief problems with American command and control. MacArthur did not command Halsey's Third Fleet with its fast carriers and battleships. Halsey worked for Nimitz. Under a compromised solution worked out with Nimitz, MacArthur provided Halsey with operational direction. In Nimitz's instructions to Halsey to support MacArthur, he indicated that if the Third Fleet got the opportunity to destroy the Japanese fleet, Halsey should take it. Indeed, in MacArthur's Operational Instructions No. 70, he also tasked Halsey "with containing or destroying the Japanese fleet."[54] Following his own aggressive instincts and interpretation of his orders, Halsey took the bait and chased the Japanese carriers far to the north. Halsey thus uncovered the invasion fleet in pursuit of the classic Mahanian decisive battle. Clearly, he believed that sea control could only be achieved once the enemy's main fleet was defeated. The problem was that Ozawa's depleted carrier force was not the main naval element on which the enemy depended for success.

Three months before the Battle of Leyte Gulf, Admiral Spruance faced the same decision. His mission was to cover Adm. Richmond K. Turner's Joint Expeditionary Force headed for the invasion of Saipan. The Japanese Combined Fleet sortied looking for the decisive battle with the U.S. Fleet. In the ensuing Battle of the Philippine Sea, Spruance's carrier aircraft savaged the Japanese fleet and naval aircraft in what became known as the "Marianas Turkey Shoot." After a brief pursuit, however, Spruance let the crippled Japanese fleet get away while he returned to support the invasion of Saipan. For most of the naval war in the Pacific, sea control for the U.S. Navy became less a search for the decisive naval encounter envisioned by Mahan than a maritime strategy of attrition in which seapower, airpower, and landpower were mutually leveraged to advance on Japan. Unlike Spruance, Halsey leaned heavily on his more aggressive instincts. It was a significant risk that ended up conforming to the Japanese plan.

Only good luck and hard fighting saved MacArthur's forces crammed into Leyte Gulf. Halsey initially detected Kurita's Center Force and struck hard on October 23 and 24, sinking the superbattleship *Musashi*. Kurita turned away, and Halsey decided to pursue the Northern Force. Kurita turned around, however, and resumed course toward Leyte. MacArthur's naval component commander, Vice Admiral Kinkaid, moved his naval fire-support group of six old battle-

ships to intercept. The Japanese Southern Force ran a gauntlet of torpedo boats and destroyers in the Surigao Strait only to be destroyed
by Kinkaid's Seventh Fleet battleships. Kurita's Center Force showed
up in the vicinity of Leyte Gulf on the morning of October 25. The
escort carriers of the Seventh Fleet put up such a valiant fight that
Kurita believed he was under attack by Halsey's powerful carrier
force. Kurita withdrew leaving the Japanese army and land-based air
forces in the Philippines to their fate. The same day, Halsey's carriers
found and crushed the Japanese Northern Force, sinking all four enemy carriers. This naval battle eliminated Japanese seapower as a
factor not only in the campaign for the Philippines but for the rest of
the war. MacArthur now had only to deal with Japanese landpower
and airpower to win the campaign.

The air battle for the Philippines began well before Krueger's
troops waded ashore. The operational strikes delivered by Halsey's
fast carriers long before the invasion not only alerted the Japanese;
they also triggered the inevitable process of attrition that would determine the battle for air superiority. The Japanese made a maximum
effort in the air as well as on the sea and ground. The Fourth Air Army
and the Imperial Navy's land-based air force mustered nearly 400
aircraft of all types initially to oppose the landings.[55] On October 24,
they began a coordinated and sustained effort, sending 150 to 200 aircraft to attack MacArthur's forces. With U.S. naval aircraft involved
in sea fights, Krueger desperately tried to establish airfields ashore to
bring forward fighters from Kenney's Fifth Air Force. The Japanese
mounted more than 250 sorties on October 25–26, but by the next
day, U.S. fighters started operating out of the airstrip at Tacloban.
Soon the Japanese began to limit their air activity to piecemeal raids
at dusk and dawn.

The Japanese rushed air reinforcements to the Philippines, but
Kenney's Fifth Fighter Command rose to the challenge. Beginning on
October 27 and for the next five weeks, American pilots shot down
314 enemy aircraft while losing only sixteen of their own aircraft
in aerial combat.[56] Beset by declining air strength and pilot quality,
the Japanese introduced the kamikaze, suicide attacks by volunteer
naval pilots. The small number of kamikaze strikes, though effective,
could not be decisive. This method of attack, however, suggested the
still lethal potential of Japanese airpower. By January 1945, the Americans had won air superiority over the southern Philippines. Vice

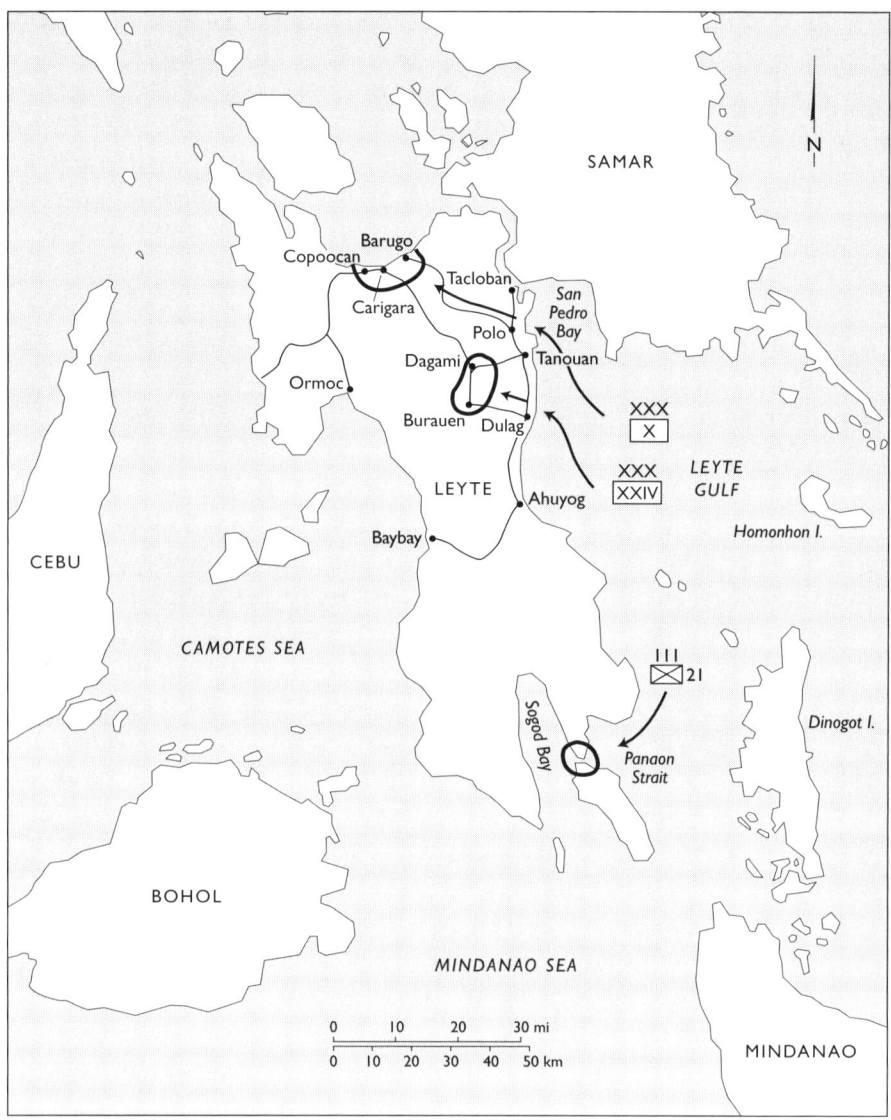

Map 7. Sixth Army objectives in Leyte operation. Source: Chamberlin Papers, Box 6, USAMHI.

Adm. Shigeru Fukudome, commander of the Japanese land-based naval air force, admitted "by the early part of January, I had lost practically all of my planes, my air force had been practically wiped out."[57] The delay in winning the air battle, however, was costly in terms of the land battle. While American airpower was weak in the early days of the operation, the Japanese reinforced their garrison in keeping with their plan to fight the decisive battle on Leyte. Between October 23 and December 11, the Japanese moved 45,000 men and 10,000 tons of supplies to Leyte aboard destroyers, barges, and small ships of all sizes.[58]

Krueger's Sixth Army drove north through the Leyte Valley to the Carigara Bay area. (See map 7.) The Japanese Sixteenth Division fought delaying actions while the Thirty-fifth Army absorbed reinforcements arriving through Ormoc on the west coast. Tropical storms, rainfall, and the subsequent mud slowed American operations, particularly efforts to build airstrips. To get at Ormoc, Krueger pushed one corps over the tough central mountain ridge along with a division up from the south on the west coast of Leyte. American forces used amphibious as well as overland routes to break into the Ormoc Valley on the Japanese right flank.

To make an amphibious landing in the Ormoc area, MacArthur borrowed another army division from Admiral Nimitz's Central Pacific resources. In what amounted to a double envelopment of the Japanese Thirty-fifth Army, Krueger sent another corps along the east coast to the north also by amphibious and overland routes to secure Limon on the Japanese left flank. By Christmas Day, after eight weeks of tough fighting, Japanese resistance collapsed. Yamashita's effort to defeat the U.S. effort to take Leyte failed. His efforts now shifted to the defense of Luzon. Although the Japanese lost Leyte, they made the Americans pay a high price. The Sixth Army suffered upwards of 15,500 casualties. Although Japanese losses are difficult to determine, the Sixth Army counted 56,263 killed and 389 captured.[59]

With Leyte in hand, MacArthur moved on to Operation Mike I, the invasion of Luzon. Without control of the sea, it was difficult for the Japanese to provide any significant ground reinforcements or supplies for the defense of Luzon, but they could still fly aerial reinforcements from the home islands and Formosa. Although much weakened, Japanese airpower remained a serious threat for the invasion fleet. MacArthur coordinated a good deal of the available airpower in the Pacific to provide operational fires for the Luzon invasion. Halsey's carrier force and the Twentieth Air Force with B-29s based in the Marianas and China reached out to Formosa, the Ryukyus, and the South China coast to destroy enemy air and naval forces. Kenney's Fifth Air Force needed to provide close air support for the invasion, but a combination of factors inhibited Kenney's ability to accomplish his mission. Airfield construction on Leyte continued to be hampered by bad weather, poor soil, and enemy activity. It became apparent that Leyte would never become the air base as originally planned.

As early as September, a staff study anticipated the potential need for a contingency plan to take Mindoro, the island just south of Luzon, as an advance air base to support future operations. In November, MacArthur decided Mindoro must be taken to extend Allied air cover. Sixth Army troops landed on Mindoro on December 15, and within five days, army engineers had the first airfield in operation. On January 9, Krueger's Sixth Army, under cover of Kenney's Fifth Air Force and the guns of Kinkaid's Seventh Fleet, came ashore at Lingayen Gulf on Luzon.

MacArthur directed Krueger's forces to land in the gulf as the most direct route to the Central Plains and Manila. Manila was the political prize closest to MacArthur's heart, but militarily Clark Air Field was the key to extending Allied airpower in the Pacific. The Japanese commander, Lieutenant General Yamashita, fully anticipated the American landing at Lingayen Gulf, but there was little he could do about it. Facing U.S. naval and air superiority, Yamashita fashioned a defense plan that maximized his ground force's ability to delay and attrite the American army. Yamashita divided his 275,000 troops between three mountain strongholds. The Shobu Group with 152,000 troops under Yamashita's direct command defended the mountainous region east and northeast of Lingayen Gulf. The Kembu Group with 30,000 troops defended the mountainous country overlooking Clark Field. Finally, the Shimbu Group with 80,000 troops defended all of southern Luzon, but was concentrated in the mountains east and northeast of Manila.[60]

The Sixth Army landed at Lingayen Gulf with two corps abreast, four divisions simultaneously assaulting the beachhead. Krueger turned one corps south along the Central Plains toward Clark Air Field and Manila while his other corps continued to attack to the north and east protecting his flank. Since Yamashita had already decided not to oppose the Americans in the Central Plains, Krueger made rapid progress moving south toward Manila. By January 28, the XIV Corps under Maj. Gen. Oscar W. Griswold (CGSS 1925, AWC 1929) occupied the airfield. To assist Krueger's southern drive on Manila, Eichelberger's Eighth Army made two amphibious landings north and south of Manila. On January 29, the IX Corps, commanded by Maj. Gen. Charles P. Hall (CGSS 1925, AWC 1930), landed north of Manila to isolate Japanese forces and prevent any withdrawal into the Bataan Peninsula. Meeting little initial resistance, the IX Corps continued south and east to assist in securing Manila Bay. Using amphibious and airborne operations, the Eighth Army also sent the Eleventh Airborne Division to strike south of Manila on January 31. Closing in from the north and the south, Krueger positioned his forces to retake Manila.

MacArthur's intelligence chief, Maj. Gen. Charles Willoughby, predicted that the Japanese would not fight for Manila. Krueger's Sixth Army intelligence disagreed. In fact, Yamashita did not want to fight for Manila, but problems with Japanese command and control led to a brutal fight and destruction of the capital. Rear Adm. Mitsuji Iwabuchi commanded the Manila Naval Defense Force numbering some 16,000 naval troops. Prior to leaving Manila, Iwabuchi's senior commander transferred operational command of the force to the Shimbu Group under Yamashita's original defense scheme. The Shimbu Group commander wanted the naval troops to expedite the evacuation of Manila and execute limited demolitions to delay the Americans. In staff conferences between the two forces, the naval officers announced their determination to defend Manila to the last, contrary to Yamashita's orders.[61] As a result, Krueger's forces had to fight their way into Manila block by blood-splattered block. MacArthur would not permit bombing, but the tank and artillery fire proved quite as destructive. It took three divisions and four weeks to secure Manila, at a cost of 6,500 U.S. casualties and virtually all of the 16,000 Japanese killed.[62]

From March to August the Sixth Army turned to reducing the remaining Japanese mountain strongholds on Luzon. Beginning in February, MacArthur launched the Victor series of operations to liberate the rest of the Philippines. The Eighth Army moved to clear the central Philippines and from April to July conducted operations to eliminate the Japanese garrison on Mindanao.[63] In every phase of the campaign, Filipino guerrillas and reconstituted army units provided invaluable intelligence and additional combat power, particularly in the final reduction of isolated Japanese forces. Still, Yamashita executed his defense plan with the usual Japanese tactical fanaticism. Long after Yamashita's forces lost any operational significance, they fought on in the northeast mountains of Luzon. Yamashita finally surrendered at the end of the war with some 50,000 troops still under his command.

Assessment

The Philippine campaign was the largest and longest in the Pacific theater of war. It absorbed two U.S. armies with a total of sixteen divisions as well as the bulk of the naval and air power in the Pacific when required. The campaign achieved its strategic and operational objectives. It severed the Japanese line of communication to the vital resources necessary to continue the war. Politically, it redeemed the American pledge to defend and liberate the Filipino people. Operationally, it provided additional bases for the advance on the Japanese home islands. But most significant of all was Japan's decision to offer decisive battle to retain the Philippines. The determination of the Japanese to mass air, land, and sea forces to wage a decisive battle gave the Americans the opportunity to crush Japanese combat power not only in the Philippines, but in the wider Pacific theater as well.

The destruction of Japanese naval and air forces gave the United States tremendous operational advantages. The Japanese had few, if any, operational cards left to play. Their losses in this campaign reduced them to only tactical expedients in delaying the Allied advance on the home islands. Tenacious defense on land and increasingly fanatical and desperate attempts in making their remaining airpower more effective constituted their remaining options. The defense of the Philippines had taken 380,000 Japanese troops out of the war and brought about the decisive defeat of the Imperial Japanese Navy. The American Sixth and Eighth armies suffered 47,000 battle casualties.[64]

When interviewed after the war, senior Japanese officers involved in the fight for the Philippines pointed out Japanese problems in intelligence, logistics, and command and control. Col. Takio Shindo, Yamashita's intelligence officer, remarked: "In my opinion, American intelligence was so far superior that a comparison is useless. It seemed to me that we were fighting our battles blindfolded, while the enemy seemed to have ten times the intelligence we possessed."[65] For one example, the U.S. move to bypass Mindanao and strike directly at Leyte took the Japanese by surprise. The ability of the Americans to conduct multiple amphibious operations at various points throughout the campaign constantly caught the enemy off guard. The ability of U.S. forces to integrate and use Filipino sources of information proved useful at both tactical and operational levels.[66]

U.S. intelligence certainly made its share of mistakes. General Willoughby, MacArthur's intelligence officer, routinely underestimated Japanese strength, particularly on Luzon. Krueger's intelligence section, however, usually painted a more accurate picture of enemy strength and intentions.[67] From the beginning of the war, American efforts in cryptographic intelligence decoding Japanese message traffic paid tremendous dividends at both the strategic and operational levels.[68] Unlike the Japanese, the Americans devoted considerable resources and effort to focus on integrating intelligence into operations at all levels.

Similarly, Americans' expertise in logistics organization and management far surpassed the opponent. Without secure lines of communication the Japanese faced tremendous difficulties. Control of the seas and air superiority allowed the Americans to isolate Japanese garrisons from sources of supply and reinforcement. Still, the Japanese logistics system was inefficient and poorly organized. There was no operational logistics organization; each service attempted to manage supply. There was no single logistical authority in Manila. Yamashita did not control his own supply organizations until after January 1945.[69] Remembering the lack of organization in Manila, his chief of staff recalled that supplies and equipment "were piled in an unsystematic helter-skelter way."[70] The Americans constantly attacked road and rail networks to disrupt enemy transportation. Regardless, the Japanese division did not even possess sufficient transportation to help sustain its own supply needs.[71]

The Americans also faced significant challenges in sustaining large expeditionary forces over lines of communication stretching all the way back to the United States. Unlike the Japanese, the Americans devoted considerable effort and attention to logistics planning and organization. There was no provision for joint logistics in the Southwest Pacific theater. Supply chains ran through service channels, but MacArthur's headquarters established priorities, while supervising the planning and control of logistics operations. The United States Army Services of Supply (USASOS) under Maj. Gen. James L. Frink (CGSS 1926, AWC 1929) handled logistics for the army and air forces throughout SWPA. Before beginning the campaign, MacArthur created an Army Service Command (ASCOM) and put his chief engineer, Maj. Gen. Hugh J. Casey, in charge. This organization provided logistic services and handled base construction for the Sixth Army until the theater service of supply could take over. It tasked a single commander with responsibility for arguably the most important mission during the campaign: the construction of air and naval bases.

Perhaps the greatest American operational advantage lay in the combining of air, sea, and land power. After the war, the chief of staff of the Japanese Thirty-fifth Army regretted the lack of joint command in the Philippines: "A great many of [Japan's] strategic plans failed in the battle of Leyte due to mix up in the commands of the air and land forces. A joint command would, undoubtedly, have integrated the units into a closer coordinating force, which would have helped the situation tremendously."[72] The lack of a single unified theater command did put the U.S. campaign at risk in the naval battle in Leyte Gulf, but the level of cooperation between Nimitz, Halsey, and MacArthur was still remarkable. Nimitz willingly offered MacArthur additional army and even marine reinforcements. Nimitz loaned MacArthur an entire marine air wing that contributed 180 aircraft and almost one-third of the Fifth Air Force sorties on Luzon. This was the first time marine aircraft provided close air support to army ground forces.[73] General Kenney functioned as a joint air component commander.

Even more impressive, MacArthur's component commanders—Kenney, Kinkaid, and Krueger—all practiced a degree of jointness noticeably lacking in the opposition. MacArthur's headquarters coordinated the planning for Leyte in a series of conferences from July 20 to August 6. For the invasion of Luzon, MacArthur specifically charged Krueger with coordinating ground, air, and naval plans. Krueger created a planning group headed by his chief of staff at Hollandia (present-day Jayapura). Staff conferences between all the services on Hollandia and Leyte hammered out the final details and briefed MacArthur in November. Likewise, in the conduct of operations, the Americans demonstrated a high degree of flexibility in supporting operations of each of the services.

Historians have rightly pointed to American mistakes in operational maneuver. Some have questioned MacArthur's failure to make an early assault on the town of Ormoc, the point of entry for Japanese reinforcements on Leyte, and indeed, his insistence on liberating the rest of the Philippines after securing sufficient bases on

Luzon.[74] The Japanese also committed significant operational mistakes in ground, air, and naval operations. Yamashita, the Japanese ground commander, conducted a tenacious and capable tactical defense on Luzon, but operational failures in logistics, intelligence, and most critically in command and control handicapped the Japanese. Senior Japanese officers were less impressed with American tactical prowess, but one Japanese general provided the best explanation for American victory in this observation: "As compared to our relatively small-scale operations, the enemy's big-scale operations enabled them to take our positions with comparative ease. Their minor operations were just as well planned in detail as their major ones, all operations utilizing the vast coordinated striking power of their combined land, air, and naval forces."[75] In his return to the Philippines, MacArthur proved that American operational art could project, conduct, and sustain joint operations sufficiently to beat a determined enemy. The Philippine campaign was an impressive military achievement, but the best expression of modern American operational art can be found in the last contest, the battle for Okinawa.

OPERATION ICEBERG

American operational art reached its peak by 1945. In Europe, the Allies developed combined staffs that proved capable of conducting and sustaining large-scale operations. In the Pacific, the American military fully developed the promise of the interwar studies on joint expeditionary warfare. By the end of the war, American operational art excelled in joint planning, organization, and logistics.

Theater strategic planning in the Pacific in 1944 focused on breaking into the China-Formosa-Luzon area. The Joint Staff plan to defeat Japan following the proposed seizure of Formosa called for "concurrent advances through the Ryukyus, Bonins, and Southeast China coast for the purpose of intensifying the blockade and air bombardment of Japan."[76] These sequenced major operations were to lead to an invasion of the home islands to capture the industrial heart of Japan located on the Tokyo plain. After the Joint Chiefs made the decision to invade Luzon rather than Formosa, the Joint Planning Staff remained committed to the move into the Ryukyu Islands as the logical next step in maintaining unremitting pressure on Japan. In the same directive ordering MacArthur to take Luzon, the Joint Chiefs directed Nimitz to occupy one or more positions in the Ryukyu Islands by March 1, 1945.[77]

The Ryukyu chain consists of a number of islands between the Japanese home island of Kyushu and Taiwan, then the Japanese colony of Formosa off the China coast. American occupation of key positions in this island chain would place American airpower and seapower approximately 400 miles of Kyushu and provide excellent staging bases for the invasion of the home islands. Okinawa, the largest island in the Ryukyus, was the obvious key objective. In November 1944, the Joint War Plans Committee in Washington, D.C., developed a plan for this major operation. The JWPC plan called for as many as five phases to seize six islands in the Ryukyu chain. The Joint Staff Planners believed Okinawa "to be strongly fortified," possessing few good beaches and offering terrain as difficult to assault and easy to defend as any yet encountered in the war against Japan.[78] The planners wanted to take two local islands prior to assaulting Okinawa to clear the approaches and provide land-based air cover for the invasion. This plan called for a two-division assault on the west coast of Okinawa and a subsequent one-division assault at Yonabaru on the southeast coast.[79] The JCS referred the plan to Admiral Nimitz for review.

In January 1945, Nimitz's joint staff submitted their own outline plan, code-named Iceberg, to the JPS in Washington. This plan called for three phases: (1) seize the southern part of Okinawa to establish air bases and develop the port of Naha for the Navy; (2) secure the remainder of Okinawa and Ie Shima (now Iejima), a small island to the north; and (3) exploit the initial lodgments by securing additional islands as necessary. Unlike the JPS plan, Nimitz's plan envisioned a single three-division assault over the Hagushi beach in the west. This assault would drive across the narrow isthmus, cutting the island in two and isolating the northern and southern ends of the island. The main effort would then push south to seize and develop the most useful part of Okinawa. The Joint Staff Planners reviewed Nimitz's plan and concluded that it was "suitable and feasible" but questioned the plans for the initial assault.[80] Once approved, the Central Pacific outline plan for Operation Iceberg became the basis for coordination and detailed planning.

While Nimitz's subordinates tackled the details, his theater staff worked out the coordination between the other Pacific theater com-

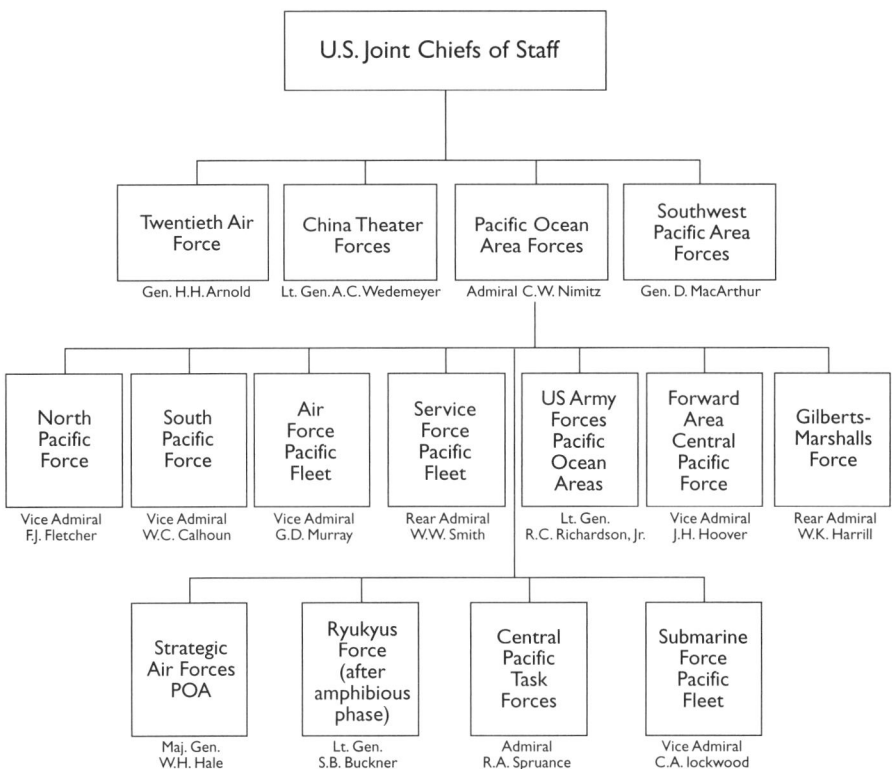

Figure 14. Theater organization for Iceberg. Source: Roy E. Appleman et al., *Okinawa: The Last Battle*, 20.

manders. The JCS directive mandating the invasion of the Ryukyus further directed the support of MacArthur's SWPA, Wedemeyer's China theater forces, and the strategic Twentieth Air Force. This coordination aimed at maximizing the airpower available to provide operational fires to shape and set the conditions for success in Iceberg.[81] MacArthur quickly agreed to order Kenney's Fifth Air Force to hit Formosa, but coordination of B-29 bomber support from the Twentieth Air Force proved more difficult. (See figure 14.)

Gen. Hap Arnold organized the Twentieth Air Force in April 1944 specifically to take advantage of the long-range capabilities of the B-29. Without a unified command in the Pacific, and in order to ensure the bombers would be used in a strategic role striking directly at the Japanese homeland, Arnold retained command even though still a member of the Joint Chiefs.[82] In establishing the Twentieth Air

Force as a strategic asset, the JCS mandated that the Pacific theater commanders might direct the employment of the bombers "only in the event of a tactical or strategic emergency."[83] Faithful to the interwar doctrine of strategic bombing, Arnold and his commander on the ground, Maj. Gen. Curtis LeMay, hated to use the B-29 against anything other than strategic targets. On March 7, 1945, representatives from Nimitz's headquarters and LeMay's XX Bomber Command met to draw up the final plan for B-29 support. The plan targeted permanent installations in the Japanese home islands. Reluctantly, Arnold approved the plan, but he made sure LeMay understood that his primary interest "was to ensure the success of Iceberg at minimum cost and casualties."[84] Essentially, Nimitz gained control of the strategic bombers for five weeks. Nimitz did not have quite the struggle Eisenhower had in Europe in gaining control of strategic airpower to support Operation Overlord. The fact suggests that Hap Arnold enjoyed greater flexibility than his British counterparts.

Joint Organization

Nimitz and his subordinate commanders made more use of joint command and control in the organization they devised to execute Iceberg than was the case in any previous American operation in World War II. Beginning with his own joint staff, Nimitz as theater commander of the expeditionary troops, made sure that such cooperation remained key to the operation. Nimitz formed three joint task forces to execute Operation Iceberg. Adm. Raymond A. Spruance (NWC 1927) commanded two of them, known collectively as the Central Pacific Task Forces, which included the Covering Force and Vice Adm. Richmond K. Turner's Joint Expeditionary Force. The Covering Force, personally commanded by Admiral Spruance, included an American and a British fast carrier force; Turner's command included the actual assault forces with the amphibious shipping and local covering force. The third joint force was formed around the Tenth Army under Lt. Gen. Simon Bolivar Buckner, Jr. (CGSS 1925, AWC 1929). The Tenth Army served as headquarters for the expeditionary troops under the direction of Vice Admiral Turner.

The Tenth Army's mission to seize Okinawa resulted from the decision to bypass Formosa, which the Tenth Army had originally been formed to secure. Its commander, General Buckner, lacked combat experience and had been prepared for this important assignment only through his military education. The son of a Confederate general, Buckner saw no combat in World War I and spent most of the interwar years as either a student or an instructor at both the GCSS and the Army War College. He spent the first three years of World War II as the army commander in Alaska. Despite his lack of combat experience, Buckner rose steadily through the ranks and received orders in June 1944 to report to Hawaii to organize and take charge of the Tenth Army. Having at last gotten his chance to command an army in combat, Buckner took the unusual but completely professional step of recommending against the Formosa operation on September 26, 1944, after reviewing the troop requirements.

The resulting CINCPOA study of other projected operations determined that it was beyond the ability of the Marine Corps alone to take Okinawa; it required an army-size force to sustain and conduct such a large-scale operation. The new mission now fell to Buckner's Tenth Army. From the beginning, however, it was clear that the Marine Corps would play a significant role in capturing the island. The requirement to coordinate and plan army and marine tactical operations supported by the navy drove Buckner to organize a joint staff at the army level.

Buckner's entire Tenth Army staff, therefore, was joint. Although an army officer headed each staff section, senior marine and navy officers were present in every one. The G-3 operations section included six army, seven marine, and five navy officers. The G-4 logistics section included seven Army, three Marine and five Navy officers.[85] In addition to organizing a joint staff, Buckner commanded his own ground, sea, and air components. (See figure 15.) The Army XXIV Corps commanded by Maj. Gen. J. R. Hodge (CGSS 1934, AWC 1935) and the Third Amphibious Corps (U.S. Marines) commanded by Maj. Gen. Roy S. Geiger (CGSS 1925, AWC 1929) made up Buckner's ground component. Buckner had known General Geiger since they had been students together at the Army General Staff School. Buckner insisted that Geiger, a marine, assume command of the Tenth Army in the event he became a casualty.[86]

Marine Maj. Gen. Francis P. Mulcahy (CGSS 1929, AWC 1936) commanded the Tenth Army's tactical air force. Mulcahy functioned as a joint air component commander. His tactical air force grew to include nine marine fighter squadrons, ten army fighter squadrons, and sixteen army bomber squadrons.[87] General Mulcahy organized

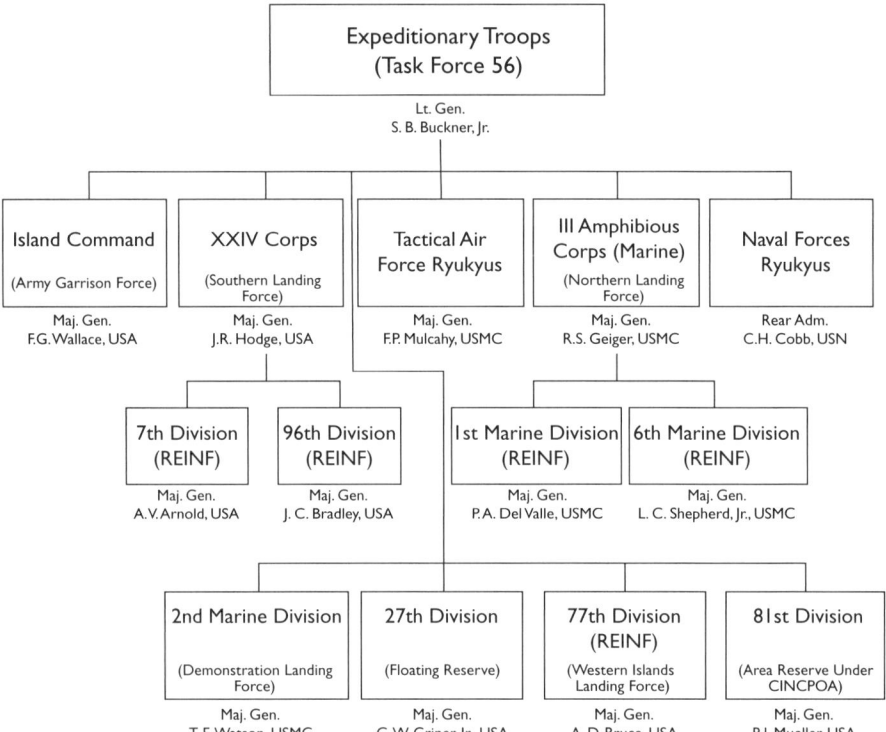

Figure 15. Organization of the Tenth Army. Source: Roy E. Appleman et al.,
***Okinawa: The Last Battle*, 24.**

his tactical air force into an air defense command, an antisubmarine unit, a photographic unit, a bomber command, and an air support control unit. He centralized control of the air assets "in order to maintain the inherent flexibility of airpower and to permit the employment of its whole weight against selected areas."[88] With air, naval, and ground components in theater, Buckner was able to organize and staff the Tenth Army like a modern joint task force.

The Tenth Army's primary mission was to secure Okinawa in order to build bases. In a fashion similar to MacArthur's use of a separate organization to sustain and develop bases, Buckner created the Island Command (ISCOM) under U.S. Army Maj. Gen. Fred C. Wallace (CGSS 1931, AWC 1936). Unlike MacArthur's logistics command, the Tenth Army's Island Command took joint organization to an unprecedented level.

Joint Logistics

Joint logistics reached their highest level of development in World War II in the Central Pacific. On March 7, 1943, the army and navy agreed on supply distribution in order to avoid the duplication of effort. As a result, the services published the Joint Logistics Plan. This plan entrusted joint commanders with full responsibility for all logistical services and directed that a single service would provide supplies common to both services whenever possible.[89] Nimitz announced a basic charter for joint logistics in the Pacific in September 1943 in the form of a basic supply policy for advance bases.[90] Shared responsibilities in the Central Pacific meant that the army provided subsistence and ammunition, while the navy provided fuel, regardless of service needs. Each service supplied and maintained service-specific equipment and requisitioned such items through their own service channels. The joint logistics staffs at the various echelons of command supervised the entire effort, reviewing requirements, assigning priorities, and developing plans and policies for base development.[91] By the time of Operation Iceberg, the Central Pacific logistics organization was expert in sustaining expeditionary warfare. The joint organization and operation of the Tenth Army's Island Command represented the most modern logistics operation of the war.

General Wallace's Island Command constituted a thoroughly joint organization. Its staff supervised a diverse and massive multi-service organization that included combat, engineer, antiaircraft artillery, military government, communications, and supply units. (See figure 16.) Eventually, Island Command grew to more than 154,000 troops. It provided administrative and logistic support to all elements of the Tenth Army regardless of service. In addition, Buckner charged it with constructing bases, establishing military government, and defending the island. To build bases, Island Command eventually controlled thirty-eight Army and naval construction battalions. At times, Island Command took operational control of the U.S. Twenty-seventh Infantry Division and marine amphibious reconnaissance units to eliminate Japanese resistance in rear areas.[92]

By the end of Operation Iceberg, Island Command had built eighteen airstrips, reconstructed 164 miles of road, supplied well over 183,000 troops from all services throughout the battle, and provided military government to nearly 200,000 Okinawans.[93] This

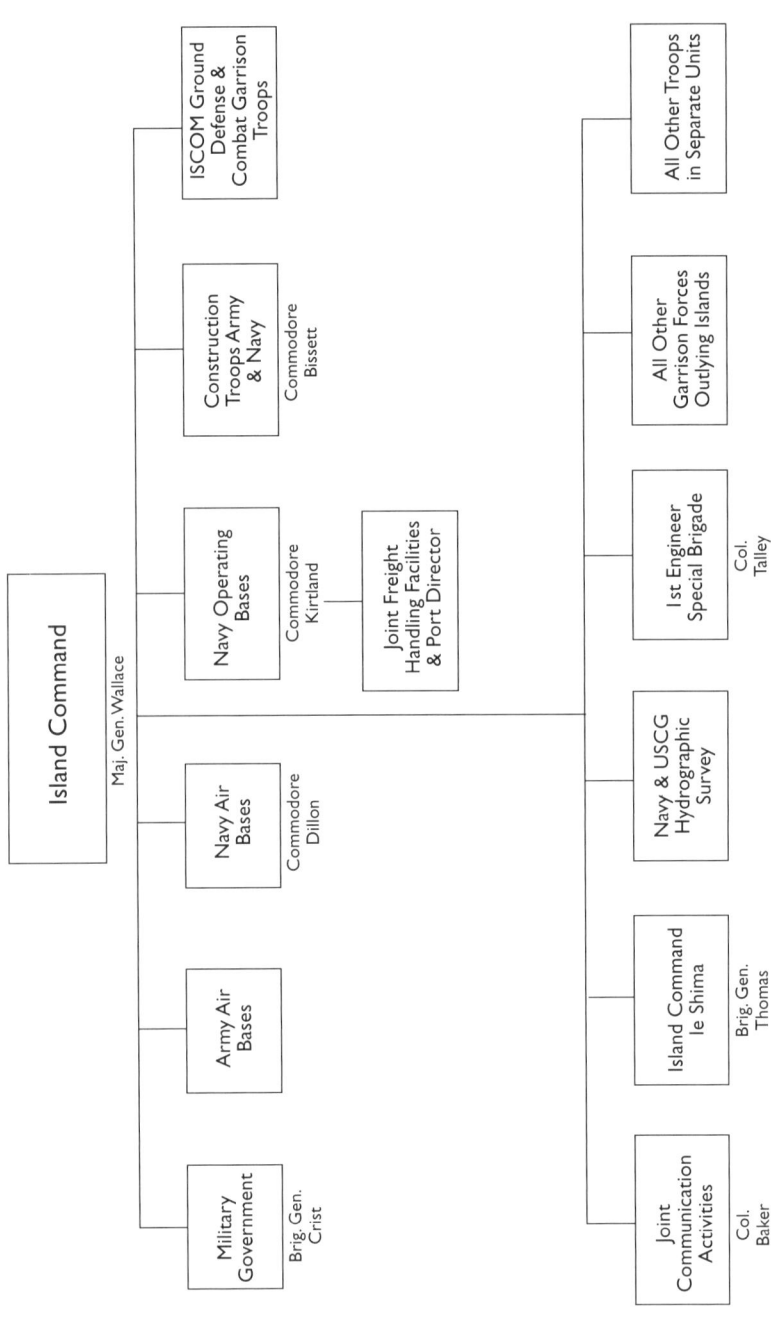

Figure 16. Organization of Island Command. Source: "Action Report Island Command Okinawa 13 December 1944–30 June 1945," USAMHI.

significant logistics achievement was possible not just because of the resources available but because of the time, attention, and expertise afforded to logistics by the U.S. military. Expertise in sustaining expeditionary warfare made success possible, but it did not guarantee victory. Victory on Okinawa came only through the combination of air, sea, and land power that by 1945 was a hallmark of American operational art.

The Battle for Okinawa

MacArthur's extended operations to take Luzon delayed Operation Iceberg by a month from its planned March 1, 1945, target date. Once additional shipping became available, Nimitz began collecting troops and resources from all over the Pacific. Now it was MacArthur's turn to provide troops and aircraft to support the Central Pacific offensive. The coordinated air strikes to neutralize Japanese airpower began in February and intensified in March. MacArthur's Fifth Air Force struck Formosa almost daily in the latter month. The Twentieth Air Force hit airfields and aircraft plants in the Japanese home islands. Nimitz's carrier force raided the Japanese home islands and Okinawa.

Buckner's plan for the invasion closely followed the outline plan Nimitz had submitted to the Joint Planning Staff. In the first phase, Tenth Army troops would seize several of the Kerama Islands fifteen miles west of Okinawa. These small islands would provide anchorages allowing navy ships to resupply and also artillery positions to support the invasion. Buckner's scheme of maneuver for the main invasion called for a two-corps assault, four divisions abreast, on the west coast of Okinawa. As in the Central Pacific plan, the assaulting troops would effectively cut the island in two, then sweep north and south to secure the objective. In the second phase, the Tenth Army would take Ie Shima, a large island to the north of Okinawa, and then exploit success by seizing and developing additional bases.

Initially, everything went according to plan—even better than expected. Beginning on March 26, the Kerama Islands fell to U.S. forces with little resistance. Meanwhile, the naval covering force pounded Okinawa for seven days. On Easter Sunday, April 1, 1945, the Tenth Army stormed ashore at Hagushi beach on the western coast. Simultaneously, a U.S. Marine division feinted a landing against Okinawa's southeast coast to pin down enemy reserves.[94] Surprisingly, there was

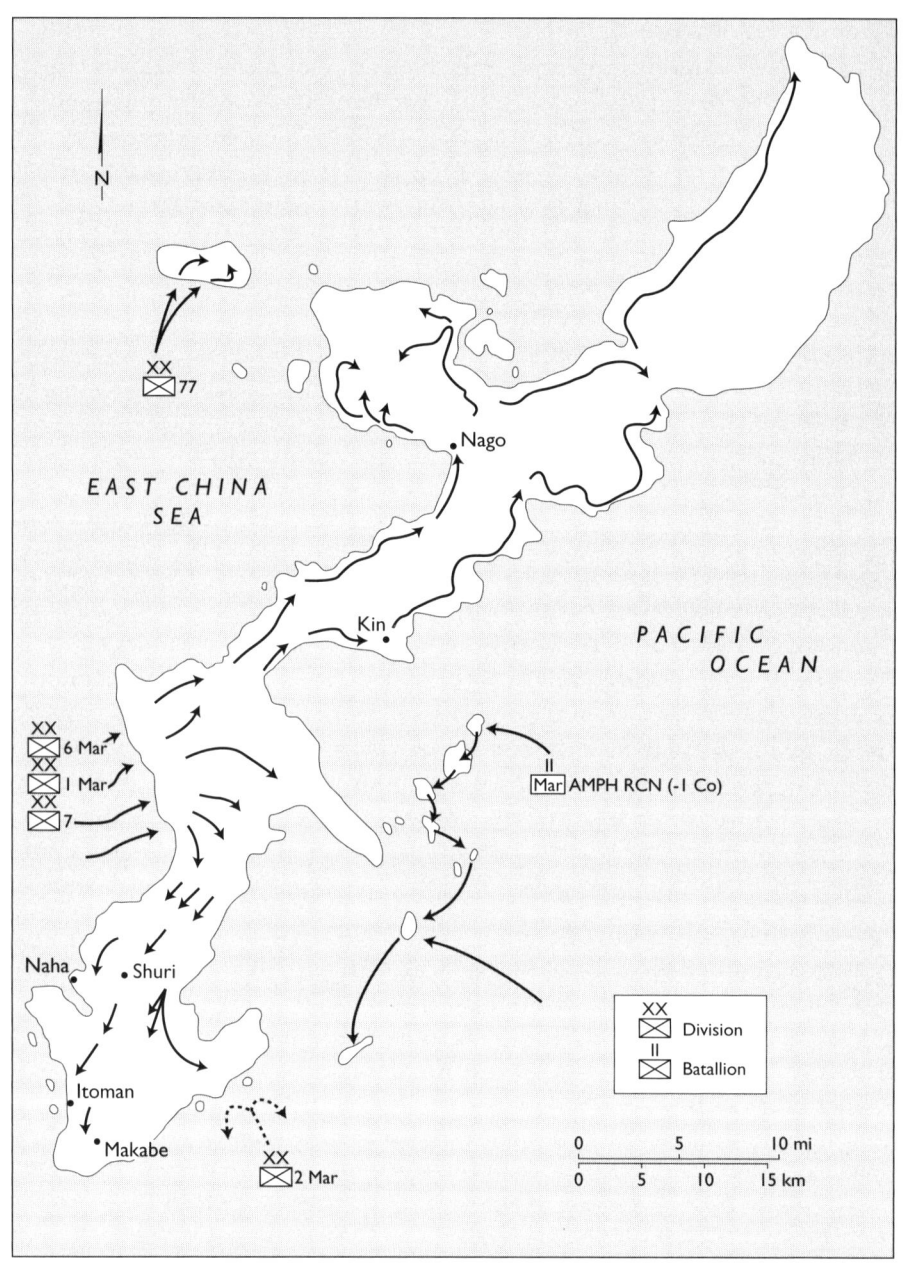

Map 8. Operation Iceberg. Adapted from Roy E. Appleman et al., *Okinawa:*
The Last Battle, 30.

little resistance. Within two days, American troops reached the east coast, cutting the island in two. The Third Amphibious Corps (Marine) turned north and the Army XXIV Corps turned south. The question on everybody's mind was: Where are the Japanese? (See map 8.)

There were plenty of Japanese on the island ready to fight, but not on the beaches. Lt. Gen. Mitsuru Ushijima, commander of the Thirty-second Army, decided instead to defend only in the southern part of Okinawa, thus denying the Americans the most valuable part of the island. Recognizing that he could only delay and attrite American forces, Ushijima concentrated his forces in the south occupying an extensive area of fortifications. For the defense of Okinawa, Ushijima had two divisions and a mixed independent brigade available. A naval base force converted to infantry and Okinawan conscripts reinforced Ushijima's army to a total troop strength of upwards of 100,000. The Thirty-second Army was well supplied with the best-trained artillery units in the greatest concentration that the Americans would face in the Pacific.[95] U.S. intelligence at all levels anticipated the main Japanese effort would be in the south, but enemy strength was underestimated at 66,000 troops. Nor did the Americans expect Japanese efforts to mobilize the local population.[96] From the time Buckner's troops landed, he suspected the enemy intended to make his stand in the southern part of the island.[97] When the XXIV Corps turned south, his suspicions were confirmed within a week.

Early progress resulted in a merging of the first and second phases of Iceberg. The marines cleared the northern area of the island by April 18, and Buckner committed the Seventy-seventh Infantry Division to take Ie Shima. The XXIV Corps pushing south ran into the first series of defensive lines that the Japanese manned in depth. Fighting its way through the outposts, the corps attack stalled by April 9 in the Japanese defensive zone in the Shuri area. The Tenth Army after-action report described the Shuri defenses thus: "the rough, jumbled ridgelines were defended from trenches, and from a vast assortment of caves used as pillboxes, elaborate multi-storied underground fortifications and gun emplacements, some of them concrete, gouged out of the ridges and hills connected by tunnels that usually opened on the reverse slopes."[98] This intricate defense, manned by a fanatically determined enemy, portended a slow and costly American advance.

Determined to resist any further advance toward the home islands, the Imperial High Command decided to launch coordinated air and naval suicide attacks against the American forces on Okinawa. The Imperial Navy sent its largest remaining battleship, the *Yamato*, with one light cruiser and eight destroyers, to attack the U.S. invasion fleet. Warned by its submarines, the American carrier covering force sank all but four Japanese destroyers by air on April 7, long before the *Yamato* task force could reach Okinawan waters.

The last vestige of Japanese airpower in the form of kamikaze attacks proved much more dangerous. On April 6, more than 700 Japanese aircraft attacked the American invasion fleet and forces ashore. On April 12–13, the Japanese launched 185 kamikaze attacks, and two days later launched another 165 suicide planes. The Americans defended themselves with naval carrier aircraft, Buckner's tactical air force with marine and army fighter squadrons, fourteen army antiaircraft battalions, and the guns of the Fifth Fleet. This jointly orchestrated force took a heavy toll on the Japanese attackers. Still, Japanese air attacks sank 20 American ships and damaged another 157 ships by the end of April.[99] Such losses made the navy impatient for success on the ground. Nimitz flew to Okinawa, and on April 23 he met with General Buckner. After remarking on the need to speed up operations, Buckner reminded Nimitz that the land battle was army business. Nimitz remarked, "Yes, but ground though it may be, I'm losing a ship and half a day. So if this line isn't moving within five days, we'll get someone here to move it so we can all get out from under these stupid air attacks."[100] Unfortunately for Buckner's troops, the Japanese fought with the same tenacity in the caves, trenches, and fortifications on the island.

Buckner ordered Hodge's XXIV Corps to mass for a concentrated thrust on April 19. A lengthy artillery preparation provided by twenty-seven artillery battalions preceded the ground assault as the XXIV Corps pushed all three of its army divisions into the attack. After five days, Hodge's troops had failed to break through, but they did force the Japanese to withdraw to the next series of defensive positions in the Shuri Line. Buckner now reconsidered his tactical and operational options. Maj. Gen. Andrew D. Bruce (CGSS 1933, AWC 1936, NWC 1937), commanding the Seventy-seventh Division, recommended another amphibious assault on the southeastern coast of Okinawa to get behind the Japanese lines.[101] Bruce's division had made a similar move during the Philippine campaign to seize the town of Ormoc, the point of entry for Japanese reinforcements on Leyte. Buckner's staff believed that another amphibious operation over the beaches near Minatoga was logistically infeasible and recommended against the operation. Buckner agreed. That judgment left Buckner with only one other option—maneuvering the entire Tenth Army (including the Marine Third Amphibious Corps) into position for a frontal assault that would rely on massive American firepower to swamp the Japanese defenses.

Before Buckner could bring the full weight of the Tenth Army to bear, however, the Japanese counterattacked. Some aggressive officers on Ushijima's staff argued for an all-out attack to defeat the Americans before Japanese combat strength drained away. After a debate, Ushijima agreed to this gambit on May 2. The Japanese did their best to put together a combined offensive, coordinating the attack with a major strike by kamikaze planes and suicide boats against American shipping. On the ground, the Japanese Twenty-fourth Division made the main effort supported by tanks and artillery. The Americans crushed the poorly coordinated attack soon after it began on the night of May 3. At a cost of more than 5,000 troops killed, the Japanese succeeded only in disclosing the positions of their artillery. Ushijima quickly returned to the tenacious static defense that required the Americans to dig his men out cave by cave.[102]

Buckner reorganized his front lines by incorporating the Third Amphibious Corps into his southern front. On May 11, the Tenth Army launched a two-corps assault against the Shuri Line. Buckner's tactics sought to reduce casualties "by a gradual and systematic destruction of their works."[103] These "blowtorch and corkscrew" tactics relied on firepower and explosives to destroy Japanese defensive positions. The army used its new tank flamethrowers as well as satchel charges, and both aerial and artillery-delivered ordnance to annihilate the Japanese. The Tenth Army engineers even resorted to using 1,000-gallon water distributors to pump gasoline into the caves to clear them with massive explosions.[104] Finally, after two weeks of fighting in poor weather against fierce resistance, the Tenth Army penetrated both flanks of the enemy defensive zone. Ushijima withdrew to the hills on the southern tip of the island.

Offensive operations resumed on June 1 and eventually enveloped the last Japanese position within three weeks. The Japanese commander and most of his staff committed suicide. General Buckner also did not survive the battle. Japanese artillery fire killed the

U.S. commander after he came to the front to observe an attack on June 18. Buckner became one of the 49,151 Americans to fall in the battle. That figure included 12,250 killed or missing. The navy paid an unusually high price with 9,731 casualties, 4,907 of them killed or missing. Okinawa cost the U.S. Navy thirty-six ships sunk and 368 damaged. The Americans also lost 763 planes from all services. The Tenth Army estimated Japanese losses at more than 7,400 captured and 107,539 counted dead, plus an additional 23,764 dead buried in caves.[105] In addition to the destruction of the Thirty-second Army, the Japanese lost 7,800 planes and sixteen ships.[106]

With Okinawa secured, the Allies owned an ideal staging area for the invasion of Japan. The Americans intended to pack the island with planes, troops, and ships for the final assault on the empire. The bitter fight for Okinawa also convinced American leadership that any invasion of the home islands would encounter the same kind of fanatical Japanese resistance. The projected cost in American casualties justified the use of any weapon that might make an invasion unnecessary.[107] On August 6, a single B-29 from the Twentieth Air Force dropped an atomic bomb on Hiroshima. Three days later, Nagasaki suffered the same fate. On August 10, the Japanese offered to surrender. The eighty-one-day fight for Okinawa turned out to be the last battle of World War II.

Assessment

Iceberg was not an independent campaign but a major operation intended to enable the ultimate invasion of the home islands. The campaigns directed by the JCS in Washington and carefully planned in detail and executed by the theater commanders synchronized these major operations in the war against Japan. These campaigns consisting of major operations bridged tactics and strategy to achieve strategic objectives in the Pacific theater. This was operational art of a high order. No other nation on earth could project, conduct, or sustain such large-scale operations across such distances by 1945. In the operational functions of intelligence, logistics, and command and control, the United States clearly surpassed its enemies. In the matter of operational maneuver, however, historians continue to take American commanders to task for their failure to provide the best solution.

Buckner's decision against a flanking amphibious assault at Okinawa received criticism at the time, and later as historians examined his performance. His mass frontal assault seemed unimaginative and became a source of controversy. On May 29, 1945, Homer Bigart soundly criticized Buckner's ultraconservative tactics in an article in the *New York Herald Tribune*. Another columnist at the time, David Lawrence, picked up this line of attack and labeled Okinawa a military fiasco. Historians have proved no less critical. In a brief discussion of the operation in A *War to Be Won*, Williamson Murray and Allan Millett condemn Buckner's "flawed generalship," declaring him unfit for command because of his lack of combat experience.[108]

Buckner's staff recommended against the second amphibious operation as logistically insupportable. This is the same conclusion reached by the Joint Staff's Logistical Committee when it reviewed the original JPS plan for an assault over these same beaches.[109] It was for similar reasons that the Minatoga beaches had been rejected in the Tenth Army plan. Later operations confirmed the staff's assessment when the Seventy-seventh Division occupied the Minatoga beach area toward the end of the battle and discovered the unit could not be supplied over the shore. Instead, the division had to be supplemented by overland supply.[110] More to the point, Buckner's primary mission was to build bases for the American advance on Japan, not to kill Japanese. Buckner's methodical approach was specifically designed to minimize friendly casualties while achieving his operational mission. In his own words, "We didn't need to rush forward, because we had secured enough airfields to execute our development mission."[111]

The Tenth Army faced a competent, determined enemy in a heavily fortified defense in depth. The American ground forces faced the greatest concentration of enemy artillery in the Pacific war. They faced an enemy unaffected by maneuver and determined to fight to the death. The great lesson of Okinawa is that even at the operational level, if the enemy is determined to fight to the death, he must be accommodated. No amount of large-scale artful maneuver can avoid a fight to the finish. Fanaticism may insist on a war of attrition. In this case, operational art may only serve to best position friendly forces to fight such a war with all possible advantage.

The battle for Okinawa showcased far more than Japanese fanaticism; it highlighted the American ability to project, conduct, and sustain large-scale expeditionary warfare. The bitter experience of war forced the U.S. Army and Navy to bring to reality the studies of the interwar period in which joint staffs and organization could synchronize air, sea, and land power. The development of the American staff system, which treated operations, logistics, and intelligence equally, made possible the application of American power at the operational level of war.

The most valid criticism of American operational art in the Pacific remains the apparent failure to appoint a single theater commander. The overall lack of unity of command encouraged service rivalry and resulted in some duplication of important efforts. Nevertheless, by keeping strategic and operational direction vested in the Joint Chiefs, the senior American leadership balanced the requirements for a global war. More important, this system, though not perfect, ensured that each decision resulted from considerable discussion, debate, and compromise. Though this runs counter to the military maxim of unity of command, it ensured full discussion of all strategic and operational issues and latitude for creativity born from competition. The system provided sound decisions and operational agility, and it directed the dual drive across the Pacific that kept the Japanese off balance and uncertain as to the main Allied effort. Fortunately, the Americans had sufficient resources to afford this theater strategy.

Like the strategy employed in Europe, this was a broad-front approach, using all available resources in a continuous concentric attack to overwhelm the enemy. This undoubtedly played to the American advantage in industrial capacity, but all this potential power could be brought to bear only through the intelligent application of operational art. The intellectual preparation of the interwar period made possible the evolution of American operational art in the Pacific. In the end, the American ability to focus, leverage, and sustain air, sea, and land power won the war in the Pacific and made it the best expression of American operational art during World War II.

8

LESSONS AND LEGACY

B ritish historian Richard Overy tackles a very fundamental question in his book *Why the Allies Won*: Why did the Allies win World War II? As he points out, the outcome was certainly not preordained. He conducts a broad examination of issues that include economic production, technology, military reform, even morality. In the end, he concludes that "the Allies won the Second World War because they turned their economic strength into effective fighting power, and turned the moral energies of their people into an effective will to win."[1] Economic production, technology, and a righteous cause that sustains public will may provide the conditions for victory, but they cannot guarantee it. As Clausewitz noted so many years ago, war is ultimately about fighting.[2] Overy does not provide a detailed discussion of military reform but limits himself to brief narratives of the war at sea, the strategic bombing campaign, and the land campaign in Europe. Like many military historians, he does not discuss or even consider operational art. How did the Allies, and specifically the Americans, solve the problems of modern warfare and apply their advantages in resources and technology? Quite simply, the Americans prepared for it. During the interwar years, America's military leaders developed a framework for operational art that allowed them to apply U.S. economic power to project, sustain, and conduct large-scale combat operations. The Americans based this operational art on experience, theory, and strategic requirements.

World War I created a great watershed in military history by the introduction of modern warfare. The three elements of air, sea, and land power now competed in an era of mass industrial warfare. The experiences of World War I drove home the reality of modern war. The tremendous cost and sacrifice of that war illustrated the problems posed by conflict in the modern world. In the years following World War I, American military professionals studied the problems the war presented and developed solutions. Senior American officers such as Gen. John J. Pershing and Gen. George C. Marshall returned from the war in Europe determined to avoid the lack of preparedness that characterized American participation in World War I.[3] As chief of staff of the Army, Pershing pinned his hopes on the military school system, which would provide the engine of professionalism for the American army in the twentieth century.[4] With a small, budget-starved peacetime army, preparation for modern war could occur only in the military schools. The great achievement of this system of military education in the interwar period was the intellectual preparation of the senior leadership that would fight and win the next world war.

Contribution of the Military School System

The Leavenworth schools and the Army War College made a tremendous contribution to Allied victory in World War II. Specifically, the senior military schools developed a problem-solving process, a modern staff system, and a framework for modern operational art. The Staff School at Leavenworth drilled into its student officers the staff study as a means for problem solving. It provided format and process in thinking through and solving complex problems. American staffs at all levels from the Joint Staff in Washington, D.C., to the tactical staffs at the division level, used the same military decision-making process for the entire range of problems from strategic to operational to tactical. With few units to command or train, the army emphasized professional education to prepare officers. Lt. Gen. Robert Eichelberger, commander of the Eighth Army in World War II, remembered, "I kept being sent back to school. In 1925, at the age of thirty-nine, I became a student at the Command and General Staff School at Fort Leavenworth. I learned more at thirty-nine than I ever learned at twenty-one."[5]

Both the Command and General Staff School and the Army War College favored an American staff system that treated the intelligence, logistics, and operations sections with equal importance. The curriculum at both schools devoted almost equal time to each staff area, demanding the same level of study and expertise in each area regardless of an officer's branch. The operators and tacticians studied logistics and intelligence along with the experts in those areas. This became a critical part of the American concept of operational art. It made senior commanders and staff officers at least competent and sensitive to the operational functional areas of intelligence and logistics. Maj. Gen. Ernest N. Harmon (CGSS 1933, AWC 1934), wartime commander of the First and then Second Armored Divisions, reflected in his autobiography: "A military historian recently asked me how the United States, indifferent and even contemptuous of the military in peacetime, had been able to produce a group of generals, proficient enough to lead armies successfully against German might. I am now convinced that the intensive and imaginative training at the Command and General Staff College had a great deal to do with it."[6]

The American military devoted a great deal of time and attention to intelligence. From the strategic to the tactical level, U.S. commanders placed a priority on intelligence gathering and analysis alike. At the strategic and operational levels, code breaking and signals intelligence contributed a great deal to Allied victory. In the European theater, the British shared the results of their code-breaking program, Ultra, which decrypted secret messages intercepted from the Germans. In the Pacific theater, the Americans benefited from their own decryption program, Magic, the code name for signal intelligence gained from the Japanese. Both intelligence activities yielded a good deal of useful operational intelligence. Intelligence at the theater and operational levels served not just to predict enemy actions but to help understand how operational maneuver might best defeat him. In the U.S. Army's standardized five-paragraph tactical field order, separate paragraphs addressed both intelligence and logistics. During the war, planners expanded the paragraphs on intelligence and logistics into annexes and even separate plans in the campaign and operations plans.[7] The intelligence annexes and plans provided not only information on the enemy but intelligence collection and counterintelligence plans.

Deception became integral to intelligence planning. Allied planners understood the value of deception in shaping the conditions necessary for operational success. Operation Fortitude, the deception plan to convince the Germans that the Allies would assault along the Pas de Calais region rather than Normandy, illustrated the potential value of operational deception. The U.S. Joint Staff also considered extensive deception measures against Japan at the theater level.[8] Commanders and staff routinely planned feints and maneuvers to deceive enemy commanders in order to fix enemy reserves or obtain tactical surprise, such as in MacArthur's plan to invade the Philippines or in Buckner's plan to feint along the eastern coast of Okinawa. American commanders used intelligence in order to shape their own operations and influence enemy reactions.

In contrast, the Axis powers rarely approached the Allies' sophistication or success in using operational intelligence. As noted in the German staff system, intelligence simply did not enjoy the same professional emphasis as operations.[9] Similarly, the Japanese staff system did not value intelligence to the same degree as operations. Col. Hiromichi Yahara, the operations officer for the Japanese Thirty-second Army defending Okinawa, admitted during his interrogation that the "unfortunate attitude that intelligence work belonged properly only to the officers incompetent for operations work prevailed even in the highest echelons."[10] The disdain or at least lack of emphasis in the Axis staff systems for intelligence also extended to logistics.

Professional advancement in Axis armies depended on demonstrated success as a field commander, namely tactical or operational expertise. In the American army, officers could reach high rank by commanding logistic organizations. At the national level, Marshall reorganized the U.S. Army in 1942 into the Army Ground Forces, Army Air Forces, and Army Services of Supply. Four-star general Brehon Somervell headed the Services of Supply, later redesignated as Army Service Forces. This organization took responsibility for procurement, inventory, and distribution of all equipment and supply. This essentially created an army logistics command coequal with the army commands charged with raising and training ground and air forces. Likewise at the operational level, each of the theaters created their own subordinate logistics commands. Lt. Gen. John C. H. Lee (AWC 1932) commanded Eisenhower's European Services of Supply. General Eisenhower, who served as both the Supreme Allied Commander and commander of the European Theater of Operations U.S. Army (ETOUSA), Eisenhower combined the Services of Supply and ETOUSA headquarters in January 1944. Although he remained commander of the U.S. ETO, he delegated control of ETOUSA to Lee primarily as an administrative and logistics headquarters. Eisenhower put Lee in command of the Communications Zone, responsible for all U.S. European logistics.[11]

In the Southwest Pacific, MacArthur created his own theater Army Services of Supply Command. In the Pacific Ocean Areas theater, the Service Force Pacific Fleet Command provided logistic support to the fleet throughout the Pacific. Nimitz relied on his army component, U.S. Army Forces POA, to provide for army needs. Both MacArthur and Nimitz used subordinate component logistic commands in their joint task forces. Commanders and planners treated these component logistic commands, such as Army Service Command in the Philippine campaign or Island Command in the battle for Okinawa, as separate components equal in importance to the subordinate combatant components.

Over time, logistics as a planning function became ingrained in all levels of army staffing. The Root reforms in 1903 created the Army General Staff on the German model. The original General Staff included only three divisions: administration, military intelligence, and military education and technical matters. In 1918, in the midst of the world war, Gen. Peyton C. March, Army Chief of Staff, finally subordinated the old War Department bureaus to the General Staff. He reorganized the staff into operations, intelligence, war plans, and a fourth section responsible for purchasing, storage, and traffic. After General Pershing became the chief of staff, he reorganized the general staff along AEF lines, establishing G-1, G-2, G-3, and G-4 staff sections.[12] This General Staff organization at the national level differed from that of the German army, which continued to invest responsibility for supply with the War Ministry. Pershing extended this new staff system throughout the U.S. Army. The Leavenworth schools and Army War College in the interwar period shaped their curriculum to prepare officers for service on senior staffs at all levels.

Perhaps more than any other aspect of war, logistics is critical to success at the operational level. When armies push 180,000 men onto a hostile shore, they need more than the spirit of the bayonet to sustain them. Logistics determines the art of the possible. There is a natural tension between logistics and maneuver in operational art: any imbalance between ends, ways, and means results in risk. The creative element in operational art comes from balancing ends, ways, and means, not just conceiving brilliant maneuvers. The Germans in World War II, whether at the operational or strategic level, frequently exceeded their means in pursuit of unclear or unachievable ends. In the desert, Rommel paid little attention to logistics and willingly accepted significant risk in his operational and tactical maneuvers.[13] Similarly on the eastern front, the Germans continually exceeded their means in the pursuit of operational victories.[14] In the Pacific, the Japanese relied more on their bushido spirit than on competent staffs.[15]

American logistical planning was not perfect. At times, lack of oversight or process produced tremendous waste, duplication of effort, and even abuse. After visiting the European theater in October 1944, Undersecretary of War Robert Patterson recommended that Eisenhower fire his chief logistician, Lt. Gen. John C. H. Lee, for all three reasons.[16] Some historians criticize Allied planners and commanders for being too sensitive to logistics, which encouraged them to be overly cautious and risk averse.[17] Even if this accusation were true, the fact remains that the Allies in general and the Americans in particular maintained a better balance between ends, ways, and means than did their enemies. No Allied operation failed after 1942 due to logistics.

All the functions of operational art—maneuver, logistics, intelligence, and command and control—come together in the art of campaign planning. In this area, the Allies again proved superior to the Axis. In the space of twenty years, the Americans emerged from their limited experience in World War I to develop a modern staff system and a framework of operational art that allowed for effective campaign planning in World War II. The faculty of the senior service colleges built this framework of operational art not just on experience, but on theory and out of necessity.

American military theory in the interwar period was largely derivative and eclectic. Jomini still provided much of the geometry of the battlefield, but American officers expanded his concepts of theaters of operations, lines of communication, and basing into a detailed organization and structure for a theater of war. Officially, at least in doctrine, U.S. military theory appeared embodied in the principles of war adopted in 1921: objective, offensive, mass, economy of force, movement, surprise, security, simplicity, and cooperation. More important, the study of Clausewitz in the senior service colleges added to American officers' operational understanding of campaign planning, focusing, and massing combat power. Theories of airpower contended between tactical, operational, and strategic tasks. Without large peacetime formations to test theories, the American army relied on academic exercises. Although their limitations were a source of concern to senior leaders, theory and exercises did provide valuable professional experience. As noted by General Marshall many years later:

> I was very much worried at the start of the Second World War for fear our—well, out of the First World War—for fear our officers were too theoretical. We didn't have an actual fleet in the water as the Navy did. We had no real Army. The officers had to get their training theoretically, and I was very much afraid that it was going to be too much theory. But afterwards I discovered that our men were so well prepared in the theoretical part, the large factors in the thing, that they were far yonder, I thought, ahead of the preparations of that nature with the British. The British had an immense advantage in tactical information because of their battle experiences, particularly in the early part of the Second World War, but when it came to the other aspects of it, it was quite the other way around.[18]

In application, American operational warfare took the form of continuous concentric pressure to overwhelm the enemy. The military's ability to bring America's tremendous economic resources to bear across intercontinental and transoceanic distances made this possible. In theory and practice, the curriculum of the senior service schools adequately recognized the three levels of war necessary to make this happen and developed a system for campaign planning to employ the forces available.

Campaign planning arranges battles and major operations to achieve strategic objectives in a theater of war or a theater of operations. From their experience in World War I, the Americans adopted a modified French general staff system and developed an operational art to solve the problems of modern warfare. The essential problem facing the American military in any modern conflict was the requirement to raise sufficient forces and to project, conduct, and sustain them in large-scale operations to achieve strategic objectives. To solve the problem of raising forces and providing resources, the faculty devoted a good deal of time and curriculum at the Army War College to mobilization. In 1924, the War Department established a separate school, the Army Industrial College, to deal with this complex and fundamental issue. The key elements of American operational art as developed in the interwar school system centered on theater structure, jointness, phasing, and establishing a firm connection to both tactical and strategic objectives.

The exercises and curriculum at the Army War College anticipated war in the Pacific and to some degree global war. Interwar instruction and doctrine provided a theater structure and an organization to deal with it. The doctrine for large-unit operations established the geometry of global warfare with the adoption of theaters of war and theaters of operations. Although command and control of this theater structure officially provided options for unity of command or cooperation, the faculty and students recognized the advantages found in unity of command. Joint command and staffing remained a preferred student solution throughout the period. American officers accepted tactical phasing in World War I as necessary for coordinating artillery with infantry advancing across the battlefield. As they contemplated an advance across the Pacific, operational phasing seemed equally obvious. A series of major operations to seize advance bases for either a retaking of the Philippines or the direct defeat of Japan became a routine part of U.S. planning.

After serving as a corps commander in World War II and later as Chief of Staff of the Army, Gen. J. Lawton Collins (CGSS 1926, AWC 1938) reflected on the values of his year at the Army War College. "These studies were of value to the students," he remarked, "in that they learned what constituted a war plan, the scope of military intelligence that would be required, the logistical support that would have to be provided, the combined Army, Navy, and Air operations that would have to be planned and the political and economic factors that would have to be considered."[19]

Russell F. Weigley has claimed that American military thought before World War II "neglected operational art to focus instead on strategy and tactics."[20] Clearly, the Army War College studied strategy and the Staff School taught tactics, but both studied and exercised large-unit operations. Precisely because the military school system dealt with large-unit operations as well as strategy and tactics, the curriculum bridged the gap with operational art. The national military planning system adopted during the interwar years started with strategic objectives and then detailed the military resources and operations necessary to achieve them. In the absence of any stated national objectives coordinated through the executive branch, the planners simply assumed them. Whether in exercises or actual war plans, the officers recognized the Clausewitzian notion that war must serve a political purpose.

The ability of American planners to tie strategy, operations, and tactics together was an important feature of American operational art. Again, the Axis powers frequently failed to grasp this fundamental concept. Both the Japanese and the Germans excelled at the tactical art, but each often pursued operational victories seemingly without clear and achievable strategic objectives. It was not simply a case of flawed strategy, but the mistaken conviction that all could be made right with a quick and decisive battlefield victory. The very militarism of these societies made them overly reliant on the pursuit of military victory divorced from any realistic or achievable strategic goals.[21]

Martin Van Creveld has criticized the American military school system during the interwar period, comparing it unfavorably with the German Kriegsakademie. He charges that no U.S. institute of advanced military learning relied on competitive exams for admittance. He asserts that American military schools "did not offer [a] comprehensive, systematic, integrated three year course on military history, art and science such as formed the core of the German *Kriegsakademie*." Finally, he concludes that no American school "served as an instrument for screening and promoting officers on their way to top commands."[22] On almost every count, Van Creveld is wrong. Although the army did not select officers for either the Command and General Staff School or the Army War College using competitive examinations, selection was nonetheless competitive and viewed as critical for promotion. Eisenhower even temporarily changed branches to ensure his admission to the CGSS.[23] True, the American army did have a more egalitarian view of military higher education. Its policies in the interwar period were crafted to increase, not limit, attendance. The army intended to produce competent, if not brilliant, staff officers and commanders in sufficient numbers to manage and lead the mass American conscript army in time of war.

As previously noted, history formed a large part of the curriculum for the senior army schools. At Leavenworth, the faculty devoted 15–17 percent of the curriculum annually to the study of history. Lectures and staff rides at both institutions used history in a fundamental way to educate students. Lieutenant General Wedemeyer recollected that "military history was an important and for most students, an enjoyable feature of the [Leavenworth] course."[24] At the War College, in particular, history was the vehicle through which the students studied the problems of modern warfare. In the twenties, the critical examination undertaken there of the campaigns of World War I made clear the complexity and scale of large-unit operations. Determined to do better if there was a next time, the faculty and students studied and compared U.S. performance with that of both allies and enemies. The faculty also used history to teach command and leadership. A course description from Leavenworth announced: "Emphasis is placed in the course of instruction at these schools on the use of historical illustrations to exemplify strategical and tactical principles and doctrines."[25]

Attendance at Leavenworth and the Army War College, if not a prerequisite for high command in the U.S. Army, certainly made attaining that status more likely. All officers commanding American army groups or armies in World War II attended both the CGSS and the AWC. Of the thirty-four generals who commanded corps, all but one graduated from Leavenworth and twenty-nine also graduated from the War College.[26] The school system educated and trained not only the senior leadership, but the countless staff officers who planned and supervised large-scale operations. The Staff School and War College educated an entire generation of senior leaders and planners in strategy, operational art, and tactics using history and exercises. History allowed the students to study the problems of past campaigns, to reflect on their lessons, and draw conclusions. Eben Swift's original applicatory method employing exercises and staff rides ensured that students developed and tested solutions to realistic and current problems. Perhaps even more than theory or experience, strategic requirements drove American officers to develop a framework for modern operational art.

The army's requirement to defend the Philippines ensured that every War College class from 1920 to 1940 exercised War Plan Orange, the plan for war with Japan. As Williamson Murray has pointed out, successful peacetime military innovation often requires specificity.[27] The need for advance bases in War Plan Orange encouraged the navy to develop carrier aviation, and the Marine Corps to develop a doctrine for amphibious operations. It also required the Army to project, conduct, and sustain large-scale expeditionary operations. This forced the army to consider the requirements for modern joint warfare—how to combine air, land, and sea forces across great distances. The solutions included joint staffs, unity of command, and leveraging air, sea, and land power in phased joint campaigns. These solutions developed at the War College remained in the curriculum but did not find their way into approved doctrine. Without the actual pressures of war, the navy and army could never have agreed on developing a true joint doctrine that might infringe on service autonomy. Once faced with the reality of war, the bureaucratic inertia of the peacetime armed forces gave way to the lessons learned and exercised in the academic freedom of the senior service colleges.

In all the discussions during the interwar years about how to fight and win the next war, the debate over the role of airpower remained the most controversial. At Leavenworth and the War College, the army insisted that airpower should be harnessed to tactical and operational tasks. The Army Air Corps, sensing the new power and potential of aviation, advocated a strategic role for airpower. America's faith in airpower well suited its economic strength, love of technology, and the great distances over which the United States had to fight. Gen. Hap Arnold, General Spaatz, and General Eaker got their chance to demonstrate the strategic reach and impact of airpower. The success of strategic bombing in World War II has been debated. The Strategic Bombing Survey conducted after the war in Europe suggested it was decisive but emphasized the resilient nature of the German economy and will.[28] In the Pacific, the survey documented the decline of the Japanese economy but noted the difficulty in determining the causes between the physical destruction and the impact of declining imports as a result of submarine interdiction of Japanese sea routes.[29] The surrender of Japan following the dropping of the atomic bombs suggested the decisive nature of airpower, but the survey asserted Japan would have surrendered by November 1945 regardless.[30] Strategic airpower, certainly in Europe, may have made its greatest contribution at the operational level. The deep raids into Germany challenged the Luftwaffe to a showdown, which resulted in an attritional struggle the outcome of which was Allied air superiority over the battlefield.

The impact of airpower at the operational level during World War II drove planning. As anticipated during the interwar years, airpower could successfully isolate the battlefield and provide distant operational fires to enhance friendly maneuver while restricting that by the enemy. In the Pacific, the need to extend Allied airpower by either land-based or carrier aviation dictated the scheme of maneuver. In Europe, air supremacy made Operation Overlord possible by delaying German reinforcements into the invasion area and allowing the Allies time to build up combat power in the beachhead, secure their foothold, and then move inland. Senior American commanders noted the importance of airpower as one of the great lessons of World War II. Perhaps they learned that lesson too well.

THE OPERATIONAL LESSONS OF WORLD WAR II

The United States military gathered many lessons from World War II that would shape American operational art in the coming years. Modern warfare meant joint warfare. General Eisenhower believed that "war is waged in three elements but there is no separate land, air, or naval war. Unless all assets in all elements are efficiently combined and coordinated against a properly selected, common objective, their maximum potential power cannot be realized."[31]

In January 1946, fifty carefully selected officers reported to Washington, D.C. These men represented every branch of the armed forces, and they had seen extensive action in World War II's different theaters. They convened as a board to review joint operations in the war. The Joint Chiefs of Staff directed the board to recommend joint doctrine based on wartime experience.[32] The resulting study contains the best summary of the operational lessons that the American military learned from World War II. The study began with a list of principles that emphasized unity of command, joint staffing, and the operational functions of intelligence and logistics. The first principle asserted that "unified command is required for the effective coordinated employment of land, sea and air forces and will be the normal form of command in overseas theaters of operation."[33] The study highlighted the requirements for joint staffing, intelligence, and logistics as key principles.

The study discussed the responsibilities of the theater commander in planning and conducting operations. The theater commander prepares a campaign plan, if required, followed by a tentative operations plan. The operations plan becomes an operations order when so directed by the theater commander. The commander then "exercises command of these forces through the senior commanders of the respective services and through the commanders of joint task forces which he may constitute."[34] The study reflected a mature understanding of the nature of joint warfare. In the conduct of operations "all of the strategic operations of air, ground, and naval forces within the theater, whether they be a combination of air-ground, air-naval, air-ground-naval or independent service operations are joint in nature in that they mutually support each other directly or indirectly and ultimately affect the success or failure of the theater commander's mission."[35]

The study devoted considerable attention to the organization and planning for joint intelligence and logistics. The study suggested staff organizations that included a planning cell in every staff section. The chapter on intelligence included specific recommendations regarding staff organization, requirements, and procedures for intelligence, counterintelligence, estimates, and plans. The longest chapter dealt with logistics. The responsibilities of the theater commander for logistics included directing all transportation services, controlling the flow of shipping, establishing joint logistics organizations, and determining which service should provide common supplies to other services.[36]

The study also addressed the role of seapower and airpower in the joint force. Seapower's general function remained sea control—gaining and maintaining command of vital sea areas. Seapower protected vital sea lanes to the United States while denying the seas to enemy commerce. Finally, seapower assisted land and air forces.[37] The study reflected the wartime compromise on the strategic and operational role of airpower. Air superiority remained the priority mission for U.S. air forces. Airpower was centralized under the theater commander as a component force, but the study recommended that strategic air forces should be made available for his use only "when this action will contribute more to the total prosecution of the war than would the continuance of the strategic air offensive."[38] Based on experiences in both theaters of war, the air force built upon a new prestige sufficient to establish airpower as an independent and fully coequal service along with the army and navy. The obvious and important role of airpower in World War II was not simply the proud claim of Hap Arnold and the air barons. Eisenhower, the most respected army general to emerge from the war, noted, "Foremost among the military lessons was the extraordinary and growing influence of the airplane in the waging of war."[39]

The advent of the atomic bomb finally offered real promise that airpower might produce decisive strategic results by itself, but the consequences of such potential destruction realistically offered only deterrence, not victory. If deterrence failed in lesser confrontations requiring the use of force, the value of atomic weapons became almost irrelevant. The atomic bomb did usher in a new era of limited warfare—an era in which faith in airpower proved overly optimistic. President Eisenhower's New Look defense policy emphasizing atomic deterrence left the army scrambling to find strategic relevance. In the limited wars that followed World War II, Americans consistently overrated the impact of airpower. The tremendous impact of airpower proved less decisive in the limited wars of the second half of the twentieth century.[40]

Under the shadow of airpower and the resumption of peacetime interservice rivalry, the services unfortunately ignored or forgot many of the operational lessons of World War II. This led to an inevitable decline in American operational art after 1945. The lessons so meticulously documented by the board of officers on joint operations after World War II remained in the curriculum of the postwar staff colleges, but they did not find their way into doctrine. For example, the preface to the study used in the Armed Forces Staff College for 1950 noted "the Joint Chiefs of Staff have not accepted the study as an expression of approved joint doctrine but have authorized merely its tentative use for instructional purposes."[41] Just as in the interwar years, the more forward-looking operational concepts might be discussed and exercised in the academic freedom allowed in the school system, but the peacetime services reverted to more constrained institutional behavior when it came to official service doctrine, which remained more narrowly focused. Without war as a forcing function, jointness faded into the bureaucratic interservice rivalry over budgets and service goals. No true joint service culture emerged in the postwar years. Interest in joint doctrine did not emerge until after the passage of the Goldwater-Nichols Act in 1986. This act increased the power of the Chairman of the Joint Chiefs at the expense of the service chiefs.[42]

American operational art declined after 1945. The stalemate in Korea and the failure in Vietnam seemed to suggest that while technology gave the United States an unparalleled advantage at the tactical level, at the operational and strategic level military performance was somehow flawed. After 1973, this led the army to a rediscovery of operational art. The military may study national or grand strategy, but that is frequently seen as the province of civilian policymakers advised by senior military leaders. In contrast, operational art is the business of professional military leaders. Faced with the problem of a massive Soviet military threat and the recent experience of defeat in Vietnam, the U.S. armed forces embarked on an intense introspective review of military theory and doctrine. In 1982, the U.S. Army published *Field Manual 100-5*, which named and described operational art as a distinct level of war.[43] With this official recognition, operational art became a permanent part of U.S. military doctrine. Operational art was not new in 1982; American officers simply rediscovered it. Known as large-scale operations, campaigning, or a type of military strategy, operational art had been part of an earlier renaissance rooted in the professional military education of the interwar period that found decisive expression in World War II.

Perhaps the key lesson that can be gained from World War II is the importance of professional education. The army's commitment to professional military education after World War I proved to be the best preparation available to it for its second global war. It enabled the military leadership to prepare for the conflict and allowed them the intellectual flexibility to adapt to the challenges of modern global war. Even in the midst of a global war, the military did not abandon its commitment to military education. Wartime conditions abbreviated the time available, but not the necessity for education and training. Even though the Army War College closed its doors in 1940, the Staff School remained open to provide short staff courses for officers at the division level. Leavenworth ran a total of twenty-seven classes and graduated 1,800 officers during the war years.[44] On November 3, 1942, Gen. Hap Arnold submitted to the Joint Chiefs a plan to establish a school for joint staffs.[45]

A year later, the Army and Navy Staff College (ANSCOL) opened in Washington, D.C. This new institution instructed senior field-grade officers in the employment of airpower in combined operations, amphibious operations, joint logistics, joint intelligence, and joint communications.[46] The JCS chose Maj. Gen. John L. DeWitt, former commandant of the Army War College, to organize and run the school. The course of instruction lasted for twenty-one weeks, a considerable length of time for highly qualified officers to devote to professional education in a time of war. This commitment to education extended into the postwar years.

The JCS charged General DeWitt with developing a postwar plan for joint education. He submitted his report in June 1945. It recommended that ANSCOL be made a permanent part of the postwar school system and that joint education be provided at the junior, mid, and senior levels of an officer's career. The services also submitted their own reports on postwar military education. After a good deal of discussion, a joint school system emerged. The JCS redesignated the Army and Navy Staff College as the National War College in 1946.[47] General Eisenhower as Army Chief of Staff arranged the donation of the old Army War College site at Fort McNair in Washington, D.C., to house the new college. Additional initiatives in joint education resulted in the establishment of the Armed Forces Staff College for midlevel officers at Norfolk, Virginia, in 1946. The Army Industrial College became the Joint Industrial College of the Armed Forces.

The JCS intended the National War College to become the pinnacle of professional military education. It accepted students from all the services as well as agencies of the federal government. For a while, it seemed as if there would be no need to reopen the Army War College. The Army Command and Staff College at Fort Leavenworth expanded its instruction to provide the necessary education at the brigade and division level.[48] It soon became evident, however, particularly after the outbreak of war in Korea, that the National War College could not graduate enough army officers to meet the service's needs. Even more importantly, the National War College's curriculum focused on national strategy, not large-scale military operations. To fill the gap between strategy and tactics, the army decided to reopen the Army War College at Leavenworth in 1950 and move it to Carlisle Barracks, Pennsylvania, the next year. Guided by the vision of then Army Chief of Staff, J. Lawton Collins, the Army War College returned to the study not only of large-unit operations but of national strategy.[49] The military's commitment to professional education remained as strong in the postwar period as it had been in the prewar period, but expanded to include joint education.

Legacy

U.S. Army officers in the interwar period determined that developing a modern operational art to project, conduct, and sustain large-scale operations was a strategic imperative. Any meaningful participation in a future war would involve expeditionary warfare over vast distances on an intercontinental scale. In an era of mass industrial warfare, modern American operational art played to American strengths and military tradition—the republic's great economic strength and ability to raise forces sufficient to provide continuous concentric pressure to overwhelm enemies.

In the final analysis, the armed forces' commitment to professional military education made this possible. The interwar military school system provided officers the invaluable opportunity to study their profession and develop into competent planners and leaders. The origins of modern American operational art lay in the classrooms of the interwar military schools. Here American military professionals interpreted their experience of World War I and adapted the lessons to the requirements of modern war. They stuck it out in the lean years of indifference, low pay, and little chance for promotion. Ultimately, it was owing to the quality of these professional officers, who proved flexible and adaptable amid the experience of war, and their commitment to duty that America survived its most formidable military challenge of the twentieth century.

Notes

AFHRA   Air Force Historical Research Agency
ASC   Army Staff College
AWC   Army War College
CARL   Combined Arms Research Library, Fort Leavenworth, Kans.
CGSS   Command and General Staff School
NARA   National Archives and Records Administration
NWC   Naval War College
USAMHI   United States Army Military History Institute Collection, Carlisle Barracks, Pa.

INTRODUCTION

1. George C. Marshall, quoted in *Times-Herald* (Washington, D.C.), March 3, 1942, 1; emphasis added.

2. A search of the Army War College and U.S. Army Military History Institute databases revealed 131 titles dealing with operational art: 98 student papers, 17 contemporary theoretical works, and only 16 books by historians. Most of the historical works are on Soviet operational art.

3. B. J. C. McKercher and Michael A. Hennessy, eds., *The Operational Art: Developments in the Theories of War* (Westport, Conn.: Praeger Publishers, 1996), 1.

4. John English, "The Operational Art: Developments in the Theories of War," in McKercher and Hennessy, *Operational Art*, 8.

5. See Richard W. Harrison, *The Russian Way of War: Operational Art, 1904–1940* (Lawrence: University Press of Kansas, 2001), 152–216.

6. See Shimon Naveh, *In Pursuit of Military Excellence: The Evolution of Operational Theory* (London: Frank Cass, 1997), 7–23.

7. Russell F. Weigley, "From the Normandy Beaches to the Falaise-Argentan Pocket," *Military Review* 70 (September 1990): 45.

8. David E. Johnson, *Fast Tanks and Heavy Bombers: Innovation in the U.S. Army 1917–1945* (Ithaca, N.Y.: Cornell University Press, 1998), 229.

9. William O. Odom, *After the Trenches: The Transformation of U.S. Army Doctrine, 1918–1939* (College Station: Texas A&M University Press, 1999), 9.

10. Joint operations refer to operations that involve two or more services. For example, army, air, and naval forces conduct joint operations to leverage each service's unique capabilities. Combined operations refer to co-ordinated military actions between allies.

11. See English, "Operational Art," 8–14.

12. Hermann Foertsch, *The Art of Modern Warfare* (New York: Veritas Press, 1940), 20.

13. See Jacob Kipp, "Two Views of Warsaw: The Russian Civil War and Soviet Operational Art, 1920–1932," in McKercher and Hennessy, *Operational Art*; David M. Glantz, *The Nature of Soviet Operational Art, 1920–1932* (Fort Leavenworth, Kans.: U.S. Army Combined Arms Center, 1985), and Naveh, *In Pursuit of Military Excellence*.

14. English, "Operational Art," 16; see also Richard M. Swain, "Filling the Void: Operational Art and the U.S. Army," in McKercher and Hennessy, *Operational Art*; and Naveh, *In Pursuit of Military Excellence*.

15. After World War II, the U.S. Army shifted its doctrinal focus from conventional to nuclear, to counterinsurgency, and then back again to conventional operations. See Robert A. Doughty, *The Evolution of U.S. Army Tactical Doctrine, 1946–76*, Leavenworth Papers (Fort Leavenworth, Kans.: Combat Studies Institute, 1979), 1–18.

16. U.S. Department of Defense, *Joint Publication 3.0: Doctrine for Joint Operations* (Washington, D.C., 2008), iv–3.

17. Ibid.

18. Carl von Clausewitz, *On War*, ed. and trans. Michael Howard and Peter Paret (Princeton, N.J.: Princeton University Press, 1976), 595.

19. *Doctrine for Joint Operations*, iii–1.

20. Joint Chiefs of Staff, *Joint Publication 1: Joint Warfare of the Armed Forces of the United States* (Washington, D.C.: Joint Chiefs of Staff, 2007), i.

1. THE ROOTS OF OPERATIONAL ART

1. Quoted in Richard A. Preston and Sydney Wise, *Men in Arms* (New York: Holt, Rinehart and Winston, 1979), 15.

2. Although there are many definitions of strategy, the use of force to achieve political objectives is common to most current definitions. See Colin S. Gray, *Modern Strategy* (New York: Oxford University Press, 1999), 1–3.

3. Peter Connolly, *Greece and Rome at War* (London: Greenhill Books, 1998), 38–41.

4. Geoffrey Parker, *The Military Revolution: Military Innovation and the Rise of the West, 1500–1800* (London: Cambridge University Press, 1996), 24.

5. In the case of Spain, Ferdinand and Isabella conquered Granada in 1492 with no more than 20,000 men. Their grandson, Charles V, commanded perhaps 100,000 against the Turks in Hungary in 1592. Charles VIII of France invaded Italy in 1494 with 18,000 men. By the time of the Battle of Malpaquet in 1709, Louis XIV of France fielded an army of 112,000 men. Ibid., 1, 24.

6. Maurice de Saxe, *Reveries upon the Art of War*, ed. and trans. Thomas R. Phillips (Harrisburg, Pa.: Military Service Publishing Company, 1944), 121.

7. Azar Gat, *A History of Military Thought from the Enlightenment to the Cold War* (London: Oxford University Press, 2001), 75.

8. Ibid., 82.

9. Hew Strachan, *European Armies and the Conduct of War* (London: Unwin & Hyman, 1983), 43.

10. Louis-Alexandre Berthier to Gouvion Saint-Cyr, 2 September 1805, in *Napoleon on the Art of War*, ed. and trans. Jay Luvaas (New York: Free Press, 1999), 90–91.

11. Henri de Jomini, *The Art of War*, trans. G. H. Mendell and W. P. Craighill (Philadelphia: J. B. Lippincott & Co., 1862), 69.

12. Ibid., 2.

13. Ibid., 178.

14. Bernard Brodie, "The Continuing Relevance of *On War*," in Clausewitz, *On War*, 53.

15. Wallace P. Franz, "Two Letters on Strategy: Clausewitz' Contribution to the Operational Level of War," in *Clausewitz and Modern Strategy*, ed. Michael I. Handel (London: Frank Cass, 1986), 172.

16. Clausewitz, *On War*, 379.

17. Ibid., 281.

18. Ibid., 280.

19. J. F. C. Fuller, *The Foundations of the Science of War* (London: Hutchinson & Co., 1926), 108.

20. Clausewitz, *On War*, 177.

21. Carl von Clausewitz, "Unfinished Note, Presumably Written in 1830," in Clausewitz, *On War*, 70.

22. William B. Skelton, *An American Profession of Arms: The Army Officer Corps, 1784–1861* (Lawrence: University Press of Kansas, 1996), xiii.

23. James J. Schneider, "The Loose Marble—and the Origins of Operational Art," *Parameters* 19 (March 1989): 90.

24. Ibid., 92.

25. Arden Bucholz, *Moltke, Schlieffen and Prussian War Planning* (Oxford: Berg Publishers, 1991), 9.

26. Helmuth von Moltke, *Moltke on the Art of War: Selected Writings*, ed. Daniel J. Hughes, trans. Harry Bell and Daniel J. Hughes (Novato, Calif.: Presidio Press, 1993), 47.

27. Ibid., 125.

28. John A. English, *Marching through Chaos: The Descent of Armies in Theory and Practice* (Westport, Conn.: Praeger Publishers, 1996), 56.

29. Baron von Fretag-Loringhoven, *Generalship in the World War* (Berlin: E. S. Mittler & Sohn, 1920), 34.

30. Moltke, "Instructions for Large Unit Commanders," in *Moltke on the Art of War*, 214.

31. Fuller, *Foundations of the Science of War*, 108.

32. Ibid., 107.

33. B. H. Liddell Hart, *Strategy of the Indirect Approach* (London: Faber and Faber, 1941), 31.

34. Jacob Kipp, *Mass, Mobility, and the Red Army's Road to Operational Art, 1918–1936* (Fort Leavenworth, Kans.: Command and General Staff College, 1987), 17.

35. Quoted in David M. Glantz, *Soviet Military Operational Art: In Pursuit of Deep Battle* (London: Frank Cass, 1991), 23.

36. For Tukhachevsky's views on successive and deep operations, see United States Army War College, *New Problems in Warfare* (Carlisle Barracks, Pa.: U.S. Army War College, 1983), 4–6, 16, 17, 42–44; from an Art of War colloquium, this text was prepared by the Department of Doctrine, Planning, and Operations.

37. David Glantz, "The Nature of Operational Art," *Parameters* 15 (Spring 1985): 6, 7.

2. OPERATIONAL ART IN THE AMERICAN ARMY BEFORE 1919

1. Michael Howard, *Studies in War and Peace* (New York: Viking Press, 1970), 99.

2. Williamson Murray and MacGregor Knox, "Thinking about Revolutions in Warfare," in *The Dynamics of Military Revolution, 1300–2050*, ed. MacGregor Knox and Williamson Murray (Cambridge: Cambridge University Press, 2001), 11.

3. Russell F. Weigley, *History of the United States Army*, rev. ed. (Bloomington: Indiana University Press, 1984), 598.

4. "The growth of armies has kept pace with that of civil business and the armies of today must be handled by the same methods that are pursued by great commercial corporations." Capt. W. D. Connor, lecture on Organization and Duties of the Staff, November 23, 1904, Infantry and Cavalry School, United States Army Military History Institute Collection, Carlisle Barracks, Pa. (hereafter cited as USAMHI), File UB 220. See also Allan R. Millett, *The General: Robert L. Bullard and Officership in the United States Army, 1881–1925* (Westport, Conn.: Greenwood Press, 1975), 3–10.

5. U.S. War Department, "The Report of the Secretary of War for 1899," in *Five Years of the War Department Reports Following the War with Spain, 1899–1903* (Washington, D.C.: Government Printing Office, 1904), 58.

6. As late as 1898, congressional debates on permanent expansion of the regular army reflected prejudice against the regular army. "One congressman struck the chords that pointed out that citizen soldiers were more idealistic and an all-round better sort than Regular Army 'hirelings,' and, above all,

they 'do not menace our liberties.'" Edward M. Coffman, *The Regulars: The American Army, 1898–1941* (Cambridge: Harvard University Press, 2004), 5.

7. Michael Howard, *War in European History* (London: Oxford University Press, 1976), 100.

8. Bronsart von Schellendorff, *The Duties of the General Staff*, 4th ed., trans. for the General Staff (London: Harrison and Sons, 1905), 2.

9. Ibid., 42.

10. William T. Sherman, *Memoirs of General W. T. Sherman* (1885; reprint, New York: Literary Classics of the United States, 1990), 2:893.

11. Timothy K. Nenninger, *The Leavenworth Schools and the Old Army: Education, Professionalism, and the Officer Corps of the United States Army, 1881–1918* (Westport, Conn.: Greenwood Press, 1978), 85.

12. U.S. War Department, "Report of the Secretary of War for 1901," in *Five Years of the War Department Reports*, 162.

13. Coffman, *The Regulars*, 55.

14. Brian Bond, *The Pursuit of Victory from Napoleon to Saddam Hussein* (Oxford: Oxford University Press, 1996), 3.

15. See Antulio J. Echevarria II, *After Clausewitz: German Military Thinkers before the Great War* (Lawrence: University Press of Kansas, 2000), 182–212.

16. Arthur L. Wagner, *Strategy: A Lecture Delivered by Colonel Arthur L. Wagner, to the Officers of the Regular Army and National Guard at the Maneuvers at West Point, Kentucky, and at Fort Riley, Kansas, 1903* (Kansas City, Mo.: Hudson-Kimberly, 1904), 3–5.

17. *Operations of War* went through seven editions, the last published in 1922. It was still the main text at the Indian Staff College just prior to World War I. Strahan, *European Armies and the Conduct of War*, 68.

18. Arthur L. Wagner et al., *Strategical Operations: Illustrated by Great Campaigns in Europe and America* (Fort Leavenworth, Kans.: United States Army Infantry and Cavalry School, 1897), preface.

19. Wagner, *Strategy*, 43.

20. Arthur L. Wagner, *Organization and Tactics* (Kansas City, Mo.: Hudson-Kimberly, 1897), 1.

21. T. R. Brereton, *Educating the Army: Arthur L. Wagner and Reform, 1875–1905* (Lincoln: University of Nebraska Press, 2000), 121.

22. John Bigelow, *The Principles of Strategy* (Philadelphia: J. B. Lippincott & Co., 1894), 6.

23. Nenninger, *Leavenworth Schools*, 46, 47.

24. Brereton, *Educating the Army*, 121.

25. Harry P. Ball, *Of Responsible Command: A History of the U.S. Army War College*, 2nd ed. (Carlisle Barracks, Pa.: Alumni Association of the United States Army War College, 1994), 141.

26. William Carter, "The Training of Officers," *United Services Journal* 2 (1902): 341.

27. Steven T. Ross, *American War Plans, 1890–1939* (London: Frank Cass, 2002), 8.

28. Army War College Session 1910–1911 Record of Work, 4 vols., USAMHI, 1:11.

29. "Another function which is now performed to a very slight degree, and which is of very great importance, should be performed by the proposed War College acting in cooperation with the existing Naval War College, that is the union of the Army and Navy in the collection and utilization of information, studying and formulating plans for defense and attack, and the testing and selection of material of war." U.S. War Department, "Report for 1899," in *Five Years of the War Department Reports*, 66.

30. Ball, *Of Responsible Command*, 116.

31. U.S. Army War College (hereafter cited as AWC), Memorandum for AWC, Course for 1909–1910, Oct. 30, 1909, USAMHI.

32. General Orders No. 107, July 20, 1903, listed as an appendix to U.S. War Department, "Report of the Secretary of War for 1903," in *Five Years of the War Department Reports*, 334.

33. U.S. War Department, *Field Service Regulations, United States Army 1914* (Washington, D.C.: Government Printing Office, 1914), 10.

34. In a lecture at First Army Headquarters, AEF, on December 18, 1918, Brig. Gen. Hugh A. Drum, chief of staff of the First Army, observed, "Prior to this war, our military students have limited their thoughts to divisions. In fact, at home, in the beginning of our organizations for this war, the same mistake was made, the authorities did not seem to grasp the composition of an Army and could not expand their view to a larger unit than a division." Drum Papers, Box 14, USAMHI.

35. AWC, Course of 1909–1910 Assignment of Map Problems, Map Maneuvers, and Rides, Officers of the Permanent Personnel, in "Outline of Course," USAMHI.

36. Record of Map Maneuver 17, April 12–29, 1915, USAMHI.

37. "AWC Curriculum for 1916–17, Part I: Problems and Exercises," USAMHI.

38. George C. Marshall, who served as the G-3 of the First Army, wrote after the war: "No one of us had a definite conception of the character of the war, and certainly none of us understood the method in which the staffs of the Allied armies functioned. In light of later experience, some of the questions asked and ideas proposed now seem ludicrous. Today it is inconceivable that we should have found ourselves committed to a war while yet in such a complete state of unpreparedness." George C. Marshall, *Memoirs of My Services in the World War, 1917–1918* (Boston: Houghton Mifflin, 1976), 8.

39. Millett, *The General*, 311.

40. Maj. Gen. Fox Connor, "G-3, G.H.Q., A.E.F., and Its Major Problems," lecture delivered to the AWC, March 21, 1933, 3, Richards Papers, Box 13, USAMHI.

41. *The United States Army in the World War, 1917–1919* (Washington, D.C.: Center of Military History, 1991), 12:4.

42. Letter of Instruction from the Secretary of War to Maj. Gen. Pershing, May 26, 1917, quoted in full in John J. Pershing, *My Experiences in the World War* (New York: Frederick A. Stokes Company, 1931), 1:38.

43. Connor, "G-3, G.H.Q., A.E.F."

44. "General Organization Project HQ AEF July 10, 1917," *United States Army in the World War*, 1:93.

45. C-in-C, AEF, "Report on Organization HQ AEF," February 12, 1919, *United States Army in the World War*, 1:144.

46. Historical Branch, War Plans Division, General Staff, *Organization of the Services of Supply: American Expeditionary Forces* (Washington, D.C.: Government Printing Office, 1921), 22–23.

47. "Report of the Chief of Staff, GHQ, AEF," June 30, 1919, in *United States Army in the World War*, 12:91, 92.

48. "Lecture Delivered by Brig. Gen. H. A. Drum, Chief of Staff, First Army," December 18, 1918, Drum Papers, Box 14, USAMHI.

49. "A Strategical Study on the Employment of the A.E.F. against the Imperial German Government," Drum Papers, Box 15, USAMHI.

50. Ibid.

51. Railroads, Exhibit B to "Strategical Study," Drum Papers, 3.

52. Drum, "Lecture," December 18, 1918.

53. Pershing, *My Experiences*, 1:103

54. Ibid., 356.

55. "Joint Note with Recommendations of Secretary of War," March 28, 1918, *United States Army in the World War*, 2:261. Later in a meeting at Abbeville on May 1 and 2 the Allies agreed that the British would provide additional shipping to increase the number of infantry sent to Europe while American shipping would concentrate on the necessary combat and service support units.

56. "Minutes of Conference Held at the Hotel De Ville, Beauvais, on April 3, 1918," in *United States Army in the World War*, 2:277.

57. Pershing, *My Experiences*, 1:376.

58. Ferdinand Foch to John J. Pershing, July 22, 1918, in *United States Army in the World War*, 2:543.

59. "Memorandum to the Commanders in Chief of the Allied Armies," July 24, 1918, in *United States Army in the World War*, 2:551.

60. Marshall, *Memoirs of My Services*, 120–27.

61. Drum, "Lecture," December 18, 1918, 4.

62. Pershing, *My Experiences*, 2:260.

63. "Instructions for the Reduction of the St-Mihiel Salient, September 2, 1918," in *United States Army in the World War*, 2:177, 178.

64. Field Orders No. 9, September 7, 1918, Drum Papers, Box 14, USAMHI.

65. Pershing, *My Experiences*, 2:263.

66. Erich Ludendorff, *Ludendorff's Own Story, August 1914–November 1918* (New York: Harper & Brothers, 1919), 2:361.

67. Ibid., 270.

68. Marshall, *Memoirs of My Services*, 137, 138.

69. Ibid., 149.

70. Ibid.

71. Brig. Gen. Hugh Drum, chief of staff of the First Army, described the army's task as "a major operation between the Meuse and the Argonne, having for its objective the taking of the Hindenburg position, with a following development in the direction of Buzancy and Stonne and an outflanking of

the enemy's positions on the Vouzieres-Rethel Line from the east." "Summary of Operations First Army," Drum Papers, Box 14, USAMHI.

72. Paul F. Braim, *The Test of Battle: The American Expeditionary Forces in the Meuse-Argonne Campaign* (Newark: University of Delaware Press, 1987), 96.

73. Pershing, *My Experiences*, 2:292.

74. Field Orders No. 20, September 20, 1918, Drum Papers, Annex 3, Box 14, USAMHI.

75. By comparison, Pershing's First Army in August 1918 was eight times larger than Meade's Army of the Potomac in May 1864. Grant's Overland campaign during the Civil War lasted forty days and cost 55,000 casualties. Mark Grimsley, *And Keep Moving On: The Virginia Campaign, May–June 1864* (Lincoln: University of Nebraska Press, 2002), 224.

76. Tim Travers, *The Killing Ground: The British Army, the Western Front and the Emergence of Modern War, 1900–1918* (Boston: Allen & Unwin, 1987), 86.

77. Holger H. Herwig, quoting Ludendorff, "The Dynamics of Necessity: German Military Policy during the First World War," in Allan R. Millett and Williamson Murray, eds., *Military Effectiveness: The First World War* (Boston: Allen & Unwin, 1988), 99. Ludendorff further stressed his tactical approach in his memoirs: "I favored the center attack; but I was influenced by the time factor and by tactical considerations, first among them being the weakness of the enemy, tactics had to be considered before purely strategical objects, which it is futile to pursue unless tactical success is possible." Ludendorff, *Ludendorff's Own Story*, 2:221.

78. See Robert A. Doughty, *The Seeds of Disaster: The Development of French Army Doctrine, 1919–1939* (Hamden, Conn.: Archon Books, 1985).

3. LANDPOWER

1. Weigley, *History of the United States Army*, 599.

2. U.S. War Department, *Report of the Secretary of War to the President for 1922* (Washington, D.C.: Government Printing Office, 1922), 13.

3. U.S. War Department, *Report of the Secretary of War to the President for 1923* (Washington, D.C.: Government Printing Office, 1923), 4.

4. The National Defense Act of 1920 authorized a regular army of 280,000 officers and men. Anticipating this large force, the army commissioned 5,229 officers. Subsequently, Congress never appropriated the money to support this force and the army, in consequence, shrank considerably. This group of officers commissioned in 1920 became known as the "hump." The sheer size of this block of officers, combined with promotion by seniority and the mandatory retirement age of sixty-four, resulted in a slowdown in promotions. Even though commissioned before the "hump," Wedemeyer's class was outranked by the many veterans joining the regular army officer corps in 1920. Coffman, *The Regulars*, 239, 240.

5. United States Army, American Expeditionary Force, "Superior Board on Organization and Tactics," 1919, 6, USAMHI.

6. Ibid., 5.

7. Ibid., 7.

8. Ibid., 18.

9. Ibid., 19.

10. The National Defense Act of 1920 confirmed the organization of the army into regular, reserve, and National Guard components. The regular army was charged with preparing the other components for war. The country was divided into nine corps areas under three army headquarters. Each corps area contained one Regular, two National Guard, and three Organized Reserve divisions. The act also created branch chiefs for infantry, cavalry, coast artillery, and field artillery and established the Air Service, Chemical Warfare Service, and Finance Department as new branches of the service.

11. "Report of the Chief of Staff," included in *Report of the Secretary of War to the President for 1922,* 119.

12. General Service School, "Manuscript for Training Regulations No. 15" (Field Service Regulations) (Fort Leavenworth, Kans.: 1922), 17, CGSC File, Box 4, USAMHI.

13. Ibid., 144.

14. Ibid., 137.

15. Ibid., 145.

16. Ibid., 146, 147.

17. Ibid., 137–141.

18. Ibid., iii.

19. See Odom, *After the Trenches,* 35, 36.

20. Boyd L. Dastrup, *The U.S. Army Command and General Staff College: A Centennial History* (Manhattan, Kans.: Sunflower University Press, 1982), 60–65.

21. United States Army, *Instruction Circular No. 1* (Fort Leavenworth, Kans.: General Service Schools Press, 1925–26), Combined Arms Research Library, Fort Leavenworth, Kans. (hereafter cited as CARL), 14, 15.

22. Carlo D'Este, *Eisenhower: A Soldier's Life* (New York: Henry Holt & Co., 2002), 178.

23. Dwight Eisenhower, "On the Command and General Staff School," in Daniel D. Holt and James W. Leyerzapf, eds., *Eisenhower: The Prewar Diaries and Selected Papers, 1905–1941* (Baltimore: Johns Hopkins University Press, 1998), 43–58.

24. Peter J. Schifferle, "Anticipating Armageddon: The Leavenworth Schools and U.S. Army Military Effectiveness, 1919 to 1945" (Ph.D. diss., University of Kansas, 2002), 107.

25. Compare William K. Naylor, *Principles of Strategy: With Historical Illustrations* (Fort Leavenworth, Kans.: General Service Schools Press, 1921), 153, with Victor Derrecagaix, *Modern War,* vol. 1: *Strategy,* trans. C. W. Foster (Washington, D.C.: James J. Chapman, 1888), 6.

26. Naylor, *Principles of Strategy,* 150.

27. Ibid., 49.

28. Clausewitz, *On War,* 90.

29. Ibid., 595, 596.

30. Ibid.

31. Ibid., 528.

32. Naylor, *Principles of Strategy*, 105.

33. Ibid.

34. In 1921, the War Department announced nine principles of war: objective, offensive, mass, economy of force, movement, surprise, security, simplicity, and cooperation. The British theorist, soldier, and historian J. F. C. Fuller originally developed the principles of war.

35. Oliver P. Robinson, "Course in Strategy," lecture delivered to the Command and General Staff School, May 10, 1926, 2 (bound vol.), CARL.

36. Ibid., 11.

37. Oliver P. Robinson, *The Fundamentals of Military Strategy* (Washington, D.C.: United States Infantry Association, 1928), 2.

38. In World War II, Leonard T. Gerow would command V Corps and the Fifteenth Army. Geoffrey Keyes, John Millikin, and Walton H. Walker would command the II, III, and XX Corps, respectively.

39. Oliver P. Robinson, "Principle of the Objective," lecture delivered to the Command and General Staff School, May 14, 1926, 10 (bound vol.), CARL.

40. Ibid., 11, 12.

41. Oliver P. Robinson, "The Principle of the Offensive," lecture delivered to the Command and General Staff School, May 23, 1926, 24, 25 (bound vol.), CARL.

42. D' Este, *Eisenhower*, 168.

43. Robinson, "Principle of the Offensive,"2.

44. The U.S. Army consistently encouraged officers to read Clausewitz. *On War* was placed on the Leavenworth reading list throughout the interwar period as especially recommended. The War Department instituted a voluntary army-wide reading program for officers in 1928. The officers were encouraged to read seven to eight books a year from a published list available at post libraries. From its inception to 1941, the reading program listed all three volumes of *On War*. Memorandum, Maj. Gen. William Connor to the Adjutant General, "Subject: Reading Course for Officers," March 8, 1928, and War Department Bulletin No. 44, "Reading Course for Officers," January 15, 1941, AWC Curricular File, Box 1-105, Copy No. 1 1-82 TAG, Faculty Comments, USAMHI.

45. U.S. Army, *Tactical and Strategical Studies, Corps and Army* (Fort Leavenworth, Kans.: General Service Schools Press, 1922), 14.

46. U.S. Army, *Tactical and Strategical Studies* (Fort Leavenworth, Kans.: General Service Schools Press, 1928), 2.

47. Ibid., 5, 6.

48. U.S. Army, "Resume of Program of Instruction, General Staff School, 1921–22" (Fort Leavenworth, Kans.: General Service Schools Press, 1921), CARL.

49. See U.S. Army, "Instruction Circular No. 1" (Fort Leavenworth, Kans.: General Service Schools Press), for the years 1922–1930, CARL.

50. "Summary of Courses at the Army War College since the War," General Staff College Course 1919–1920, 1, AWC Curricular File 1-105, USAMHI.

51. "Summary of Courses," AWC Curricular File 1-105, Course 1923–1924, 7, USAMHI.

52. "Summary of Courses," Course 1928–1929, 12, AWC Curricular File 1-105, USAMHI.

53. Col. H. B. Crosby, "Orientation Lecture," delivered to the AWC, September 3, 1924, AWC Curricular File 294-2, 4, USAMHI.

54. Committee No. 10, "German Plan of 1914," February 28, 1923, War Plans Course, AWC Curricular File 254-10, 8, USAMHI.

55. Ibid., 10.

56. Committee No. 1, "War Planning in the Past," September 14, 1927, War Plans Course, AWC Curricular File 346-1, 2, USAMHI.

57. Ibid., 12.

58. Committee No. 3, "Naval War Plans of Great Britain and Germany, 1914," September 14, 1926, AWC Curricular File 336-3, 19, USAMHI.

59. Ball, *Of Responsible Command*, 211.

60. Maj. Gen. John L. Hines, "Grand Joint Army and Navy Exercise No. 3," lecture delivered to the AWC, June 26, 1925, AWC Curricular File 294-7, USAMHI.

61. Army-Navy Joint Exercise 1927, Estimate, Plans, and Orders for First Army, Black Expeditionary Force, Box 1926–27, War Plans Course, AWC Curricular File 336-1-11, 3, USAMHI.

62. Charles Keller, "The Army and Navy Joint Board and Joint Planning Committee; and the Methods of the War Plans Division War Department General Staff," lecture delivered to the AWC, September 4, 1926, AWC Curricular File 336A-4, 6, USAMHI.

63. Joint Board, "Statement of Coordination of Operations of the Army and Navy," December 1, 1926, AWC Curricular File, Box 232-5, 2, USAMHI.

64. Committee No. 8, "Recommendation for a System of High Command for Major Joint Army and Navy Expeditionary Forces," October 6, 1928, G-3 Course, AWC Curricular File 352-8B, 15, USAMHI.

65. Subcommittee No. 3, "Training System of the Navy Including the Marine Corps with a View to More Effective Cooperation between the Army and Navy in Joint Operations," October 4, 1928, G-3 Course, AWC Curricular File 352-4B, 1, 10, USAMHI.

66. Command Group No. 6, "Historical Study: German Operations on the Western Front, 1914: From the Concentration of Their Armies to Include the Battle of the Marne," March 29, 1929, Conduct of War Course, AWC Curricular File 356-3B, 2, USAMHI.

67. Ibid., 19.

68. Maj. H. S. Grier, "A Study of the Organization, Functions, and Relations of the Supply Division of the German General Staff or Its Equivalent, with Lessons Therefrom Applicable to the United States," Individual Staff Memorandum, December 1, 1926, G-4 Course, AWC Curricular File 334-22, 13, USAMHI.

69. Ibid., 11.

70. Maj. James W. Barber, "A Study of the Organization, Functions, and Relations of the Supply Division of the German General Staff or Its Equiva-

lent, with Lessons Therefrom Applicable to the United States," Individual Staff Memorandum, December 5, 1926, AWC Curricular File 334-22, 4, USAMHI.

71. "Orientation and Outline for G-4 Course," November 11, 1926, AWC Curricular File 334-A-1, USAMHI.

72. Report of Committee No. 5, "Influence of Logistics on Strategy," December 21, 1925, G-4 Course, AWC Curricular File 314-5, 17, USAMHI.

73. Ibid.

74. Brig. Gen. Fox Conner, "G-4 from a G-3 Point of View," lecture delivered to the AWC, January 6, 1925, AWC Curricular File 290-A-5, 6, USAMHI.

75. Adm. W. V. Pratt, "The Exercise of High Naval Command," lecture delivered to the AWC, April 14, 1927, bound vol., 1926–27, 1, 2, USAMHI.

76. Committee No. 1, "War and Its Principles, Methods, and Doctrines," February 27, 1928, Command Course, AWC Curricular File 347-1, 10, USAMHI.

77. Maj. C. C. McCornack, "The G-4 and Some of His Problems," lecture delivered to the AWC, November 15, 1926, G-4 Course, AWC Curricular File 334A-3, 1, USAMHI.

78. Brig. Gen. Frank Parker, "The Application of Strategy by Tactics in Combat," lecture delivered to the AWC, February 17, 1928, Command Course, AWC Curricular File 347-A, 5, USAMHI.

79. Lt. Col. A. D. Chaffin, "Strategy of the Central Powers in the World War," lecture delivered to the AWC, March 5, 1929, bound vol., 1928–29, AWC Curricular File, 10, USAMHI.

80. Committee No. 1, "Command and Organization of Large Units," February 23, 1926, Command Course, AWC Curricular File 338-1, 28, USAMHI.

81. Col. C. M. Bundel, "Orientation and Outline for the War Plans Course," lecture delivered to the AWC, April 1, 1926, AWC Curricular File 310A, WPD #14, 2, USAMHI.

82. WPD Committee No. 8, "Joint Plans, Army Plans, GHQ Plans," September 26, 1925, War Plans Course, AWC Curricular File 310-11, 1–9, USAMHI.

83. Committee No. 11, "War Plans Division," September 18, 1926, AWC Curricular File 336-11, 8, USAMHI.

84. Lt. Col. Charles Keller, "The Army and Joint Board and Joint Planning Committee," lecture delivered to the AWC, September 4, 1926, AWC Curricular File 336-4, USAMHI.

85. Memorandum for the Director, Command Division, "A Plan of Campaign and a Plan of Concentration," March 2, 1926, AWC Curricular File 316-11-12-13-14, 1, 2, USAMHI.

86. Steven T. Ross, *U.S. War Plans, 1939–1945* (Malabar, Fla.: Krieger Publishing Company, 2000), 11.

87. Map Problem No. 1, Command #35, 1924–25, AWC Curricular File 293A-35, USAMHI.

88. AWC Curricular File 1-105, Course 1924–1925, 9, USAMHI.

89. Ibid., 10.

90. Maj. Gen. W. D. Connor, "Notes on European Trip, 1929," AWC Curricular File 241-64, 7–20, USAMHI.

91. Paul Kennedy, *The Rise and Fall of the Great Powers: Economic Change and Military Conflict from 1500 to 2000* (New York: Random House, 1987), 328, 329.

92. "Report of the Chief of Staff of the Army to the Secretary of War for 1932," extract in *Annual Report of the Secretary of War* (Washington, D.C.: Government Printing Office, 1932), 54, 56.

93. Weigley, *History of the United States Army*, 417.

94. In early 1932, veterans began arriving in Washington, D.C., to support passage of a Bonus bill that would provide immediate payment of a promised bonus for their service in World War I. By May 1932, their numbers swelled to an estimated 10,000–17,000 men, who camped out in the Anacostia mud flats near the capitol, but the bill did not pass. After this failure of Congress to act, many of the veterans went home, but some 2,000 marchers remained. On July 28 violence broke out between a small group of veterans and the authorities. Secretary of War Patrick Hurley's fears of communist agitation and riots prompted orders to General MacArthur to clear the affected area. In full uniform and against the advice of his aide, Maj. Dwight D. Eisenhower, MacArthur exceeded his orders and broke up and burned the veterans' camp. D'Este, *Eisenhower*, 219–21.

95. Hans Schmidt, *Maverick Marine: General Smedley D. Butler and the Contradictions in American Military History* (Lexington: University Press of Kentucky, 1987), 236–38.

96. Maxwell D. Taylor, *Swords and Plowshares* (New York: W. W. Norton & Company, 1972), 29.

97. Ibid., 37.

98. The Franklin Roosevelt administration mandated a further cut in 1933. D'Este, *Eisenhower*, 213.

99. Robert L. Eichelberger, *Our Jungle Road to Tokyo* (Nashville, Tenn.: Battery Press, 1989), xvi.

100. Alan Schom, *The Eagle and the Rising Sun* (New York: W. W. Norton & Company, 2004), 168.

101. J. Lawton Collins, *Lightning Joe: An Autobiography* (Baton Rouge: Louisiana State University Press, 1979), 57.

102. The course provided thirteen lessons on the employment of field artillery at both corps and army levels and twelve conferences on the air corps. U.S. Army, "Schedule for 1933–1934, Second Year Class" (Fort Leavenworth, Kans.: Command and General Staff School Press, 1933), CARL.

103. U.S. Army, *Principles of Strategy for an Independent Corps or Army in a Theater of Operations* (Fort Leavenworth, Kans.: Command and General Staff School Press, 1936), 70; and Box 12, CGSC Files, USAMHI.

104. Ibid., 3.

105. Clausewitz, *On War*, 101–3, 170–74.

106. U.S. Army, *Principles of Strategy*, 37. For Clausewitz's discussion of the importance of the superiority of numbers at the decisive point, see *On War*, 194–97.

107. U.S. Army, *Principles of Strategy*, 70.

108. Ibid., 8.

109. Ibid., 28.

110. Ibid., 16.

111. Ibid.

112. Ibid.

113. Ibid., 46, 47.

114. Ibid., 18.

115. Ibid., 42.

116. Ibid., 7.

117. Ibid., 8.

118. Committee No. 7, "Trends in Tactics and Techniques," October 10, 1938, AWC Curricular File 3-1939-7, USAMHI.

119. Col. C. H. Wright, "Strategic Employment of Military Forces," lecture delivered to the Naval War College, October 21, 1937, AWC Curricular File 195-38-2k; Col. Ned B. Rehkopf, "Strategy," lecture delivered to the AWC, April 11, 1939, AWC Curricular File WP #19, 1939.

120. Col. Ned B. Rehkopf, "Orientation Lecture," delivered to AWC, September 1, 1937, AWC Curricular File Misc. #1, 1938, USAMHI.

121. Ball, *Of Responsible Command*, 228.

122. See Tentative Courses 1923–1940, AWC Curricular File 1-82/A, USAMHI.

123. Report of War Plans Group No.2, April 11, 1936, File 5-1936-19/1–12, USAMHI.

124. Ball, *Of Responsible Command*, 227.

125. Omar N. Bradley and Clay Blair, *A General's Life: An Autobiography of General of the Army Omar N. Bradley* (New York: Simon and Schuster, 1983), 74.

126. Jonathan Wainwright, who became a major general, commanded and subsequently surrendered American forces on Bataan in World War II. William F. Halsey rose to four-star admiral and commander of U.S. forces in the South Pacific.

127. Taylor, *Swords and Plowshares*, 37.

128. Ibid.

129. Maj. Gen. Fox Connor, "Organization and Function of G-3, AEF," lecture delivered to the AWC, September 18, 1931, AWC Curricular File 383-A-8, G-3 Course, 6, USAMHI.

130. Ibid., 1.

131. D'Este, *Eisenhower*, 164.

132. Ibid., 167, 168.

133. William D. Connor, "Strategy of Supply," lecture delivered to the AWC, April 29, 1937, AWC Curricular File WP #21, 1937, 4, USAMHI.

134. Summary of Courses at the Army War College since the World War, AWC Curricular File 1-105, 19, USAMHI.

135. Ibid., 5–8.

136. Lt. Col. B. Q. Jones, "Comments on the Command Post Exercise, 1938," AWC Curricular File 6-1938-11, 36, 37, USAMHI.

137. Ross, *American War Plans: 1890–1939*, 183.

138. Committee No. 2, "U.S. War Planning," September 23, 1931, AWC Curricular File 383-2, 9, USAMHI.

139. Ibid., 2.

140. Ibid., 9.

141. Committee No. 1, "Subject: Joint Plans and Army Strategic Plans," September 23, 1932, AWC Curricular File 383-2, 9, USAMHI.

142. Collins, *Lightning Joe*, 92.

143. Staff Group 2, "Subject: War Plan Red-Orange Coalition," May 4, 1937, AWC Curricular File 5-1937-20/1–9, USAMHI.

144. War Plans Group 4, "Participation with Allies (Blue, Pink, Red, Yellow vs. Orange and Carnation)," April 21, 1934, AWC Curricular File 405-24, 13, USAMHI.

145. Ibid.

146. "Directives and Organization of Groups for War Plans, March 19 to April 15, 1936," AWC Curricular File 5-1936-19/1–12, USAMHI.

147. Committee No. 1, "Subject: War Planning," February 19, 1936, AWC Curricular File 5-1936A 1-20, 50, USAMHI.

148. Committee No. 3, "Subject: Joint Army and Navy Action," September 24, 1931, AWC Curricular File 383-3, 2, USAMHI.

149. Charles Bolte would later command the Thirty-fourth Infantry Division in Italy in World War II, eventually rising to four-star rank and serving as vice chief of staff of the army before retiring in 1955. Maj. Charles L. Bolte, "Joint Operations in the American Revolution and Civil War," lecture delivered to the AWC, February 4, 1938, AWC Curricular File, Conduct of War 1938, 6, USAMHI.

150. Edward S. Miller, *War Plan Orange: The U.S. Strategy to Defeat Japan, 1897–1945* (Annapolis: Naval Institute Press, 1991), 21.

151. Allan R. Millett, "Assault from the Sea: The Development of Amphibious Warfare between the Wars: The American, British, and Japanese Experiences," in Williamson Murray and Allan R. Millet, eds., *Military Innovation in the Interwar Period* (Cambridge: Cambridge University Press, 1996), 57, 58.

152. Maj. Edward Almond, "Oral Presentation of 2nd Situation, Map Study No. 2 (Orange)," June 9, 1934, Almond Papers, Box 15, USAMHI.

153. War Plans Group No. 2, "Subject: War Plan Orange," April 11, 1936, AWC Curricular File 5-1936-19/1–12, 33, USAMHI.

154. Ibid., 45.

155. Ibid., 25.

156. Orlando Ward rose to the rank of major general and commanded the First Armored Division in North Africa.

157. Committee No. 4, "Subject: War Plan Orange Oral Presentation," May 6, 1937, 5, AWC Curricular File 5-1937-22/1, 2, 3, p. 5, USAMHI.

158. Ibid., 20.

159. Ibid., 4, 19.

160. War Plans Group No. 3, "Subject: War Plan Orange Oral Presentation," May 6, 1938, AWC Curricular File 5-1938-21/1, 73, USAMHI.

161. Ibid., 34.

162. Ibid., 85. Of the twenty-four officers in this planning group, fifteen became generals.

163. Bradley and Blair, *A General's Life*, 74.

4. AIRPOWER

1. This quote is taken from a letter the Wright brothers wrote to the United States secretary of war dated October 9, 1905, in Fred C. Kelly, *The Wright Brothers* (1943; reprint, New York: Ballantine Books, 1969), 91.

2. U.S. War Department, *Field Service Regulations 1914*, 20.

3. See Roger C. Miller, *A Preliminary to War: The 1st Aero Squadron and the Mexican Punitive Expedition of 1916* (Washington, D.C.: Air Force History and Museums Program, 2003).

4. Orville Wright to Wilbur Wright, Berlin, November 27, 1910, in Marvin W. McFarland, ed., *The Papers of Wilbur and Orville Wright* (McGraw-Hill, 1953), 2:1004.

5. James S. Corum, *The Luftwaffe: Creating the Operational Air War, 1918–1940* (Lawrence: University Press of Kansas, 1997), 21.

6. Lee Kennett, *The First Air War, 1914–18* (New York: Free Press, 1991), 20, 21.

7. Ibid., 31, 33.

8. Mason M. Patrick, *The United States in the Air* (Garden City, N.Y.: Doubleday, Doran and Co., 1928), 9.

9. William Mitchell, *Memoirs of World War I: From Start to Finish of Our Greatest War* (New York: Random House, 1960), 266.

10. Col. William Mitchell, "General Principles Underlying the Use of the Air Service in the Zone of the Advance," A.E.F., October 3, 1917, USAMHI.

11. Douhet's reputation as a theorist is based primarily on the publication of *Command of the Air* in 1921. Douhet revised the book in 1927, and in the more strident revision, he claims that airpower had made armies and navies obsolete. He famously advocated the bombing of population centers to break the enemy's will. Mitchell admitted to meeting Douhet during a trip to Europe in 1922. A translation of excerpts from Douhet's book was found in the Air Service files, and in 1923 a longer translation circulated at the Air Service headquarters. None of the early American aviation pioneers claim any direct influence from Douhet. Phillip S. Meilinger, "Guillo Douhet and the Origins of Airpower Theory," in Phillip S. Meilinger, *The Paths of Heaven: The Evolution of Airpower Theory* (Maxwell Air Force Base, Ala.: Air University Press, 1997), 33.

12. Memorandum quoted in full in Maurer Maurer, ed., *The U.S. Air Service in World War I* (Washington, D.C.: Government Printing Office, 1978), 143.

13. Patrick, *United States in the Air*, 6.

14. Mitchell, *Memoirs of World War I*, 240.

15. Edward M. Coffman, *The War to End All Wars: The American Military Experience in World War I* (University Press of Kentucky, 1998), 208.

16. Lucien H. Thayer, Donald Joseph McGee, and Roger James Bender, eds., *America's First Eagles: The Official History of the U.S. Air Service, A.E.F., 1917–1918* (San Jose, Calif.: Bender Publishing, 1983), 305.

17. Kennett, *First Air War*, 20, 21.

18. Ibid.

19. Coffman, *War to End All Wars*, 208.

20. Pershing, *My Experiences*, 2:337.

21. See Mitchell, *Memoirs of World War I*, 266; and James J. Hudson, *A Combat History of the American Air Service in World War I* (New York: Syracuse University Press, 1968), 274.

22. U.S. Army, AEF, "Superior Board on Organization and Tactics," 81.

23. William Mitchell, *Winged Defense: The Development and Possibilities of Modern Air Power—Economic and Military* (New York: G. P. Putnam's Sons, 1925), 223.

24. Robert T. Finney, *History of the Air Corps Tactical School, 1920–1940* (Washington, D.C.: Air Force History and Museums Program, 1998), 38.

25. Robert F. Futrell, *Ideas, Concepts, vol. 1: Doctrine: Basic Thinking in the United States Air Force, 1907–1960* (Maxwell Air Force Base, Ala.: Air University Press, 1989), 41.

26. Ibid.

27. U.S. War Department, *Training Regulations No. 440-15: Fundamental Principles for the Employment of the Air Service* (Washington, D.C.: War Department, 1926), 2.

28. Ibid., 8, 9.

29. Ibid., 14.

30. Maj. Oscar Westover, "Tactics and Techniques of Air Corps Bombardment Aviation," lecture delivered to the Command and General Staff School, November 19, 1928, CGSC Curricular Archives, CARL.

31. Command and General Staff School, 1928–29, *Tactical Principles*, One Year Course Map Problem No. 16, Series II, Box 7, CARL. Other conferences on employment of the air division in the attack covered its various missions, such as an attack on the hostile air force, withdrawal, coast defense, defense of a city, and attack and pursuit of an army. In each case, the instruction emphasized the subordination of the air force in shaping theater operations. "Conference on Air Division Attack, Illustrative Problem," May 25, 1929, CGSC Material, Box 7, USAMHI.

32. Committee No. 3, "Fundamental Principles for the Employment of Air Service," Command Course 1925–26, Command #3, AWC Curricular File 316A, USAMHI.

33. Ibid., 5.

34. Maj. H. C. Pratt, "Air Corps Organization and Employment of GHQ and Army Air Corps," lecture delivered to the AWC, November 1, 1927, AWC Curricular File 343A-7, USAMHI.

35. William C. Sherman, *Air Warfare* (New York: Ronald Press, 1926), 217.

36. Ibid., 98.

37. Futrell, *Ideas, Concepts, Doctrine*, 70.

38. Ibid., 77.

39. Ibid.

40. U.S. War Department, *Training Regulations No. 440-15: Employment of the Air Forces of the Army* (Washington, D.C., 1935), 5–6.

41. Ibid., 6.

42. Ibid., 4.

43. Finney, *Air Corps Tactical School*, 43.

44. Air Corps Tactical School (ACTS), Air Corps Tactical Notes, November 22, 1937, 248.214, Air Force Historical Research Agency (hereafter cited as AFHRA).

45. ACTS, *Bombardment Aviation*, January 1, 1938, 248.101-9, AFHRA.

46. ACTS, "Employment of Combined Air Forces," 12, 248.101-7A, 1925–26, AFHRA.

47. ACTS, "Introduction, History, Development, Organization, and Basic Principles of Attack Aviation," 1, 248.2208A, 1938–39, AFHRA.

48. ACTS, Committee No. 2 on the Course in Attack Aviation, 1934–35, 4, 248.2204B, AFHRA.

49. ACTS, "Lecture by Maj. George Kenney: Observation Aviation," November 22, 1929, 248.2601-4, 1929–30, AFHRA.

50. ACTS, "Illustrative Problem," 248.2604-11, 1934–35, AFHRA.

51. ACTS, GHQ Memorandum 41, Intelligence, November 30, 1936, 248.2142, 1936, AFHRA.

52. Haywood S. Hansell, Jr., *The Air Plan That Defeated Hitler* (New York: Arno Press, 1980), 50.

53. ACTS, "Lecture on Logistics," 2, n.d., 248.6008A-1, 1938–39, AFHRA.

54. ACTS, Memorandum No. 4, November 2, 1935, 248.125, 1935–36, AFHRA.

55. ACTS, Maj. James C. Shively, "General System of Supply and Maintenance of an Air Force in an Air Theater of Operations," 248.602-12, 1937–1938, AFHRA.

56. ACTS, Capt. L. S. Kuter, "Conference on Operations against Naval Objectives," March 2, 1938, 248.2207A-27, AFHRA.

57. ACTS, Capt. Harold George, "Air Force Introductory Lecture," 5, 248.2014A, 1932–33, AFHRA.

58. Ibid.

59. *Field Service Regulations, United States Army 1923* (Washington, D.C.: Government Printing Office, 1923), 77.

60. George, "Air Force Introductory Lecture," 24.

61. Ibid.

62. ACTS, "Lecture on Air Force Objectives: National Economic Structure," April 14, 1936, 248.2017A-11, AFHRA.

63. ACTS, "Air Force Course: Lecture on Definitions," September 24, 1935, 15, 248.2017A, 1935–36, AFHRA.

64. ACTS, Bombardment Aviation, Illustrative Problem No. 6, 248.2205A, 1935–36, AFHRA.

65. ACTS, Lecture on Bombardment Aviation, December 13, 1930, 26, 248.101-9, 1930, AFHRA.

66. Conduct of War: Part II Map Maneuver, Group 2, Red Coalition, May 20, 1938, AWC Curricular File 6-1938-9B, 2, USAMHI.

67. Ibid.

68. ACTS, *Air Warfare—Tentative*, February 1, 1938, 1, AWC Curricular File 97-124/A, USAMHI.

69. Ibid., 65.

70. Ibid., 36.

71. The development of the naval air arm certainly testified to the navy's recognition of airpower. In a 1930 lecture to the AWC, USN Capt. William D. Pulleston unequivocally stated: "The effect of the air force on joint operations will probably be enormous; it would be a rash undertaking indeed to attempt a landing on a hostile shore without air superiority. Operating against an invading expedition, air superiority will probably be decisive." "Lecture on the Probable Future Trend of Joint Operations," delivered to the AWC, April 8, 1930, AWC Curricular File 366-A-2A, 10, USAMHI.

72. Roger J. Spiller, ed., s.v., "Arnold, Henry Harley," *Dictionary of American Military Biography* (Westport, Conn.: Greenwood Press, 1984), 1:43.

73. Richard G. Davis, *Carl A. Spaatz and the Air War in Europe* (Washington, D.C.: Center for Air Force History, 1993), 33.

74. H. H. Arnold and Ira C. Eaker, *Winged Warfare* (New York: Harper & Brothers, 1941), 125.

75. Ibid., 123.

76. Spiller, s.v., "Eaker, Ira Clarence," *Dictionary of American Military Biography*, 1:293.

77. Davis, *Spaatz*, 31–33.

78. Thomas H. Greer, *The Development of Air Doctrine in the Army Air Arm, 1917–1941* (Washington, D.C.: Office of Air Force History, 1985), 66.

79. Spiller, s.v., "Kenney, George Churchill," *Dictionary of American Military Biography*, 2:554.

80. Phillip S. Meilinger, *Airmen and Air Theory* (Maxwell Air Force Base, Ala.: Air University Press, 1997), 115.

5. SEAPOWER

1. Stephen B. Luce, "Naval Administration, III," *U.S. Naval Institute Proceedings* 29, no 4 (December 1903): 820.

2. John B. Hattendorf, B. Mitchell Simpson III, and John R. Wadleigh, *Sailors and Scholars: The Centennial History of the U.S. Naval War College* (Newport, R.I.: Naval War College Press, 1984), 17.

3. Ronald H. Spector, *Professors of War: The Naval War College and the Development of the Naval Profession* (Honolulu, Hawaii: University Press of the Pacific, 2005), 19.

4. Ibid., 27.

5. Ibid., 43.

6. Hattendorf, Simpson, and Wadleigh, *Sailors and Scholars*, 38.

7. Alfred T. Mahan, *The Influence of Sea Power upon History, 1660–1783* (Boston: Little, Brown, 1890), 88.

8. Ibid., 8.

9. Alfred T. Mahan, *Naval Strategy: Compared and Contrasted with the Principles and Practice of Military Operations on Land* (Boston: Little, Brown, 1911), 199.

10. Ibid., 107.

11. Philip A. Crowl, "Alfred Thayer Mahan: The Naval Historian," in Peter Paret, ed., *Makers of Modern Strategy from Machiavelli to the Nuclear Age* (Princeton, N.J.: Princeton University Press, 1986), 473.

12. Sadao Asada, *From Mahan to Pearl Harbor: The Imperial Japanese Navy and the United States* (Annapolis: Naval Institute Press, 2006), 8.

13. Eric J. Grove, introduction to Julian S. Corbett, *Some Principles of Maritime Strategy* (Annapolis: Naval Institute Press, 1988), xvi.

14. Julian S. Corbett, *Some Principles of Maritime Strategy* (London: Longmans, Green and Co., 1911), 14.

15. Ibid., 25.

16. Ibid., 87.

17. Ibid., 90.

18. Ibid., 13.

19. Ibid., 159.

20. John Gooch, "Mahan and Corbett," in Colin S. Gray and Roger W. Barnett, eds., *Seapower and Strategy* (Annapolis: Naval Institute Press, 1989), 43.

21. Hattendorf, Simpson, and Wadleigh, *Sailors and Scholars*, 57.

22. Spector, *Professors of War*, 118.

23. Capt. William McCarty Little, "The Strategic Naval War Game or Chart Maneuver," June 1912, RG 4, Box 1, 21, Naval War College Archives (hereafter cited as NWC Archives).

24. Spector, *Professors of War*, 95.

25. David F. Trask, *The War with Spain in 1898* (New York: Macmillan Publishing Co., 1981), 88.

26. George W. Baer, *One Hundred Years of Sea Power: The U.S. Navy, 1890–1990* (Stanford, Calif.: Stanford University Press, 1994), 33.

27. German basic naval plan, quoted in Capt. De Witt Blamer, "Lecture at the Fleet War College Sessions," August 17–19, [1922?], RG 4, Box 12, NWC Archives.

28. William S. Sims, *The Victory at Sea* (Garden City, N.J.: Doubleday, 1920), 9.

29. Capt. Dudley W. Knox, "American Naval Participation in the Great War (with Special Reference to the European Theater of Operations)," in House of Representatives, Committee on Naval Affairs, *Hearings before Committee on Naval Affairs of the House of Representatives on Sundry Legislation Affecting the Naval Establishment*, 70th Cong., 1st sess., 1927–28.

30. Allan R. Millett and Peter Maslowski, *For the Common Defense: A Military History of the United States of America* (New York: Free Press, 1994), 357.

31. Hattendorf, Simpson, and Wadleigh, *Sailors and Scholars*, 90.

32. Knox, "American Naval Participation."

33. Committee No. 5, "The Basic War Plan of the U.S. in the World War," April 12, 1923, RG 4, Box 17, NWC Archives.

34. Knox, "American Naval Participation."

35. Rear Adm. William S. Sims, "Opening Address," June 1919, RG 4, Box 12, NWC Archives.

36. Ibid.

37. "Lecture on the Organization of the Office of the Chief of Naval Operations," June 1933, RG 4, Box 63, NWC Archives.

38. Capt. Ernest J. King, future wartime CNO, participated on the board for naval education in April 1920 that recommended a progressive education "paralleling the Army course at Fort Leavenworth, Kansas." Hattendorf, Simpson, and Wadleigh, *Sailors and Scholars*, 129.

39. Adm. W. V. Pratt, "The Aspects of Higher Command," August 30, 1929, RG 14, Box 4, 24, NWC Archives.

40. Ibid., 3.

41. Ibid.

42. Jack Sweetman, ed., *The Great Admirals: Command at Sea, 1587–1945* (Annapolis: Naval Institute Press, 1997), 416.

43. David W. Hogan, *Command Post at War: First Army Headquarters in Europe, 1943–1945* (Washington, D.C.: Center of Military History, 2000), 59.

44. The navy established the Civil Engineer Corps in 1867 to design and construct all navy facilities ashore. Hattendorf, Simpson, and Wadleigh, *Sailors and Scholars*, 133.

45. Thomas A. Buell, *Spruance and the Naval War College* (Newport, R.I.: Naval War College, 1971), 22.

46. Memo on NWC Course, 1927–28, April 29, 1927, RG 4, Box 33, 1, NWC Archives.

47. Capt. Raymond Spruance, "The Nature of Naval Warfare," July 7, 1937, RG 14, Box 5, 20, NWC Archives.

48. *Prospectus of the Naval War College Courses Senior and Junior 1937–38*, Richards Papers, Box 15, USAMHI.

49. Thomas B. Buell, *The Quiet Warrior: A Biography of Admiral Raymond A. Spruance* (Boston: Little, Brown, 1971), 72.

50. Capt. Richmond K. Turner, "The Strategic Employment of the Fleet," October 28, 1937, RG 14, Box 5, 4, NWC Archives.

51. Ibid., 17.

52. Ibid.

53. Spruance, "Nature of Naval Warfare," 13.

54. Ibid., 16.

55. Ibid., 17.

56. Ibid., 19.

57. Ibid., 22.

58. In 1946, Admiral Spruance became president of the Naval War College. One of his first official acts was to "replace *Sound Military Decision* with a permanent, Navy-wide naval operations planning manual." Buell, *Spruance and the Naval War College*, 32.

59. U.S. Naval War College, *Sound Military Decision* (Annapolis, 1942), 195.

60. Ibid., 196.

61. Ibid., 95.

62. Ibid., 156.

63. Ibid., 12.

64. Ernest J. King and Walter Muir Whitehill, *Fleet Admiral King: A Naval Record* (New York: W. W. Norton & Company, 1952), 239.

65. Hattendorf, Simpson, and Wadleigh, *Sailors and Scholars*, 145.

66. Ibid.

67. Thomas Wildenberg, *Gray Steel and Black Oil: Fast Tankers and Replenishment at Sea in the U.S. Navy, 1912–1995* (Annapolis: Naval Institute Press, 1996), 29.

68. "Operations Problem III-1935-Sr. BLUE," RG 4, Box 69, 1, NWC Archives.

69. Ibid.

70. Baer, *One Hundred Years of Sea Power*, 136.

71. *Employment of Aviation in Naval Warfare*, September 1937, Richards Papers, Box 15, 4, USAMHI.

72. Thomas B. Buell, *Master of Sea Power: A Biography of Fleet Admiral Ernest J. King* (Boston: Little, Brown, 1980), 62.

73. "Operations Problem III-1935-Sr. Blue: Solution by a Member of the Staff," RG 4, Box 69, 52, NWC Archives.

74. Craig C. Felker, *Testing American Sea Power: U.S. Navy Strategic Exercises, 1923–1940* (College Station: Texas A&M University Press, 2007), 63.

75. Thomas C. Hone and Trent Hone, *Battleline: The United States Navy, 1919–1939* (Annapolis: Naval Institute Press, 2006), 111.

76. Miller, *War Plan Orange*, 320.

77. See Felker, *Testing American Sea Power*; and Robert L. O'Connell, *Sacred Vessels: The Cult of the Battleship and the Rise of the U.S. Navy* (Boulder, Colo.: Westview Press, 1991).

78. Baer, *One Hundred Years of Sea Power*, 139.

79. Capt. T. T. Craven, "Aviation," August 5, 1919, RG 15, Box 3, 21, NWC Archives.

80. Rear Adm. Bradley A. Fiske, "Lecture to the Naval War College," September 12, 1919, RG 15, Box 3, 6, 7, NWC Archives.

81. Ibid., 93.

82. Lt. Cmdr. J. F. Moloney, "Aircraft and Their Operations in Connection with the War College Course," December 3, 1933, RG 15, Box 7, 5, NWC Archives.

83. Baer, *One Hundred Years of Sea Power*, 140.

84. Felker, *Testing American Sea Power*, 35.

85. Ibid., 58, 59.

86. "Analysis of Operations Problem III-1935-SR," RG 4, Box 70, 4, NWC Archives.

87. Capt. Richmond K. Turner, "The Employment of Aviation in Naval Warfare," September 1937, Richards Papers, Box 15, 4, USAMHI.

88. Ibid., 17.

89. Allan R. Millett, *Simper Fidelis: The History of the United States Marine Corps* (New York: Macmillan Publishing Co., 1980), 325.

90. Maj. Earl Ellis, "Advanced Base Operations in Micronesia," 1921, Historical Amphibious File 165, 1, Marine Corps University Archives, Quantico, Va.

91. Maj. Gen. John A. Lejeune, "Lecture to the Naval War College," December 14, 1923, RG 15, Box 3, 14, NWC Archives.

92. Kenneth Clifford, *Amphibious Warfare Development in Britain and America from 1920–1940* (Laurens, N.Y.: Edgewood Press, 1983), 95.

93. Ibid., 99.

94. U.S. Joint Army and Navy Board, "Joint Action of the Army and Navy" (Washington, D.C.: Government Printing Office, 1927), 2.

95. William Felix Atwater, "United States Army and Navy Development of Joint Landing Operations, 1898–1942" (Ph.D. diss., Duke University, 1986), 91.

96. Ibid., 107.

97. Col. E. B. Miller, "The Marine Corps: Its Mission, Organization, Powers and Limitations, with Special Reference to Advanced Bases in Support of the Fleet," January 16, 1932, RG 14, Box 4, 1, NWC Archives.

98. Capt. G. B. Wright, "Naval Strategy," August 6, 1934, RG 14, Box 4, 30, NWC Archives.

6. THE EUROPEAN THEATER OF WAR

1. Just six months after the Japanese attack on Pearl Harbor, the United States was engaged in offensive operations in the Pacific, and within eleven months after declaring war against the Axis, it was conducting major offensive operations in the European war—Operation Torch. This global commitment of U.S. forces so soon after the commencement of hostilities far surpassed the seventeen months it took to cobble together an American army and commit it to action in France in World War I.

2. Maurice Matloff and Edwin M. Snell, *Strategic Planning for Coalition Warfare, 1941–1942*, United States Army in World War II (Washington, D.C.: Center of Military History, 1953), 101.

3. Each mention of officers in chapters 6 and 7 includes their interwar education, using the following abbreviations: CGSS (Command and General Staff School), ASC (Army Staff College), AWC (Army War College), and NWC (Naval War College).

4. Dwight D. Eisenhower, *Crusade in Europe* (New York: Doubleday & Co., 1948), 31.

5. Ibid., 36.

6. Thomas Handy, who succeeded Eisenhower as chief of the OPD, remained as Marshall's chief planner and rose to the rank of four-star general. Lyman Lemnitzer served primarily as a staff officer throughout most of World War II. Following the war, he attained the rank of four-star general holding a variety of positions and eventually becoming chairman of the Joint Chiefs of

Staff. Matthew B. Ridgway commanded the Eighty-second Airborne Division in World War II. Following the war, he became a four-star general commanding UN forces during the Korean War and Army Chief of Staff. Wedemeyer replaced Gen. Joseph Stillwell as commander of U.S. forces in China and retired as a lieutenant general.

7. Dwight D. Eisenhower to George C. Marshall, memorandum, February 1942, in Alfred D. Chandler, Jr., ed., *The Papers of Dwight David Eisenhower: The War Years* (Baltimore: Johns Hopkins Press, 1970), 1:146. This memorandum was officially presented to Marshall on March 25, 1942.

8. Clausewitz, *On War*, 596.

9. Quoted in Winston S. Churchill, *The Hinge of Fate*, vol. 4 of *The Second World War* (Boston: Houghton Mifflin, 1950), 434.

10. See Basil H. Liddell Hart, *The British Way in Warfare* (London: Faber and Faber, 1932).

11. Michael D. Pearlman, *Warmaking and American Democracy: The Struggle over Military Strategy, 1700 to the Present* (Lawrence: University Press of Kansas, 1999), 233. The president was careful not to insist on an invasion date that would precede the election, but he did urge Eisenhower to begin operations as soon as militarily practical. In fact, the invasion took place five days after the election. The Democrats lost forty-four seats in the House of Representatives and nine seats in the Senate. See Eric Larrabee, *Commander in Chief: Franklin Delano Roosevelt, His Lieutenants, and Their War* (New York: Harper & Row, 1987), 139.

12. Matloff and Snell, *Strategic Planning for Coalition Warfare*, 268.

13. Quoted in Churchill, *Hinge of Fate*, 442.

14. Wedemeyer remembered, "Sir Alan Brooke, the British Chief of Staff, did not have a very high opinion of the American military leaders' knowledge or ability. He thought we lacked experience and couldn't be expected to evolve sound strategical concepts." Albert C. Wedemeyer, *Wedemeyer Reports!* (New York: Henry Holt & Co., 1958), 132.

15. Dwight D. Eisenhower to George C. Marshall, August 10, 1942, in Chandler, *Papers of Dwight David Eisenhower*, 1:456, 457.

16. Dwight D. Eisenhower to Lord Hastings L. Ismay, August 6, 1942, in ibid., 446.

17. King readily agreed to the principle of unity of command, undoubtedly with one eye on the eventual command arrangements in the Pacific. Eisenhower, *Crusade in Europe*, 51.

18. For an account of Eisenhower's review of Anderson's instructions, see Dwight D. Eisenhower to Hastings L. Ismay, October 10, 1942, in Chandler, *Papers of Dwight David Eisenhower*, 1:602. The actual instructions are quoted in George F. Howe, *Northwest Africa: Seizing the Initiative in the West*, United States Army in World War II (Washington, D.C.: Center of Military History, 1985), 36.

19. Eisenhower, *Crusade in Europe*, 78.

20. Dwight D. Eisenhower to George C. Marshall, August 15, 1942, in Chandler, *Papers of Dwight David Eisenhower*, 1:469.

21. Ibid.

22. Allied Force Headquarters G-2 Intelligence Report, September 11, 1942, RG 407.3, Box 24349, File 478, National Archives and Records Administration (hereafter cited as NARA).

23. French politics were complicated. Britain recognized Charles de Gaulle as leader of the Free French, and although the United States was willing to provide support to de Gaulle, the U.S. government maintained diplomatic relations with Vichy France. Hoping to solicit French military support in North Africa, the Allies secretly brought Gen. Henri Geraud from France to North Africa. Eisenhower also sent his deputy Mark Clark in late October to meet with sympathetic French military leaders at Cherchel seventy-five miles west of Algiers. Regardless of this clandestine meeting, as anticipated the French navy and portions of the army resisted the Allied landings.

24. Howe, *Northwest Africa*, 16.

25. Due to losses at the Battles of Coral Sea and Midway, and the struggle for Guadalcanal, the U.S. Navy had only one operational carrier, the *Ranger*, assigned to the Atlantic Fleet in October 1942. The British carrier allocated to Torch was torpedoed and sunk in the Mediterranean that summer.

26. Dwight D. Eisenhower to George C. Marshall, August 15, 1942, in Chandler, *Papers of Dwight David Eisenhower*, 1:469.

27. Eisenhower also believed that only by landing at Casablanca could the Allies provide the appearance of overwhelming force that would convince the French not to fight. Throughout his correspondence, he remained most concerned about the potential Spanish and Axis threat to Gibraltar. Clearly in his mind and in those of the American planners, Gibraltar was a decisive point on which success of the operation depended. Dwight D. Eisenhower to Combined Chiefs, August 23, 1942, in ibid., 488.

28. Dwight D. Eisenhower to George C. Marshall, November 2, 1942, in ibid., 652.

29. Dwight D. Eisenhower to George C. Marshall, August 9, 1942, in ibid., 454.

30. Dwight D. Eisenhower to Thomas T. Handy, August 13, 1942, in ibid., 462.

31. Harry C. Butcher, *My Three Years with Eisenhower* (New York: Simon and Schuster, 1946), 85.

32. See Churchill, *Hinge of Fate*, 530–38.

33. Mark Clark, *Calculated Risk* (New York: Harper & Brothers, 1950), 51.

34. Lemnitzer added: "One of the turning points in my military experience was going to Leavenworth, because up until that time I was an artillery officer. There I comprehended and was exposed to the operations of the combined arms, the army as a whole. I could not have asked for any better preparation for all of this in that period of service which included schools, teaching, and practical experience." Gen. Lyman L. Lemnitzer, interviewed by Lt. Col. Walter J. Blickston, *U.S. Army Military History Institute Senior Officer Oral History Program*, December 18, 1972, 24, USAMHI.

35. Complete plan Operation Torch, October 8, 1942, in RG 407.3, Box 24349, File 477, NARA.

36. Clark, *Calculated Risk*, 62.

37. For Operation Reservist and Terminal see Howe, *Northwest Africa*, 202, 241.

38. Dwight D. Eisenhower to Thomas T. Handy, December 7, 1942, in Chandler, *Papers of Dwight David Eisenhower*, 2:814.

39. Annex 2 to Outline Plan Operation TORCH, September 29, 1942, RG 407.3, Box 24351, File 492, NARA.

40. L. James Binder, *Lemnitzer: A Soldier for His Time* (London: Brassey's, 1997), 92, 93.

41. Dwight D. Eisenhower to William D. Connor, March 22, 1943, in Chandler, *Papers of Dwight David Eisenhower*, 2:1051.

42. Howe, *Northwest Africa*, 258.

43. Eisenhower, *Crusade in Europe*, 156.

44. Ibid., 117.

45. Dwight D. Eisenhower to Winston S. Churchill, December 5, 1942, in Chandler, *Papers of Dwight David Eisenhower*, 1:802.

46. Dwight D. Eisenhower to George C. Marshall, August 9, 1942, in ibid., 454.

47. F. H. Hinsley et al., *British Intelligence in the Second World War: Its Influence on Strategy and Operations*, 2 vols. (New York: Cambridge University Press, 1981), 2:466.

48. See Dwight D. Eisenhower to Ernest J. King, January 35, 1943, in Chandler, *Papers of Dwight David Eisenhower*, 2:920; and Dwight D. Eisenhower to George C. Marshall, February 4, 1943, in ibid., 937.

49. Eisenhower, *Crusade in Europe*, 148.

50. Ibid., 158.

51. Gen. Sir Harold Alexander commanded Allied ground forces, Air Chief Marshal Sir Arthur Tedder commanded the air forces, while Adm. Andrew Cunningham commanded the naval forces.

52. Arthur Bryant, *The Turn of the Tide, 1939–1943: A History of the War Years Based on the Diaries of Field Marshal Lord Alanbrooke, Chief of the Imperial General Staff* (Garden City, N.Y.: Doubleday & Co., 1957), 455.

53. Dwight D. Eisenhower to George C. Marshall, February 8, 1943, in Chandler, *Papers of Dwight David Eisenhower*, 2:943, 944.

54. Dwight D. Eisenhower to Carl Spaatz, memorandum, October 13, 1942, in ibid., 1:616.

55. Dwight D. Eisenhower to the Combined Chiefs of Staff, February 3, 1943, in ibid., 935.

56. Davis, *Spaatz*, 183.

57. By the end of the campaign, the Allies were finally in a position to isolate Axis forces in North Africa. Air commanders designed Operation Flax specifically to interdict enemy air and maritime transport in an effort to prevent Axis withdrawal or reinforcement. In March and April, 41.5 percent of Axis seaborne cargoes failed to reach Tunisia. Ibid., 190.

58. Dwight D. Eisenhower to Thomas T. Handy, December 7, 1942, in Chandler, *Papers of Dwight David Eisenhower*, 1:811. Although it contains little assessment of the North African campaign at the operational level, for a good tactical analysis of U.S. doctrine and effectiveness, see Peter Mansoor,

The G.I. Offensive in Europe: The Triumph of American Infantry Division, 1941–1945 (Lawrence: University Press of Kansas, 1999), 98; and Michael D. Doubler, *Closing with the Enemy: How GI's Fought the War in Europe, 1944–1945* (Lawrence: University Press of Kansas, 1994), 12–14.

59. Dwight D. Eisenhower to George C. Marshall, memorandum, March 25, 1942, in Chandler, *Papers of Dwight David Eisenhower*, 1:205.

60. OPD to the Chief of Staff, memorandum, Operations in Western Europe, RG 407.3, Entry 427, Box 24325, File 308, NARA.

61. Ibid.

62. H. H. Arnold to the Chief of Staff, memorandum, March 30, 1942, RG 407.3, Entry 427, Box 24325, File 308, NARA.

63. Davis, *Spaatz*, 75.

64. General Morgan informed the staff: "If you will remember he had a really small board of selected officers who dealt with major decisions on broad lines, the day to day work of the war being delegated completely to commanders of army groups. This is what I have in mind." COSSAC Meetings Digest of Decisions, April 17, 1943, RG 407.3, Entry 427, Box 24234, File 296, NARA.

65. Ibid., COSSAC Meeting, August 27, 1943.

66. "Digest of Operation Overlord," in Gordon A. Harrison, *Cross-Channel Attack*, United States Army in World War II (Washington, D.C.: Center of Military History, 1951), appendix A, 453.

67. Ibid., 79.

68. Winston Churchill, *Closing the Ring*, vol. 6 of *The Second World War* (Boston: Houghton Mifflin, 1950), 85.

69. Eisenhower, *Crusade in Europe*, 217.

70. General Eisenhower served as commander in chief, Allied Forces Mediterranean theater of operations, from December 10, 1942, to January 8, 1944. Ray Cline, *Washington Command Post: The Operations Division*, United States Army in World War II (Washington, D.C.: Center of Military History, 1990), 376.

71. See Dwight D. Eisenhower to the Combined Chiefs of Staff, January 23, 1944, and Memorandum for Diary, February 7, 1944, both in Chandler, *Papers of Dwight David Eisenhower*, 3:1673, 1711.

72. George C. Marshall to Dwight D. Eisenhower, February 10, 1944, in Larry I. Bland and Sharon Ritenour Stevens, eds., *The Papers of George Catlett Marshall*, 4 vols. (Baltimore: Johns Hopkins Press, 1996), 4:282.

73. Dwight D. Eisenhower to George C. Marshall, February 19, 1944, in Chandler, *Papers of Dwight David Eisenhower*, 3:1558.

74. The Supreme Commander established the First Allied Airborne Army on August 8, 1944, which consisted of the British Airborne Corps and the U.S. XVIII Airborne Corps as well as the U.S. IX Troop Carrier Command and 38 Group, RAF. The First Allied Airborne Army planned several operations but executed only two: Market-Garden and Varsity. Market-Garden, conducted in September 1944, failed to secure crossings over the Rhine in the Netherlands. Operation Varsity in March 1945 provided the airborne component to Montgomery's large-scale crossing of the northern Rhine. Eisenhower consistently used this theater reserve to weight the main effort in the north.

See John J. Abbatiello, "The First Allied Airborne Army in Operation Varsity: Applying the Lessons of Arnhem" (master's thesis, King's College, London, 1995), 1–5.

75. Dwight D. Eisenhower to Sir Trafford Leigh-Mallory, May 30, 1944, in Chandler, *Papers of Dwight David Eisenhower*, 3:1895.

76. Dwight D. Eisenhower to Carl Spaatz, January 5, 1944, in ibid., 1654.

77. Dwight D. Eisenhower, memorandum for Butcher Diary, March 22, 1944, in ibid., 1784.

78. Dwight D. Eisenhower to the Combined Chiefs of Staff, memorandum, January 23, 1944, in ibid., 3:1675.

79. Letter from the British chiefs to Field Marshall Dill, cited in ibid., 3:1707.

80. Churchill convinced the Allies to continue operations in the Mediterranean after the fall of North Africa. On July 9, 1943, the Allies invaded Sicily. After two months of hard fighting, they captured the island and positioned their forces for the invasion of Italy. On July 24, the Italian High Command overthrew Mussolini and sought an armistice with the Allies. Hitler quickly ordered the occupation of Italy and moved additional forces south to Italy. On September 9, 1943, Lt. Gen. Mark Clark's U.S. Fifth Army landed at Salerno, on the Italian mainland, beginning a long and tough fight up the Italian Peninsula. By October 1943, the Allies had consolidated southern Italy. The Allies committed two armies to the Italian campaign: the British Eighth and the U.S. Fifth. Throughout the rest of 1943 and the first six months of 1944, the Germans stalemated the Allies in terrain ideally suited for defense.

81. Dwight D. Eisenhower, memorandum for Butcher diary, August 7–17, 1944, in Chandler, *Papers of Dwight David Eisenhower*, 4:2057.

82. *Report by the Supreme Commander to the Combined Chiefs of Staff on the Operations in Europe of the Allied Expeditionary Force, 6 June 1944 to 8 May 1945* (Washington, D.C.: Government Printing Office, n.d.), vi.

83. Eisenhower, *Crusade in Europe*, 229.

84. Figures for serviceable landing craft are from Samuel Eliot Morison, *The Invasion of France and Germany, 1944–45*, vol. 2 of *History of the United States Naval Operations in World War II* (Boston: Little, Brown, 1957), 57. Figures for the combat forces are from *Report by the Supreme Commander*, 8–10.

85. Harrison, *Cross-Channel Attack*, 266.

86. Ibid., 261.

87. Gunther Blumentritt, "Report of the Chief of Staff," in David C. Isby, ed., *Fighting the Invasion: The German Army at D-Day* (London: Greenhill Books, 2000), 20. This book contains interviews of senior German commanders taken immediately after the war with regard to the invasion of France.

88. Rommel, quoted by Lt. Gen. Hans Speidel in "Ideas and Views of Generalfeldmarschall Rommel on Defense and Operations in the West in 1944," in ibid., 43.

89. Blumentritt, "Report of the Chief of Staff," 26.

90. Ibid., 22.

91. Maj. Gen. Freiherr von Gersdorff, "Preparations against the Invasion," in Isby, *Fighting the Invasion*, 35.

92. Gen. de Panzertruppen Leo Freiherr Geyr von Schweppenburg, "Preparation by Panzer Gruppe West against Invasion," in Isby, *Fighting the Invasion*, 77.

93. Ibid.

94. Dwight D. Eisenhower to Combined Chiefs, June 23, 1944, in Chandler, *Papers of Dwight David Eisenhower*, 3:1943.

95. Harrison, *Cross-Channel Attack*, 94.

96. Martin Van Creveld, *Supplying War: Logistics from Wallenstein to Patton* (London: Cambridge University Press, 1977), 215.

97. See Russell F. Weigley, "Normandy to Falaise: A Critique of Allied Operational Planning in 1944," in Michael D. Krause and R. Cody Phillips, eds., *Historical Perspectives of the Operational Art* (Washington, D.C.: Center of Military History, 2005), 408; and Williamson Murray and Allan R. Millett, *A War to Be Won: Fighting the Second World War* (Cambridge: Harvard University Press, 2000), 432.

98. *Report by the Supreme Commander*, 121.

99. J. D. Hittle, *The Military Staff* (Harrisburg, Pa.: Stackpole Co., 1961), 71.

100. Schweppenburg, "Panzer Gruppe West," 76.

7. THE PACIFIC THEATER OF WAR

1. A colorful character, Homer Lea was born in Colorado in 1876. After attending Stanford University for several years he traveled to China to offer his military services to help reform the Chinese government. An associate of Dr. Sun Yat-sen, he served as a general with Chinese reform forces. His travels through Asia convinced him to write about his perception of the Japanese threat to America. Once published, *The Valor of Ignorance* caught the attention of William Randolph Hearst, who reworked Lea's thesis using his yellow journalism approach to boosting circulation. Lea died in California in 1912. See Clare Boothe Luce's introduction to Homer Lea, *The Valor of Ignorance* (New York: Harper & Brothers, 1942).

2. Bywater assumed the Japanese would strike without warning and seize Guam and the Philippine Islands. He believed they must achieve a decisive naval battle or lose the war. He suggested U.S. strategy should focus on retaking Guam and conducting a distant blockade. He saw a great future for carrier aviation in the vast distances of the Pacific theater. See Hector C. Bywater, *Sea-power in the Pacific: A Study of the American-Japanese Naval Problem* (Boston: Houghton, Mifflin, 1921).

3. In his fictional account of a future war in the Pacific, Bywater predicted the Japanese would defeat the U.S. Asiatic Squadron, capture the Philippines, mine Hawaiian and West Coast waters, and raid Dutch Harbor in Alaska. In this account the United States eventually takes Truk (Chuuk), Jaluit, and Ponape (now Pohnpei) and threatens Yap (all islands in the Japanese mandates). A decisive naval battle takes place off Yap, and following a

bloodless air raid on Tokyo, the Japanese sue for peace. Harry C. Butcher, *The Pacific War: The Campaign of 1931–33 (Boston: Houghton Mifflin, 1925).*

4. Matloff and Snell, *Strategic Planning for Coalition Warfare,* 166.

5. George C. Marshall to Douglas S. Freeman, April 7, 1943, in Bland and Stevens, *Papers of George Catlett Marshall,* 3:355.

6. Louis Morton, *Strategy and Command: The First Two Years,* United States Army in World War II (Washington, D.C.: Office of the Chief of Military History, 1962), 244.

7. Directive to the Supreme Commander in the Southwest Pacific Area (CCS 57/1), March 30, 1942, in Morton, *Strategy and Command,* 615, appendix C.

8. Ibid., 249.

9. Directive to the Commander in Chief of the Pacific Ocean Area (CCS 57/1), March 30, 1942, in ibid., appendix D, 617.

10. Ibid., 250. Also see Ronald H. Spector, *The American War with Japan: Eagle against the Sun* (New York: Free Press, 1985), 145.

11. Gen. Thomas Handy recalled in later years: "MacArthur hated the Navy. Practically accused them of running off and leaving him, and they never forgave him. General Marshall said one time, and the Navy just threw a fit, when the question of unified command out there was raised, If we set up a unified command, the commander can't be anybody but MacArthur on the basis of pure competence alone. That was the word he used. And I think it was the truth, but it wouldn't work, they wouldn't agree to it and, after all, the United States Navy was in the war." Gen. Thomas T. Handy, interviewed by Lt. Col. Edward M. Knoff, *U.S. Army Military History Institute Senior Officer Oral History Program,* vol. 2, 1974, 35, USAMHI.

12. Marshall wrote in a memorandum to the army's senior commanders, "It is apparent that vigorous action must be taken to suppress service jealousies and suspicions." Memorandum for Higher Commanders, September 11, 1942, in Bland and Stevens, *Papers of George Catlett Marshall,* 3:355.

13. Wedemeyer, *Wedemeyer Reports!,* 175, 192.

14. Ibid.

15. Cline, *Washington Command Post,* 237

16. Directive, quoted in ibid., 240.

17. See *Records of the Joint Chiefs of Staff,* pt. 1: *1942–1945: Pacific Theater* (Frederick, Md.: University Publications of America, 1981), Reel 13, USAMHI.

18. Joint Staff Planners, Operations to Recapture the Philippine Islands, September 24, 1943, *Records of the Joint Chiefs of Staff,* pt. 1: *Pacific Theater,* Reel 12, USAMHI.

19. Cline, *Washington Command Post,* 264.

20. Ibid.

21. Samuel Milner, *The War in the Pacific: Victory in Papua,* United States Army in World War II (Washington, D.C.: Center of Military History, 1957), 23.

22. Joint Chiefs of Staff Directive: Unified Command for U.S. Joint Organizations, April 20, 1943, in Morton, *Strategy and Command,* appendix L, 642.

23. George C. Marshall to Ernest J. King, memorandum, April 10, 1944, in Bland and Stevens, *Papers of George Catlett Marshall*, 4:393.

24. The JCS directive mandating joint staffs for unified commanders "will not function in a dual capacity as joint force commander and as commander of a component of his force, unless so directed by the Joint Chiefs." King argued that the JCS established Nimitz as commander of POA knowing that he was already commander of the Pacific Fleet as well. Morton, *Strategy and Command*, 477.

25. Emmons and later Richardson commanded the Hawaiian Department. On August 14, 1943, the War Department activated U.S. Army forces in the Central Pacific Ocean Areas under General Richardson, assigning him specifically to command and administer all army forces in Nimitz's theater of operations. The War Department redesignated Richardson's command as U.S. Army Forces, Pacific Ocean Areas, on August 1, 1944. Cline, *Washington Command Post*, 378.

26. Quoted in Morton, *Strategy and Command*, 491.

27. Marshall to King, memorandum, April 10, 1944, 4:393. See also Morton, *Strategy and Command*, 498–500.

28. Joint Chiefs of Staff Memorandum, Strategic Plan for the Defeat of Japan, May 19, 1943 (JCC 287/1), in Morton, *Strategy and Command*, appendix M, 645.

29. Headquarters, SWPA, Reno III: Outline Plan for Operations of the Southwest Pacific Area 1944, USAMHI.

30. Joint Staff Planners, "Future Operations in the Pacific," March 10, 1944, *Records of the Joint Chiefs of Staff*, Reel 9, USAMHI.

31. Radio Nos. 5171 and 989 to Douglas MacArthur and Chester W. Nimitz, March 12, 1944, in Bland and Stevens, *Papers of George Catlett Marshall*, 4:336.

32. Ibid.

33. MacArthur met with President Roosevelt and Admiral Nimitz in Pearl Harbor on July 10, 1944. No decision was made at the time of the conference, and there is debate among historians about whether MacArthur actually persuaded FDR to approve his return to Luzon. Apparently, MacArthur did persuade Admiral Leahy of his point of view, which carried great weight with the other chiefs. Larrabee, *Commander in Chief*, 344.

34. Robert Ross Smith, "Luzon Versus Formosa," in Kent Roberts Greenfield, ed., *Command Decisions* (Washington, D.C.: Center of Military History, 2000), 471, 472.

35. Headquarters, SWPA, RENO III: Outline Plan.

36. Ibid.

37. Ibid.

38. Ibid.

39. *Reports of General MacArthur*, vol. 1: *The Campaigns of MacArthur in the Pacific*, prepared by his General Staff (Washington, D.C.: Center for Military History, 1994), 170.

40. Larry I. Bland, ed., *George C. Marshall Interviews and Reminiscences for Forrest C. Pogue* (Lexington, Va.: George C. Marshall Foundation, 1996), 568.

41. Headquarters SWPA, Basic Plan for Musketeer III Operations, USAMHI.

42. Ibid.

43. Ibid.

44. See Operations Instructions No. 70, September 21, 1944, and Operations Instructions No. 73, October 12, 1944, in Chamberlin Papers, Box 3, USAMHI.

45. Walter Krueger, *From Down Under to Nippon* (Nashville, Tenn.: Battery Press, 1989), 141.

46. Louis Morton, "Japan's Decision for War," in Kent Roberts Greenfield, ed., *Command Decisions* (Washington, D.C.: Center of Military History, 2000), 106.

47. Milan Vego, *The Battle for Leyte, 1944: Allied and Japanese Plans, Preparations, and Execution* (Annapolis: Naval Institute Press, 2006), 47, 48.

48. Morton, *Strategy and Command*, 235.

49. Ibid., 237.

50. Masataka Chihaya, "The Organization of the Japanese Naval General Staff Headquarters in Tokyo," in Donald M. Goldstein and Katherine V. Dillon, eds., *The Pacific War Papers: Japanese Documents of World War II* (Washington, D.C.: Potomac Books, 2004), 38, 39; and U.S. War Department, *Handbook on Japanese Military Forces* (1944; reprint, Baton Rouge: Louisiana State University Press, 1995), 10.

51. M. Hamlin Cannon, *Leyte: The Return to the Philippines*, United States Army in World War II (Washington, D.C.: Center of Military History, 1954), 50.

52. Interview with Maj. Gen. Yoshiharu, October 14, 1947, in *Interrogations of Japanese Officials on World War II* (English trans.), General Headquarters, Far East Command, USAMHI.

53. *Campaigns of MacArthur*, 370.

54. Operations Instructions No. 70, and Operations Instructions No. 73, Chamberlin Papers.

55. *Reports of General MacArthur*, vol. 2: *Japanese Operations in the Southwest Pacific Area*, prepared by his General Staff (Washington, D.C.: Center for Military History, 1994), 385.

56. The Fifth Fighter Command lost a total of 203 aircraft from bombing, accidents and other causes. Herman S. Wolk, "George C. Kenney: MacArthur's Premier Airman," in *We Shall Return! MacArthur's Commanders an and the Defeat of Japan*, ed. William M. Leary (Lexington: University Press of Kentucky, 1988), 110.

57. *Japanese Operations in the Southwest Pacific*, 237.

58. Cannon, *Leyte*, 102.

59. Ibid.

60. *Japanese Operations in the Southwest Pacific*, 451–54.

61. Robert Ross Smith, *Triumph in the Philippines*, United States Army in World War II (Washington, D.C.: Center of Military History, 1991), 242.

62. Ibid., 306.

63. Curiously, the original directive from the Joint Chiefs directed MacArthur to take Luzon; there was no explicit order authorizing him to liberate

the rest of the Philippines. In March 1945, Marshall sent MacArthur a draft of the directive outlining future final operations against Japan that included the authorization "to conduct such additional operations toward completing the liberation of the Philippines as can be mounted without prejudice to the accomplishment of the overall objective." The JCS issued the directive on April 4, 1945. George C. Marshall to Douglas MacArthur, March 29, 1945, in Bland and Stevens, *Papers of George Catlett Marshall*, 5:104.

64. This figure does not provide the whole picture of U.S. casualties. From January to June, the Sixth Army suffered 93,400 nonbattle casualties, mostly for illness. Obviously, the great majority of these were returned to duty. Smith, *Triumph in the Philippines*, 652.

65. Interview with Col. Takio Shindo, December 19, 1947, in *Interrogations of Japanese Officials*, 253.

66. Interview with Maj. Gen. Yoshiharu Tomochika, Chief of Staff, 35th Army, October 14, 1947, in ibid., 7, 13.

67. Willoughby, who initially estimated Japanese strength on Luzon at 158,000, later revised the figure in January to 195,000. Krueger's staff's initial estimate of Japanese strength was 234,500; they then raised their estimate in January to 287,000. The actual number of Japanese troops on Luzon in January 1945 was 267,000. Murray and Millett, *War to Be Won*, 495, 496.

68. Early in the war the United States broke Japan's naval and diplomatic codes. The code breakers named these intercepted and decoded messages MAGIC. The decryptions proved enormously useful in determining Japanese operations such as in the attack on Midway in 1943.

69. Smith, *Triumph in the Philippines*, 92.

70. Ibid.

71. Ibid., 91.

72. Interview with Maj. Gen. Tomochika, 28.

73. Robert Sherrod, *History of Marine Corps Aviation in World War II* (San Rafael, Calif.: Presidio Press, 1952), 311.

74. See Murray and Millett, *War to Be Won*, 502, 503; and Vego, *Battle for Leyte*, 342.

75. Interview with Maj. Gen. Tomochika, 21.

76. Joint Staff Planners, "Future Operations in the Pacific," September 23, 1944, *Records of the Joint Chiefs of Staff*, Reel 9, USAMHI.

77. Ibid., October 3, 1944.

78. Joint War Plans Committee, "Plan for the Seizure of the Ryukyus," November 6, 1944, *Records of the Joint Chiefs of Staff*, Reel 7, USAMHI.

79. Ibid.

80. Ibid., "Examination of ICEBERG," January 2, 1945.

81. Nimitz was worried about Japanese airpower. There were fifty-five airfields on Kyushu and sixty-five on Formosa. With no other major combat operations going on in the Pacific, the Japanese would be able to mass their airpower against Central Pacific forces assaulting Okinawa. E. B. Potter, *Nimitz* (Annapolis: Naval Institute Press, 1976), 368.

82. The official U.S. Air Force history notes, "None of the theater commanders—Nimitz, MacArthur, Stilwell—had shown himself an enthusiastic advocate of the type of mission for which the B-29 was being prepared, and it

was not unnatural that the AAF should be reluctant to assign permanently to those leaders its most potent bomber." Wesley Frank Craven and James Lea Cate, *The Army Air Forces in World War II*, 5 vols. (Chicago: University of Chicago Press, 1953), 5:35.

83. Ibid., 630.

84. Ibid.

85. "Tenth Army After Action Report for March 26 to June 30," September 3, 1945, USAMHI.

86. Buckner Diary entry, February 7, 1945, in Nicholas Evan Sarantakes, ed., *Seven Stars: The Okinawa Battle Diaries of Simon Bolivar Buckner, Jr., and Joseph Stilwell* (College Station: Texas A&M University Press, 2004), 19.

87. Sherrod, *Marine Corps Aviation in World War II*, 374.

88. Headquarters Tenth Army, Tentative Operations Plan 1-45, Iceberg, January 6, 1945, USAMHI.

89. Robert W. Coakley and Richard M. Leighton, *Global Logistics and Strategy: 1943–1945* (Washington, D.C.: Center of Military History, 1989), 428.

90. Ibid., 448.

91. Ibid., 449.

92. Headquarters Island Command, "Action Report Island Command Okinawa, June 30, 1945," USAMHI.

93. Roy E. Appleman et al., *Okinawa: The Last Battle*, The United States Army in World War II (Washington, D.C.: Center of Military History, 1993), 419.

94. Ibid., 74.

95. Ibid., 91.

96. "Tenth Army After Action Report for March 26 to June 30," September 3, 1945, USAMHI.

97. Buckner Diary entry, April 2, 1945, in Sarantakes, *Seven Stars*, 30.

98. "Tenth Army Action Report for March 26 to June 30."

99. Appleman et al., *Okinawa*, 102.

100. Potter, *Nimitz*, 375.

101. Buckner Diary entry, April 11, 1945, in Sarantakes, *Seven Stars*, 37.

102. For the best Japanese account of the battle, see Hiromichi Yahara, *The Battle for Okinawa*, trans. Frank B. Gibney (New York: John Wiley and Sons, 1995). Colonel Yahara served as the operations officer for the Thirty-second Army. This book also includes the original interrogation report of Colonel Yahara conducted by Frank Gibney in 1945.

103. Letter to Adele Buckner, April 14, 1945, in Sarantakes, *Seven Stars*, 39.

104. Appleman et al., *Okinawa*, 256.

105. "Tenth Army Action Report March 26–30 June 1945."

106. Appleman et al., *Okinawa*, 474.

107. Bland, *Marshall Interviews and Reminiscences*, 423.

108. Murray and Millett, *War to Be Won*, 514, 515.

109. Joint Logistics Committee, "Joint Logistical Plan for Operations in the Ryukyus," January 6, 1945, *Joint Chiefs of Staff Records*, Reel 7, USAMHI.

110. Appleman et al., *Okinawa*, 263.

111. Press conference, June 15, 1945, in Sarantakes, *Seven Stars*, 80.

8. LESSONS AND LEGACY

1. Richard Overy, *Why the Allies Won* (New York: W. W. Norton & Company, 1995), 325.

2. Clausewitz observed, "Combat is the only effective force in war; its aim is to destroy the enemy's force as a means to a further end." *On War*, 97.

3. Pershing, *My Experiences in the World War*, 1:16; and Marshall, *Memoirs of My Services*, 8.

4. Military historian Edward M. Coffman noted, "The spectacular row over air power and the development of an armored force in the postwar period were indications of change, but more basic was the new professionalism of the army. In part this was a result of the war experience, but there was a greatly increased emphasis on the schools and professional training. Although a few officers attended the Leavenworth schools and the AWC before 1917, many did not think this training essential. After the war, it was." Coffman, *War to End All Wars*, 361–62.

5. Eichelberger, *Our Jungle Road to Tokyo*, xv.

6. Ernest N. Harmon, *Combat Commander: Autobiography of a Soldier* (London: Prentice Hall International, 1970), 49.

7. In the plans for Operations Overlord, Reno, and Iceberg the intelligence and particularly the logistic annexes or separate plans frequently constituted the largest portions of the overall plans.

8. Joint Staff Planners, Directive for Deception Measures against Japan, May 3, 1944, *Records of the Joint Chiefs of Staff*, pt. 1: *Pacific Theater*, Reel 8, USAMHI.

9. Schweppenburg, "Panzer Gruppe West," 76.

10. "Prisoner of War Interrogation Report," August 6, 1945, in Yahara, *Battle for Okinawa*, 216.

11. Forrest C. Pogue, *The Supreme Command*, United States Army in World War II (Washington, D.C.: Center of Military History, 1954), 74.

12. Weigley, *History of the United States Army*, 322, 323, 379, 405.

13. Van Creveld, *Supplying War*, 201.

14. Ibid., 180.

15. Bushido refers to the way of the warrior, the samurai code that demanded loyalty and considered surrender to be the ultimate dishonor. The Imperial Army encouraged bushido in its officers. This personal philosophy often produced a death-and-glory fatalism over more rational or realistic tactical and operational decision making. Meirion Harries and Susie Harries, *Soldiers of the Sun: The Rise and Fall of the Imperial Japanese Army* (New York: Random House, 1991), 481.

16. D'Este, *Eisenhower*, 592.

17. Van Creveld, *Supplying War*, 215.

18. Bland, *Marshall Interviews and Reminiscences*, 161.

19. Collins, *Lightning Joe*, 94.

20. Weigley, "From Normandy to Falaise-Argentan," 45.

21. Bond, *Pursuit of Victory*, 154.

22. Martin Van Creveld, *The Training of Officers: From Military Professionalism to Irrelevance* (New York: Free Press, 1990), 66.

23. D'Este, *Eisenhower*, 177.

24. Wedemeyer, *Wedemeyer Reports!*, 48.

25. Command and General Staff Schools, *Historical Illustrations and References* (Fort Leavenworth, Kans.: General Service School Press, 1927), iii, Richards Papers, Box 10, USAMHI.

26. Robert H. Berlin, *U.S. Army World War II Corps Commanders: A Composite Biography* (Fort Leavenworth, Kans.: Combat Studies Institute, 1989), 10, 12.

27. Williamson Murray, "Innovation: Past and Present," in Murray and Millett, *Military Innovation*, 311.

28. *The United States Strategic Bombing Surveys: European War, Pacific War* (Maxwell Air Force Base, Ala.: Air University Press, 1987), 37–39.

29. Ibid., 90.

30. Ibid., 107.

31. Eisenhower, *Crusade in Europe*, 210.

32. Paula Reading, "History of the Army and Navy Staff College" (Washington, D.C.: National War College Library, 1970), 32.

33. *Joint Overseas Operations*, pt. 1 (1946; reprint, Norfolk, Va.: U.S. Armed Forces Staff College, 1950), 1-4.

34. Ibid., 2-4.

35. Ibid., 5-1, 5-2.

36. Ibid., 7-4.

37. Ibid., 5-7.

38. Ibid., 5-6.

39. Eisenhower, *Crusade in Europe*, 452.

40. Conrad C. Crane, *American Air Power Strategy in Korea, 1950–1953* (Lawrence: University Press of Kansas, 2000), 183; and Mark Clodfelter, *The Limits of Airpower: The American Bombing of North Vietnam* (New York: Free Press, 1989), 203.

41. *Joint Overseas Operations*, preface.

42. James R. Locher III, *Victory on the Potomac: The Goldwater-Nichols Act Unifies the Pentagon* (College Station: Texas A&M University Press, 2002), 446.

43. The best discussion of how the U.S. Army incorporated operational art as a concept into official military doctrine can be found in Swain, "Filling the Void," 147–72.

44. Ball, *Of Responsible Command*, 257.

45. Reading, "Army and Navy Staff College," 2.

46. Ibid., 9.

47. For a discussion of the development of joint education, see Joint Chiefs of Staff Historical Section, *Joint Chiefs of Staff and the Joint Education System, 1943–1986* (Washington, D.C., 1988).

48. The army renamed the Command and General Staff School as the Command and General Staff College in 1947.

49. Ball, *Of Responsible Command*, 272, 273.

Selected Bibliography

ARCHIVAL SOURCES

U.S. Army Military History Institute, Carlisle Barracks, Pennsylvania
(USAMHI)
    Mircrofilm Collection
        Records of the Joint Chiefs of Staff: Pacific Theater
    Army War College Files
    Personal Papers Collection
        Almond, Edward M. Papers
        Chamberlin, Stephen J. Papers
        Devers, Jacob Papers
        Drum, Hugh A. Papers
        Eaker, Ira C. Papers
        Krueger, Walter. Papers
        Lee, John C. H. Papers
        Richards, George J. Papers
        Simpson, William H. Papers
    Oral History Collection
        Interview transcriptions
            Handy, Thomas T.
            Lemnitzer, Lyman L.
            Simpson, William H.
U.S. Army Combined Arms Research Library, Fort Leavenworth, Kansas
(CARL)
    Command and General Staff School Files
    National Archives and Records Administration, College Park, Maryland
(NARA)

Records Group 407.3, Reports Relating to World War II and Korean War
   Combat Operations, 1940–54
      Records of the Supreme Allied Commander, Entry 427
Naval War College Archives
   Curricular Files

                      PUBLISHED PRIMARY SOURCES

Arnold, H. H., and Ira C. Eaker. *Winged Warfare*. New York: Harper & Broth-
   ers, 1941.
Bigelow, John. *The Principles of Strategy*. Philadelphia: J. B. Lippincott & Co.,
   1894.
Bland, Larry I., ed. *George C. Marshall Interviews and Reminiscences for
   Forrest C. Pogue*. Lexington, Va.: George C. Marshall Foundation, 1996.
Bland, Larry I., and Sharon Ritenour Stevens, eds. *The Papers of George Catlett
   Marshall*. 4 vols. Baltimore: Johns Hopkins University Press, 1981–96.
Bradley, Omar N., and Clay Blair. *A General's Life: Autobiography of General
   of the Army Omar N. Bradley*. New York: Simon and Schuster, 1983.
Butcher, Harry C. *My Three Years with Eisenhower*. New York: Simon and
   Schuster, 1946.
Bywater, Hector C. *Sea-power in the Pacific: A Study of the American-
   Japanese Naval Problem*. Boston: Houghton Mifflin, 1921.
——. *The Pacific War: The Campaign of 1931–33*. Boston: Houghton Mifflin,
   1925.
Chandler, Alfred G., Jr., ed. *The Papers of Dwight David Eisenhower: The
   War Years*. 5 vols. Baltimore: Johns Hopkins University Press, 1970.
Churchill, Winston S. *The Hinge of Fate*. Vol. 4 of *The Second World War*.
   Boston: Houghton Mifflin, 1950.
——. *Closing the Ring*. Vol. 6 of *The Second World War*. Boston: Houghton
   Mifflin Company, 1950.
Clark, Mark. *Calculated Risk*. New York: Harper & Brothers, 1950.
Clausewitz, Carl von. *On War*. Translated by Michael Howard and Peter
   Paret. Princeton, N.J.: Princeton University Press, 1976.
Collins, J. Lawton. *Lightning Joe: An Autobiography*. Baton Rouge: Louisiana
   State University Press, 1979.
Corbett, Julian S. *Some Principles of Maritime Strategy*. London: Longmans,
   Green and Co., 1911.
Derrecagaix, Victor. *Modern War*. Vol. 1: *Strategy*. Translated by C. W. Foster.
   Washington, D.C.: James J. Chapman, 1888.
Eichelberger, Robert L. *Our Jungle Road to Tokyo*. Nashville, Tenn.: Battery
   Press, 1989.
Eisenhower, Dwight D. *Crusade in Europe*. New York: Doubleday & Co.,
   1948.
Foertsch, Hermann. *The Art of Modern Warfare*. New York: Veritas Press,
   1940.
Fretag-Loringhoven, Hugo. *Generalship in the World War*. Translated from
   the German at the Army War College, Washington, D.C.; Berlin: E. S.
   Mitter & Sohn, 1920.

Fuller, J. F. C. *The Foundations of the Science of War*. London: Hutchinson and Co., 1926.

Goldstein, Donald M., and Katherine V. Dillon, eds. *The Pacific War Papers: Japanese Documents of World War II*. Washington, D.C.: Potomac Books, 2004.

Harmon, Ernest N. *Combat Commander: Autobiography of a Soldier*. London: Prentice Hall International, 1970.

Historical Branch, War Plans Division, General Staff. *Organization of the Services of Supply: American Expeditionary Forces*. Washington, D.C.: Government Printing Office, 1921.

War IHolt, Daniel D., and James Leyerzapf, eds. *Eisenhower: The Prewar Diaries and Selected Papers, 1905–1941*. Baltimore: Johns Hopkins University Press, 1998.

Isby, David C., ed. *Fighting the Invasion: The German Army at D-Day*. London: Greenhill Books, 2000.

Joint Chiefs of Staff. *Joint Publication 1: Joint Warfare of the Armed Forces of the United States*. Washington, D.C.: Joint Chiefs of Staff, 2007.

——. *Joint Publication 3.0: Doctrine for Joint Operations*. Washington, D.C.: Joint Chiefs of Staff, 2006, incorporating Change 1 February 2008.

Joint Chiefs of Staff Historical Section, *Joint Chiefs of Staff and the Joint Education System, 1943–1986*. Washington, D.C.: Joint Chiefs of Staff, 1988.

Joint Overseas Operations. 1946. Reprint, Norfolk, Va.: U.S. Armed Forces Staff College, 1950.

Jomini, Henri de. *The Art of War*. Translated by G. H. Mendell and W. P. Craigill. Philadelphia: J. B. Lippincott & Co., 1862.

King, Ernest J., and Walter Muir Whitehill. *Fleet Admiral King: A Naval Record*. New York: W. W. Norton & Company, 1952.

Krueger, Walter. *From Down Under to Nippon*. Nashville, Tenn.: Battery Press, 1989.

Lea, Homer. *The Valor of Ignorance*. New York: Harper & Brothers, 1942.

Liddell Hart, Basil H. *The British Way in Warfare*. London: Faber and Faber, 1932.

——. *Strategy of the Indirect Approach*. London: Faber and Faber, 1941.

Luce, Stephen B. "Naval Administration, III." *U.S. Naval Institute Proceedings* 29, no. 4 (December 1903).

Ludendorff, Erich. *Ludendorff's Own Story: August 1914–November 1918*. New York: Harper & Brothers, 1919.

Luvaas, Jay, ed. and trans. *Napoleon on the Art of War*. New York: Free Press, 1999.

Mahan, Alfred T. *The Influence of Sea Power upon History, 1660–1783*. Boston: Little, Brown, 1890.

——. *Naval Strategy: Compared and Contrasted with the Principles and Practice of Military Operations on Land*. Boston: Little, Brown, 1911.

Marshall, George C. *Memoirs of My Services in the World War, 1917–1918*. Boston: Houghton Mifflin, 1976.

McFarland, Marvin W., ed. *The Papers of Wilbur and Orville Wright*. New York: McGraw-Hill, 1953.

Mitchell, William. *Memoirs of World War I: From Start to Finish of Our Greatest War*. New York: Random House, 1960.

——. *Winged Defense: The Development and Possibilities of Modern Air Power—Economic and Military*. New York: G. P. Putnam's Sons, 1925.

Moltke, Helmuth von. *Moltke on the Art of War: Selected Writings*. Edited by Daniel J. Hughes. Translated by Harry Bell and Daniel J. Hughes. Novato, Calif.: Presidio Press, 1993.

Naylor, William K. *Principles of Strategy: With Historical Illustrations*. Fort Leavenworth, Kans.: The General Service Schools Press, 1921.

Patrick, Mason M. *The United States in the Air*. Garden City, N.J.: Doubleday, Doran and Co., 1928.

Pershing, John J. *My Experiences in the World War*. 2 vols. New York: Frederick A. Stokes Company, 1931.

Report by the Supreme Commander to the Combined Chiefs of Staff on the Operations in Europe of the Allied Expeditionary Force 6 June 1944 to 8 May 1945. Washington, D.C.: Government Printing Office, n.d.

Reports of General MacArthur. 2 vols. Prepared by his General Staff. 1950. Reprint, Washington, D.C.: Center for Military History, 1994.

Robinson, Oliver P. *The Fundamentals of Military Strategy*. Washington, D.C.: United States Infantry Association, 1928.

Sarantakes, Nicholas Evan, ed., *Seven Stars: The Okinawa Battle Diaries of Simon Bolivar Buckner, Jr., and Joseph Stilwell*. College Station: Texas A&M University Press, 2004.

Saxe, Maurice de. *Reveries upon the Art of War*. Edited and translated by Thomas R. Phillips. Harrisburg, Pa.: Military Service Publishing Company, 1944.

Schellendorff, Bronsart von. *The Duties of the General Staff.* Translated for the General Staff. London: Harrison and Sons, 1905.

Sherman, William C. *Air Warfare*. New York: Ronald Press, 1926.

Sherman, William T. *Memoirs of General W. T. Sherman*. 1875 (D. Appleton and Co.). Reprint, New York: Literary Classics of the United States, 1990.

Sims, William S. *The Victory at Sea*. Garden City, N.J.: Doubleday, 1920.

Taylor, Maxwell D. *Swords and Plowshares*. New York: W. W. Norton & Company, 1972.

United States Army. *Principles of Strategy for an Independent Corps or Army in a Theater of Operations*. Fort Leavenworth, Kans.: Command and General Staff School Press, 1936.

United States Army in the World War, 1917–1919. 17 vols. Washington, D.C.: Center of Military History, 1948.

United States Army War College. *New Problems in Warfare*. Carlisle Barracks, Pa.: U.S. Army War College, 1983.

United States Naval War College. *Sound Military Decision*. Annapolis, 1942.

United States Strategic Bombing Surveys: European War, Pacific War. Maxwell Air Force Base, Ala.: Air University Press, 1945.

United States War Department. *Annual Report of the Secretary of War*. Washington, D.C.: Government Printing Office, 1932.

——. *Field Service Regulations, United States Army 1914*. Washington, D.C.: Government Printing Office, 1914.

——. *Five Years of the War Department Reports Following the War with Spain, 1899–1903*. Washington, D.C.: Government Printing Office, 1904.

——. *Handbook on Japanese Military Forces*. Introduction by David Isby. 1944. Reprint, Baton Rouge: Louisiana State University Press, 1995.

——. *Report of the Secretary of War to the President for 1922*. Washington. D.C.: Government Printing Office, 1922.

——. *Report of the Secretary of War to the President for 1923*. Washington, D.C.: Government Printing Office, 1923.

——. *Training Regulations No. 440-15: Employment of the Air Forces of the Army*. Washington, D.C., 1935.

——. *Training Regulations No. 440-15: Fundamental Principles for the Employment of the Air Service*. Washington, D.C., 1926.

Wagner, Arthur L. *Organization and Tactics*. Kansas City, Mo.: Hudson-Kimberly, 1897.

——. *Strategy: A Lecture Delivered by Colonel Arthur L. Wagner to the Officers of the Regular Army and National Guard at the Maneuvers at West Point, Kentucky and at Fort Riley, Kansas, 1903*. Kansas City, Mo.: Kimberly Publishing Company, 1904.

Wagner, Arthur L., Eben Swift, J. T. Dickman, and A. L. Miles. *Strategical Operations: Illustrated by Great Campaigns in Europe and America*. Fort Leavenworth, Kans.: United States Army Infantry and Cavalry School, 1897.

Wedemeyer, Albert C. *Wedemeyer Reports!* New York: Henry Holt & Co., 1958.

Yahara, Hiromichi. *The Battle for Okinawa*. Translated by Frank B. Gibney. New York: John Wiley and Sons, 1995.

BOOKS, ARTICLES, AND THESES

Abbatiello, John J. "The First Allied Airborne Army in Operation Varsity: Applying the Lessons of Arnhem." Master's thesis, Kings College, London, 1995.

Appleman, Roy E., James M. Burns, Russell A. Gugeler, and John Stevens. *Okinawa: The Last Battle*. The United States Army in World War II. Washington, D.C.: Center of Military History, 1993.

Asada, Sadao. *From Mahan to Pearl Harbor: The Imperial Japanese Navy and the United States*. Annapolis: Naval Institute Press, 2006.

Baer, George W. *One Hundred Years of Sea Power: The U.S. Navy, 1890–1990*. Stanford, Calif.: Stanford University Press, 1994.

Ball, Harry P. *Of Responsible Command: A History of the U.S. Army War College*. 2nd rev. ed. Carlisle Barracks, Pa.: Alumni Association of the U.S. Army War College, 1994.

Berlin, Robert H. *U.S. Army World War II Corps Commanders: A Composite Biography*. Fort Leavenworth, Kans.: Combat Studies Institute, 1989.

Binder, L. James. *Lemnitzer: A Soldier for His Time*. London: Brassey's, 1997.

Bond, Brian. *The Pursuit of Victory from Napoleon to Saddam Hussein*. Oxford: Oxford University Press, 1996.

Braim, Paul F. *The Test of Battle: The American Expeditionary Forces in the Meuse-Argonne Campaign.* Newark: University of Delaware Press, 1987.

Brereton, T. R. *Educating the Army: Arthur L. Wagner and Reform, 1875–1905.* Lincoln: University of Nebraska Press, 2000.

Bryant, Arthur. *The Turn of the Tide, 1939–1943: A History of the War Years Based on the Diaries of Field Marshal Lord Alanbrooke, Chief of the Imperial General Staff.* Garden City, N.Y.: Doubleday & Co., 1957.

Buell, Thomas B. *Master of Sea Power: A Biography of Fleet Admiral Ernest J. King.* Boston: Little, Brown, 1980.

——. *The Quiet Warrior: A Biography of Admiral Raymond A. Spruance.* Boston: Little, Brown, 1971.

Cannon, M. Hamlin. *Leyte: The Return to the Philippines.* The United States Army in World War II. Washington, D.C.: Center of Military History, 1954.

Carter, William. "The Training of Officers." *United Services Journal* 2 (1902).

Clifford, Kenneth. *Amphibious Warfare Development in Britain and America from 1920–1940.* Laurens, N.Y.: Edgewood Press, 1983.

Cline, Ray. *Washington Command Post: The Operations Division.* The United States Army in World War II. Washington, D.C.: Center of Military History, 1990.

Clodfelter, Mark. *The Limits of Airpower: The American Bombing of North Vietnam.* New York: Free Press, 1989.

Coakley, Robert W., and Richard M. Leighton. *Global Logistics and Strategy, 1943–1945.* Washington, D.C.: Center of Military History, 1989.

Coffman, Edward M. *The Regulars: The American Army, 1898–1941.* Cambridge: Harvard University Press, 2004.

——. *The War to End All Wars: The American Military Experience in World War I.* Lexington: University Press of Kentucky, 1968.

Connolly, Peter. *Greece and Rome at War.* London: Greenhill Books, 1998.

Corum, James S. *The Luftwaffe: Creating the Operational Air War, 1918–1940.* Lawrence: University Press of Kansas, 1997.

Crane, Conrad C. *American Air Power Strategy in Korea, 1950–1953.* Lawrence: University Press of Kansas, 2000.

Dastrup, Boyd L. *The U.S. Army Command and General Staff College: A Centennial History.* Manhattan, Kans.: Sunflower University Press, 1982.

Davis, Richard G. *Carl A. Spaatz and the Air War in Europe.* Washington, D.C.: Center for Air Force History, 1993.

D'Este, Carlo. *Eisenhower: A Soldier's Life.* New York: Henry Holt & Co., 2002.

Doughty, Robert A. *The Seeds of Disaster: The Development of French Army Doctrine 1919–1939.* Hamden, Conn.: Archon Books, 1985.

Echevarria, Antulio J., II. *After Clausewitz: German Military Thinkers before the Great War.* Lawrence: University Press of Kansas, 2000.

English, John A. *Marching through Chaos: The Descent of Armies in Theory and Practice.* Westport, Conn.: Praeger Publishers, 1996.

Felker, Craig C. *Testing American Sea Power: U.S. Navy Strategic Exercises, 1923–1940.* College Station: Texas A&M University Press, 2007.

Franz, Wallace P. "Two Letters on Strategy: Clausewitz' Contribution to the Operational Level of War." In *Clausewitz and Modern Strategy*. Edited by Michael I. Handel. London: Frank Cass, 1986.

Futrell, Robert F. *Ideas, Concepts, Doctrine*, vol. 1: *Basic Thinking in the United States Air Force, 1907–1960*. Maxwell Air Force Base, Ala.: Air University Press, 1989.

Gat, Azar. *A History of Military Thought from the Enlightenment to the Cold War*. London: Oxford University Press, 2001.

Glantz, David. "The Nature of Operational Art." *Parameters* 15 (1985): 2–12.

———. *The Nature of Soviet Operational Art, 1920–1932*. Fort Leavenworth, Kans.: U.S. Army Combined Arms Center, 1985.

———. *Soviet Military Operational Art: In Pursuit of Deep Battle*. London: Frank Cass, 1991.

Gray, Colin S. *Modern Strategy*. New York: Oxford University Press, 1999.

Greenfield, Kent Roberts, ed. *Command Decisions*. Washington, D.C.: Center of Military History, 2000.

Greer, Thomas H. *The Development of Air Doctrine in the Army Air Arm, 1917–1941*. Washington, D.C.: Office of Air Force History, 1985.

Grimsley, Mark. *And Keep Moving On: The Virginia Campaign, May–June 1864*. Lincoln: University of Nebraska Press, 2002.

Harries, Meirion, and Susan Harries. *Soldiers of the Sun: The Rise and Fall of the Imperial Japanese Army*. New York: Random House, 1991.

Harrison, Richard W. *The Russian Way of War: Operational Art, 1904–1940*. Lawrence: University Press of Kansas, 2001.

Hattendorf, John B., B. Mitchell Simpson III, and John R. Wadleigh. *Sailors and Scholars: The Centennial History of the U.S. Naval War College*. Newport, R.I.: Naval War College Press, 1984.

Hinsley, F. H., C. F. G. Ransom, R. C. Knight, and E. E. Thomas. *British Intelligence in the Second World War*, vols. 1–2: *Its Influence on Strategy and Operations*. 2 vols. New York: Cambridge University Press, 1981.

Hogan, David W. *Command Post at War: First Army Headquarters in Europe, 1943–1945*. Washington, D.C.: Center of Military History, 2000.

Hone, Thomas C., and Trent Hone. *Battleline: The United States Navy, 1919–1939*. Annapolis: Naval Institute Press, 2006.

Howard, Michael. *Studies in War and Peace*. New York: Viking Press, 1970.

Howe, George F. *Northwest Africa: Seizing the Initiative in the West*. The United States Army in World War II. Washington, D.C.: Center of Military History, 1985.

Hudson, James J. *A Combat History of the American Air Service in World War I*. New York: Syracuse University Press, 1968.

Johnson, David E. *Fast Tanks and Heavy Bombers: Innovation in the U.S. Army 1917–1945*. Ithaca, N.Y.: Cornell University Press, 1998.

Kelly, Fred C. *The Wright Brothers*. 1943. Reprint, New York: Ballantine Books, 1969.

Kennedy, Paul. *The Rise and Fall of the Great Powers: Economic Change and Military Conflict from 1500 to 2000*. New York: Random House, 1987.

Kennett, Lee. *The First Air War, 1914–1918*. New York: Free Press, 1991.

Kipp, Jacob. *Mass, Mobility, and the Red Army's Road to Operational Art, 1918–1936.* Fort Leavenworth, Kans.: Command and General Staff College, 1987.

Knox, MacGregor, and Williamson Murray, eds. *The Dynamics of Military Revolution, 1300–2050.* Cambridge: Cambridge University Press, 2001.

Krause, Michael D., and R. Cody Phillips, eds. *Historical Perspectives of the Operational Art.* Washington, D.C.: United States Army Center of Military History, 2005.

Larrabee, Eric. *Commander in Chief: Franklin Delano Roosevelt, His Lieutenants, and Their War.* New York: Harper & Row, 1987.

Locher, James R., III. *Victory on the Potomac: The Goldwater-Nichols Act Unifies the Pentagon.* College Station: Texas A&M University, 2002.

Matloff, Maurice, and Edwin M. Snell. *Strategic Planning for Coalition Warfare, 1941–42.* The United States Army in World War II. Washington, D.C.: Office of the Chief of Military History, Department of the Army, 1953.

——. *Strategic Planning for Coalition Warfare, 1943–1944.* The United States Army in World War II. Washington, D.C.: Office of the Chief of Military History, Department of the Army, 1959.

Maurer, Maurer, ed. *The U.S. Air Service in World War I.* Washington, D.C.: Government Printing Office, 1978.

McKercher, B. J. C., and Michael Hennessy, eds. *The Operational Art: Developments in the Theories of War.* Westport, Conn.: Praeger Publishers, 1996.

Meilinger, Phillip S., ed. *Airmen and Air Theory.* Maxwell Air Force Base, Ala.: Air University Press, 1997.

——. *The Paths of Heaven: The Evolution of Airpower Theory.* Maxwell Air Force Base, Ala.: Air University Press, 1997.

Miller, Edward S. *War Plan Orange: The U.S. Strategy to Defeat Japan, 1897–1945.* Annapolis: Naval Institute Press, 1991.

Millett, Allan R. *The General: Robert L. Bullard and Officership in the United States Army, 1881–1925.* Westport, Conn.: Greenwood Press, 1975.

——. *Simper Fidelis: The History of the United States Marine Corps.* New York: Macmillan Publishing Co., 1980.

Millett, Allan R., and Peter Maslowski. *For the Common Defense: A Military History of the United States of America.* New York: Free Press, 1994.

Millett, Allan R. and Williamson Murray, eds. *Military Effectiveness: The First World War.* Boston: Allen & Unwin, 1988.

Milner, Samuel. *The War in the Pacific: Victory in Papua.* The United States Army in World War II. Washington, D.C.; Center of Military History, 1957.

Morison, Samuel Eliot. *The Invasion of France and Germany, 1944–45.* Vol. 11 of *History of the United States Naval Operations in World War II.* Boston: Little, Brown, 1957.

Morton, Louis. *Strategy and Command: The First Two Years.* United States Army in World War II. Washington, D.C.: Office of the Chief of Military History, Department of the Army, 1962.

Murray, Williamson, and Allan R. Millett, eds. *Military Innovation in the Interwar Period*. Cambridge: Cambridge University Press, 1996.

—— *A War to Be Won: Fighting the Second World War*. Cambridge: Harvard University Press, 2000.

Naveh, Shimon. *In Pursuit of Military Excellence: The Evolution of Operational Theory*. London: Frank Cass, 1997.

Nenninger, Timothy K. *The Leavenworth Schools and the Old Army: Education, Professionalism, and the Officer Corps of the United States Army, 1881–1918*. Westport, Conn.: Greenwood Press, 1978.

Overy, Richard. *Why the Allies Won*. New York: W. W. Norton & Company, 1995.

Parker, Geoffrey. *The Military Revolution: Military Innovation and the Rise of the West, 1500–1800*. London: Cambridge University Press, 1996.

Pearlman, Michael D. *Warmaking and American Democracy: The Struggle over Military Strategy, 1700 to the Present*. Lawrence: University Press of Kansas, 1999.

Pogue, Forrest C. *The Supreme Command*. The United States Army in World War II. Washington, D.C.: Center of Military History, 1954.

Potter, E. B. *Nimitz*. Annapolis: Naval Institute Press, 1976.

Preston, Richard A., and Sydney Wise. *Men in Arms*. New York: Holt, Rinehart and Winston, 1979.

Reading, Paula. "History of the Army and Navy Staff College." Photocopy of typescript draft. Washington, D.C.: National War College Library, 1970.

Ross, Steven T. *American War Plans, 1890–1939*. London: Frank Cass, 2002.

——. *U.S. War Plans, 1939–1945*. Malabar, Fla.: Krieger Publishing Company, 2000.

Schifferle, Peter J. "Anticipating Armageddon: The Leavenworth Schools and U.S. Army Military Effectiveness 1919 to 1945." Ph.D. diss., University of Kansas, 2002.

Schmidt, Hans. *Maverick Marine: General Smedley D. Butler and the Contradictions in American Military History*. Lexington: University Press of Kentucky, 1987.

Schneider, James J. "The Loose Marble—and the Origins of Operational Art." *Parameters* 19 (1989): 85–99.

Schom, Alan. *The Eagle and the Rising Sun*. New York: W. W. Norton & Company, 2004.

Sherrod, Robert. *History of Marine Corps Aviation in World War II*. San Rafael, Calif.: Presidio Press, 1952.

Skelton, William B. *An American Profession of Arms: The Army Officer Corps, 1784–1861*. Lawrence: University Press of Kansas, 1996.

Smith, Robert Ross. *Triumph in the Philippines*. The United States Army in World War II. Washington, D.C.: Center of Military History, 1991.

Spector, Ronald H. *Eagle against the Sun: the American War with Japan*. New York: Free Press, 1985.

——. *Professors of War: The Naval War College and the Development of the Naval Profession*. Honolulu, Hawaii: University of the Pacific Press, 2005.

Spiller, Roger J., ed. *Dictionary of American Military Biography*. 3 vols. Westport, Conn.: Greenwood Press, 1984.

Strachan, Hew. *European Armies and the Conduct of War*. London: Unwin & Hyman, 1983.

Sweetman, Jack, ed. *The Great Admirals: Command at Sea, 1587–1945*. Annapolis: Naval Institute Press, 1997.

Trask, David F. *The War with Spain in 1898*. New York: Macmillan Publishing, Co., 1981.

Travers, Tim. *The Killing Ground: The British Army, the Western Front and the Emergence of Modern War, 1900–1918*. Boston: Allen & Unwin, 1987.

Van Creveld, Martin. *Supplying War: Logistics from Wallenstein to Patton*. London: Cambridge University Press, 1977.

——. *The Training of Officers: From Military Professionalism to Irrelevance*. New York: Free Press, 1990.

Vego, Milan. *The Battle for Leyte, 1944: Allied and Japanese Plans, Preparations, and Execution*. Annapolis: Naval Institute Press, 2006.

Weigley, Russell F. "From the Normandy Beaches to the Falaise-Argentan Pocket." *Military Review* 70 (1990): 45–64.

——. *History of the United States Army*. Bloomington: Indiana University Press, 1984.

Wildenberg, Thomas. *Gray Steel and Black Oil: Fast Tankers and Replenishment at Sea in the U.S. Navy, 1912–1995*. Annapolis: Naval Institute Press, 1996.

Index